Ecology of Macrofungi

An Overview

Series: Progress in Mycological Research

- Fungi from Different Environments (2009)
- Systematics and Evolution of fungi (2011)
- Fungi from Different Substrates (2015)
- Fungi: Applications and Management Strategies (2016)
- Advances in Macrofungi: Diversity, Ecology and Biotechnology (2019)
- Advances in Macrofungi: Industrial Avenues and Prospects (2021)
- Advances in Macrofungi: Pharmaceuticals and Cosmeceuticals (2021)
- Fungal Biotechnology: Prospects and Avenues (2022)
- Bioprospects of Macrofungi: Recent Developments (2023)
- Ecology of Macrofungi: An Overview (2023)

Series: Progress in Mycological Research

Ecology of Macrofungi

An Overview

Editors

Kandikere R. Sridhar

Department of Biosciences

Mangalore University, Mangalore, India

Sunil Kumar Deshmukh

R & D Division

Greenvention Biotech Pvt. Ltd., Pune, India

CRC Press

Taylor & Francis Group

Boca Raton London New York

CRC Press is an imprint of the

Taylor & Francis Group, an **informa** business

A SCIENCE PUBLISHERS BOOK

First edition published 2024
by CRC Press
2385 NW Executive Center Drive, Suite 320, Boca Raton FL 33431

and by CRC Press
4 Park Square, Milton Park, Abingdon, Oxon, OX14 4RN

CRC Press is an imprint of Taylor & Francis Group, LLC

Library of Congress Cataloging-in-Publication Data (applied for)

ISBN: 978-1-032-55153-1 (hbk)
ISBN: 978-1-032-55155-5 (pbk)
ISBN: 978-1-003-42927-2 (ebk)

DOI: 10.1201/9781003429272

Typeset in Times New Roman
by Radiant Productions

Preface

Mycology is one of the fast-developing disciplines that has been influenced by many facets of human life similar to plants and animals (Naranjo-Ortiz and Gabaldón, 2019). Macrofungi are natural bioresources with wide distribution responsible for the decomposition of lignocellulosic materials, mutualistic association with tree species as ectomycorrhizas, source of nutrition and possess or produce bioactive compounds valuable in human health, agriculture and industries. Although a global conservative estimate of macrofungal diversity ranges from 0.14 to 1.25 million, a few thousand are described and very few of them are explored for their usefulness (Hawksworth, 2019). Evaluation of macrofungal resources, distribution, lifestyles, adaptations, substrate preferences and ecology in different ecosystems enhances our knowledge of their conservation and applications in various fields (Dighton, 2019). Macrofungal ecology is dependent on environmental variations, climatic conditions, substrates (availability and quality), disturbances, symbiotic partners and insect population (Hussain and Sher, 2021; Čejka et al., 2022). In spite of considerable developments in applied mycology in the 21st century, there are several gaps in understanding the macrofungal ecosystem services (e.g., decomposition, enhancement of soil qualities, nutrient acquisition, mutualistic association, energy flow, biogeochemical cycles, bioremediation and conservation).

The book, *Ecology of Macrofungi: An Overview* projects some of the current developments in macrofungal ecology in different habitats. Contributions of 41 researchers from 11 countries offered chapters on different facets of macrofungal ecology. This volume emphasizes four subdivisions of macrofungal ecology: (1) Importance in forest ecosystems; (2) Ectomycorrhizal associations; (3) Wood-preference of macrofungi; (4) Polymorphism in macrofungi. The first subdivision highlights the role of macrofungi in forests such as nutrient cycling, mycorrhizal functions, prevention of soil erosion, animal nutrition, parasitic macrofungi, host preference, ammonia fungi and macrofungal dynamics. The second subdivision discusses the significance of ectomycorrhizal fungi in tropical and subtropical forest ecosystems. The third subdivision mainly delivers information about the macrofungal substrate ecology, wood-rot polypores, *Pinus*-dwelling Hymenochaetaceae and indoor macrofungi followed by fourth subdivision dealing with polymorphism in woody-decaying macrofungi.

We are optimistic that the efforts made by the authors of chapters in understanding the ecology of macrofungi in different ecosystems will be valuable for readers with a wide interest in natural science. Owing to the topics dealing with macrofungal diversity, distribution, dynamics, lifestyles, ecosystem preference, substrate preference and ecosystem services draw the attention of mycologists, botanists, zoologists, ecologists, foresters, geneticists, biochemists, agronomists and field biologists.

We are indebted to the kind gesture of the contributors and reviewers for the on-time submission and meticulous evaluation. We are grateful to the CRC Press for its cooperation to fulfill several official formalities to bring out this book on time.

January 09, 2023

Kandikere R. Sridhar
Mangalore, India

Sunil K. Deshmukh
Pune, India

References

Čejka, T., Isaac, E.L., Oliach, D., Martínez-Peña, F. et al. (2022). Risk and reward of the global truffle sector under predicted climate change. Environ. Res. Lett., 27: 024001. 10.1088/1748-9326/ac47c4.

Dighton, J. (2019). The roles of macrofungi (Basidiomycotina) in terrestrial ecosystems. pp. 701–104. *In*: Sridhar, K.R. and Deshmukh, S.K. (eds.). Advances in Macrofungi: Diversity, Ecology and Biotechnology. CRC Press, Boca Raton, USA.

Hawksworth, D.L. (2019). The macrofungal resource: Extant, current utilization, future prospects and challenges. pp. 1–9. *In*: Sridhar, K.R. and Deshmukh, S.K. (eds.). Advances in Macrofungi: Diversity, Ecology and Biotechnology. CRC Press, Boca Raton, USA.

Hussain, S. and Sher, H. (2021). Ecological characterization of Morel (*Morchella* spp.) habitats: A multivariate comparison from three forest types of district Swat, Pakistan. Acta Ecol. Sinica, 41: 1–9.

Naranjo-Ortiz, M.A. and Gabaldón, T. (2019). Fungal evolution: Diversity, taxonomy and phylogeny of the fungi. Biol. Rev., 94: 2101–2137.

Contents

Preface v

List of Contributors ix

1. **Ecology of Macrofungi** 1
   Kandikere R Sridhar and *Sunil K Deshmukh*

Forest Macrofungi

2. **The Roles of Macrofungi in Humid Tropical Forests and the Effects of Disturbance** 17
   D Jean Lodge and *Sharon A Cantrell*

3. **Ecological Aspects of Ammonia Fungi in Various Vegetation Sites** 53
   Akira Suzuki and *Neale L Bougher*

4. **Ecology of Wild Mushrooms** 78
   Hoda M El-Gharabawy

5. **Ecology of Macrofungi in Southwest India** 101
   Kandikere R Sridhar

Ectomycorrhizal Fungi

6. **Mycogeography and Ecology of Ectomycorrhizal Fungi in Northern México** 123
   Fortunato Garza Ocañas, Miroslava Quiñonez Martínez, Lourdes Garza Ocañas, Artemio Carrillo Parra, Jesús García Jiménez, Gonzalo Guevara Guerrero, Ricardo Valenzuela Garza, Mario García Aranda, Javier de la Fuente, Gerardo Cuellar Rodríguez and *José Guadalupe Martínez Avalos*

7. **On the *Amanita* in Southwest India** 159
   Kandikere R Sridhar, Shivannegowda Mahadevakumar and *Namera C Karun*

Wood-Inhabiting Macrofungi

8. **Substrate Ecology of Wood-Inhabiting Basidiomycetes** 179
Ivan V Zmitrovich, Stanislav P Arefiev, Vladimir I Kapitonov, Anton G Shiryaev, Kiran R Ranadive and *Margarita A Bondartseva*

9. **Wood-Rot Polypores of Kerala, India** 222
TK Arun Kumar and *N Vinjusha*

10. **Host Preferences of *Pinus*-dwelling Hymenochaetaceae** 244
Balázs Palla, Yuan Yuan, Yu-Cheng Dai and *Viktor Papp*

11. **Occurrence and Adaptive Potential of Indoor Macrofungi** 280
Dmitry Yu Vlasov, Nadezhda V Psurtseva, Ivan V Zmitrovich, Katerina V Sazanova, Oleg N Ezhov and *Margarita A Bondartseva*

Polymorphism in Macrofungi

12. **Genetic and Morphological Polymorphism in Wood-Decaying Fungi** 311
Badalyan SM, Vladykina VD, Zhuykova EV, Diyarova DK and *Mukhin VA*

Index 333

About the Editors 335

List of Contributors

Aranda, Mario García

Mar Caspio, 8212, Loma Linda, Monterrey, N.L. C.P. 64120, México.

Arefiev, Stanislav P.

The Institute for the Development of the North – A Structural Subdivision of the Federal State Budgetary Institution of Science, The Federal Research Center Tyumen Scientific Center of the Siberian Branch of the Russian Academy of Sciences, Russia.

Arun Kumar, T.K.

Department of Botany, The Zamorin's Guruvayurappan College, Kerala 673014, India.

Avalos, José Guadalupe Martínez

Instituto de Ecología Aplicada, Universidad Autónoma de Tamaulipas, Cd. Victoria, Mexico.

Badalyan S.M.

Laboratory of Fungal Biology and Biotechnology, Institute of Biology, Yerevan State University, 1 A. Manoogian St., 0025 Yerevan, Armenia.

Bondartseva, Margarita A.

V.L. Komarov Botanical Institute, Russian Academy of Sciences, 2 Professor Popov Street, Saint Petersburg, 197376, Russia.

Bougher, Neale L.

Western Australian Herbarium, Biodiversity and Conservation Science, Department of Biodiversity, Conservation and Attractions, Western Australia, Australia.

Cantrell, Sharon A.

Department of Biology, Universidad Ana G Méndez – Gurabo, PO Box 3030, Gurabo, Puerto Rico 00778.

Dai, Yu-Cheng

Institute of Microbiology, School of Ecology and Nature Conservation, Beijing Forestry University, Beijing 100083, China.

Deshmukh, Sunil K.

R & D Division, Greenvention Biotech Pvt. Ltd., Uruli-Kanchan, Pune 412202, Maharashtra, India.

de la Fuente, Javier

Colegio de Postgraduados, km 36.5, 56230 Montecillo, Texcoco, Estado de México, Mexico.

Diyarova D.K.

Institute of Plant and Animal Ecology, Ural Branch of Russian Academy of Sciences, 202, 8 Marta Street, 620144 Ekaterinburg, Russia.

El-Gharabawy, Hoda M.

Botany and Microbiology Department, Faculty of Science, Damietta University, New Damietta, 34517, Egypt.

Ezhov, Oleg N.

N. Laverov Federal Center for Integrated Arctic Research of the Ural Branch of the Russian Academy of Sciences, 23 North Dvina Enbankment, Arkhangelsk, 163000, Russia.

Garza, Ricardo Valenzuela

Laboratorio de Micologία, Departamento de Botánica, Escuela Nacional de Ciencias Biológicas, Instituto Politécnico Nacional, Apartado Postal 63-351, 02800, México.

Guerrero, Gonzalo Guevara

Tecnológico Nacional de México, Instituto Tecnológico de Ciudad Victoria, Boulevard Emilio Portes Gil número 1301, 87010 Cd. Victoria, Tamaulipas, Mexico.

Jiménez, Jesús García

Tecnológico Nacional de México, Instituto Tecnológico de Ciudad Victoria, Boulevard Emilio Portes Gil número 1301, 87010 Cd. Victoria, Tamaulipas, Mexico.

Kapitonov, Vladimir I.

Tobolsk complex scientific station Ural Branch of the Russian Academy of Sciences, 15 Academician Osipov, Tobolsk, 626152, Russia.

Karun, Namera C.

Western Ghats Macrofungal Research Foundation, Bittangala, Virajpet, Kodagu, Karnataka, India.

Lodge, D. Jean

University of Georgia, Odum School of Ecology and Dept. of Plant Pathology, Athens, GA 30606, USA.

Mahadevakumar, Shivannegowda

Forest Pathology Department, Division of Forest Protection, Kerala Forest Research Institute, Peechi, Thrissur, Kerala, India.
Botanical Survey of India, Andaman and Nicobar Regional Centre, Haddo 744102, Port Blair, Andaman, India.

Martínez, Miroslava Quiñonez

Universidad Autónoma de Ciudad Juárez, Departamento de Ciencias Químico-Biológicas, Av. Benjamín Franklin No. 4650, Zona PRONAF, Cd. Juárez, Chihuahua C.P. 32315, México.

Mukhin V.A.

Department of Biodiversity and Bioecology, Institute of Natural Sciences and Mathematics, Ural Federal University named after the first President of Russia B.N. Yeltsin, 19 Mira Street, 620003 Ekaterinburg, Russia; Institute of Plant and Animal Ecology, Ural Branch of Russian Academy of Sciences, 202 on 8 Marta Street, 620144 Ekaterinburg, Russia.

Ocañas, Fortunato Garza

Universidad Autónoma de Nuevo León, Campus Linares, Facultad de Ciencias Forestales, Carretera Nacional km 145, Apdo. postal 41, 67700 Linares, Nuevo León, Mexico.

Ocañas, Lourdes Garza

Facultad de Medicina, Departamento de Farmacología y Toxicología, Universidad Autónoma de Nuevo León, México.

Palla, Balázs

Department of Botany, Hungarian University of Agriculture and Life Sciences, Budapest, Hungary.

Papp, Viktor

Department of Botany, Hungarian University of Agriculture and Life Sciences, Budapest, Hungary.

Parra, Artemio Carrillo

Instituto de Silvicultura e Industria, de la Madera Universidad, Juárez del Estado de Durango, México.

Psurtseva,Nadezhda V.

V.L. Komarov Botanical Institute, Russian Academy of Sciences, 2 Professor Popov Street, Saint Petersburg, 197376, Russia.

Ranadive, Kiran R.

Annasaheb Magar Mahavidyalaya, Hadapsar, Maharashtra, India.

Rodríguez, Gerardo Cuellar

Universidad Autónoma de Nuevo León, Campus Linares, Facultad de Ciencias Forestales, Carretera Nacional km 145, Apdo. postal 41, 67700 Linares, Nuevo León, Mexico.

Sazanova, Katerina V.

V.L. Komarov Botanical Institute, Russian Academy of Sciences, 2 Professor Popov Street, Saint Petersburg, 197376, Russia.

Shiryaev, Anton G.

Institute of Plant and Animal Ecology, Ural Branch of Russian Academy of Sciences, 202, 8 Marta Street, 620144 Ekaterinburg, Russia.

Sridhar, Kandikere R.

Department of Biosciences, Mangalore University, Mangalagangotri, Mangalore, Karnataka, India.

Suzuki, Akira

Faculty of Science and Engineering, Tokyo City University - Chiba University, Japan.

Vinjusha, N.

Department of Botany, The Zamorin's Guruvayurappan College, Kerala 673014, India.

Vladykina V.D.

Department of Biodiversity and Bioecology, Institute of Natural Sciences and Mathematics, Ural Federal University named after the first President of Russia B.N. Yeltsin, 19 Mira Street, 620003 Ekaterinburg, Russia.

Vlasov, Dmitry Yu

V.L. Komarov Botanical Institute, Russian Academy of Sciences, 2 Professor Popov Street, Saint Petersburg, 197376, Russia.

Yuan, Yuan

Institute of Microbiology, School of Ecology and Nature Conservation, Beijing Forestry University, Beijing 100083, China.

Zhuykova E.V.

Institute of Plant and Animal Ecology, Ural Branch of Russian Academy of Sciences, 202, 8 Marta Street, 620144 Ekaterinburg, Russia.

Zmitrovich, Ivan V.

V.L. Komarov Botanical Institute, Russian Academy of Sciences, 2 Professor Popov Street, Saint Petersburg, 197376, Russia.

1

Ecology of Macrofungi

Kandikere R Sridhar[1,*] and *Sunil K Deshmukh*[2]

1. Introduction

Fungal lineage represents a diverse group of organisms that evolved parallelly with plants and animals. Fungi are the largest communities after the class Insecta with the evolutionary history of about 1,800 million years (Kirk et al., 2008; Hawksworth, 1991). They are morphologically versatile (unicellular, filamentous and produce visible fruit bodies), have wide distribution in varied ecosystems and exhibit distinct lifestyles. Macrofungi occupies an important place in mycology owing to their uniqueness such as nutritional, antinutritional values (metabolites), mutualistic associations (plants and animals) and several ecosystem functions (e.g., nutrient recycling and environmental protection). The ecosystem services offered by macrofungi are beneficial directly as well as indirectly (Fig. 1). Considering 10% of all the fungi (2.2–3.8 million), a conservative global estimate reveals the existence of macrofungi between 0.22 and 0.38 million species (Hawksworth, 1991, 2019; Hawksworth and Lucking, 2017). Based on the plant-macrofungal ratio, Mueller et al. (2007) predict the worldwide existence of 0.053–0.11 species of macrofungi. However, so far known macrofungi will be about 3.7–6.4% (Hawksworth, 2019).

[1] Department of Biosciences, Mangalore University, Mangalagangotri, Mangalore, Karnataka, India.
[2] R & D Division, Greenvention Biotech Pvt. Ltd., Uruli-Kanchan, Pune 412202, Maharashtra, India.
* Corresponding author: kandikere@gmail.com

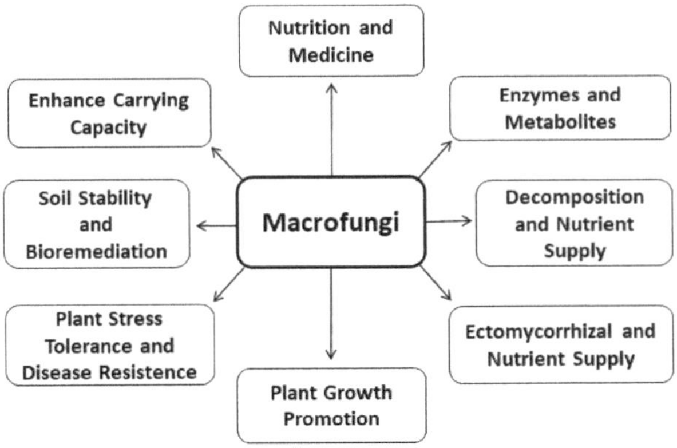

Fig. 1. Direct and indirect ecosystem services of macrofungi.

Macrofungal distribution and diversity in a landscape are dependent on the nature of a specific ecosystem and the availability of suitable substrates. They are known to prefer a wide range of substrates for growth, reproduction and perpetuation (e.g., soil, leaf detritus, woody litter and insects). Macrofungal diversity and distribution are mainly dependent on the topography, vegetation and several environmental factors (e.g., vegetation, topography, landscape, rainfall, season, temperature, light, humidity, wind speed and soil fauna) (Brown et al., 2006; Trierveiler-Pereira et al., 2013; Chen et al., 2018; Carteron et al., 2022; Hu et al., 2022) (Fig. 2). The richness of many macrofungi is depended on the availability, quantity and quality of woody substrates (Rudolf et al., 2012). Habitat degradation rather than fragmentation has severe detrimental effects on the macrofungal diversity in the Western Ghats forests of India (Brown et al., 2006). For instance, although the sacred groves exist in small areas, they possess diverse as well as versatile macrofungi owing to the uniqueness of their habitat. The coffee plantations were also supported by more diverse macrofungi, which has been supported by a recent study by Karun and Sridhar (2016) owing to the least disturbance of woody litter in coffee agroforests.

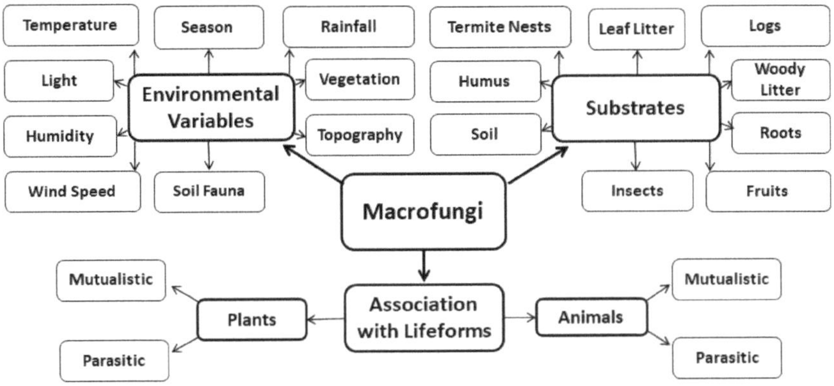

Fig. 2. Environmental variables, substrates and association of macrofungi with other lifeforms.

Based on studies in the Greater Mekong Sub-Region of China, the substrate preference explains the biogeography of macrofungi (Ye et al., 2019). Edible, non-edible and poisonous mushrooms as non-timber forest products worldwide serve as major economic resources useful in various environmental and industrial applications (e.g., forestry, food source and value-added pharmaceuticals) (Zotti et al., 2013). Macrofungi are well known for the bioremediation of several environmental pollutants owing to their powerful enzymes (e.g., pesticides, heavy metals, hydrocarbon contaminants and xenobiotics) (Rahi et al., 2021; Niego et al., 2023).

This overview provides an outlook of macrofungal ecology such as substrate preference, nutrient recycling, mutualistic associations, pathogens and impacts of disturbance based on the literature and chapters offered in this book.

2. Substrates and Nutrient Cycling

Forest ecosystems consist of mainly four major types of macrofungi (saprotrophic, ectomycorrhizal, pathogenic and parasitic) (Niego et al., 2023). The accessibility as well as the quality of substrates, are crucial for the occurrence, distribution and diversity of macrofungi (Ye et al., 2019) (Fig. 2).

The occurrence, contributions and functions of wild mushrooms in various microhabitats have been discussed by El-Gharabawy (Chapter 4). The ecology of macrofungi with special reference to the hotspot Western Ghats of India has been discussed by Sridhar (Chapter 5). Mycogeography and ecology of ectomycorrhizal (EM) fungi in Northern Mexico in comparison with the Northeast United States and Northeast Canada discussed by Ocañas et al. (Chapter 6).

2.1 *Substrates*

Studies in the Greater Mekong Sub-Region of China revealed the occurrence of 957 species of macrofungi (in 73 families, 189 genera) (Ye et al., 2019). Substrates macrofungi were classified into four major categories (soil, root, litter and rare substrates) and divided into 14 sub-categories (branches, leaves and fruits). About 50% of macrofungal species were ectomycorrhizal with roots of specific host plant species (major genus, *Russula*), 30% were saprotrophs on decaying plant litter (e.g., major genus, *Marasmius*), 15% were found in soil and 5% preferred rare substrates (e.g., insects, lichens and termite nests). Macrofungi commonly prefer single substrate that will not be influenced by the changes in vegetation as well as climate. However, some species (< 1%) (e.g., *Marasmius* aff. *maximus*) preferred multiple substrates. In the secondary forests of the Greater Mekong Sub-Region of China showed a significant correlation between preference substrates vs. taxonomic diversity (Ye et al., 2019). Such multiple substrates preferring pioneer groups of macrofungi are important communities to establish in new environments. Thus, higher substrate diversity is related to higher macrofungal diversity. Forest naturalness has been positively correlated with the diversity of wood-preferring fungi (e.g., canopy tree age, number of cut stumps and volume of dead wood) (Edman et al., 2004). Similar observations are made for specific macrofungal groups (e.g., agarics, corticioids and polypores). Macrofungi are important candidates in nutrient cycling in soil ecosystems of forests (Niego et al., 2023).

Substrate ecology of wood-inhabiting basidiomycetes with detailed emphasis on macrofungal succession, saprotrophs, enzymes of wood-rot fungi and trophic differentiation have been discussed by Zmitrovich et al. (Chapter 8). Distribution, environmental factors, substrate preferences, pathogens, medicinal species, edible species and mutualistic association of wood-rot polypores of southwest India (Kerala) have been presented by Arun Kuman and Vinjusha (Chapter 9). Global mycogeography of *Pinus*-dwelling Hymenochataceae has been discussed along with ecology, habitat preference, host preference and saprophytism by Palla et al. (Chapter 10). The biology, ecology and adaptive potential of xylotrophic indoor macrofungi in residential buildings have been elucidated by Vlasov et al. (Chapter 11). Genetic and morphological polymorphism of wood-decaying fungi, their geographical patterns and relationships with environmental factors has been studied based on morphology as well as molecular approaches by Badayan et al. (Chapter 12) to facilitate future biotechnological applications. The impact of various environmental variables on wild mushrooms and ecological role of wild mushrooms have been discussed by El-Gharabawy (Chapter 4).

2.2 *Nutrient Cycling*

Release N and P by the mycorrhizas (Ericoid and ectomycorrhizal) from the detritus (plant and faunal origin) is of high significance in terrestrial ecosystems (Read and Perez-Moreno, 2003). The EM fungi are known to efficiently drive the global carbon nutrient cycle including nitrogen and phosphorus cycles in terrestrial ecosystems (van der Heijden et al., 2015; Netherway et al., 2021; Vaario and Matusushita, 2021; Niego et al., 2023). Rahi et al. (2021) discussed the production, significance and industrial applications of white-rot fungi. Brown-rot fungi swiftly degrade cellulose and hemicellulose, but they are unable to produce lignin-degrading enzymes (Arantes and Goodell, 2014; Langer et al., 2021). They are well-known to attack the coniferous woods (Goodell et al., 2003). White-rot fungi degrade wood cellulose, hemicelluloses and lignins (Suryadi et al., 2022). Lodge and Cantrell gave a detailed account of the transfer of nutrients by EM fungi, wood decomposers, leaf decomposers and the conservation of nutrients by macrofungi (Chapter 2). Mutualism and parasitism of wild mushrooms with plants are discussed in Chapter 4.

3. Association with Plants

Mycorrhizal fungi (arbuscular and ectomycorrhizal) are ubiquitous worldwide with mutualistic association with multifunctionality links plant roots with soil. The EM fungi are mutualistically associated with various tree species in forest ecosystems.

3.1 *Ectomycorrhiazae*

The EM fungi consist of over 5000 spp. (in about 250 genera) mainly Basidiomycota and some Ascomycota (e.g., truffles) established mutualistic associations with root systems of woody plants (angiosperms and gymnosperms) (Geml, 2017). Among 1928 genera of macrofungi (Basidiomycota) up to 100 spp. are known to be ectomycorrhizal with About 2% of vascular plants (Brundrett and Tedersoo, 2018; Corrales et al., 2018; He et al., 2019). The Western Ghats of India represent up to 250 species of EM fungi with tree species/EM fungi ratio of 1:3 with the highest association with Dipterocarpaceae (Sridhar and Karun, 2019). The EM fungi are specialized to pool up and supply necessary nutrients to plants in nitrogen and phosphorus-limited habitats. The EM fungi are also known for their occupation in many natural extreme habitats (e.g., glaciers, volcanic deserts, sand dunes and mangroves) (Ghate and Sridhar, 2016a, 2016b; Kałucka and Jagodziński, 2017). Ecology and host tree species of a new *Amanita* (*A. konajensis*) occurring in the southwestern part of India has been discussed with emphasis on its edibility, ectomycorrhizal association and bioactive potential (Sridhar et al., Chapter 7).

3.2 *Dipterocarpaceae*

Tree species belonging to Dipterocarpaceae are highly valuable in sylviculture in the Western Ghats, Himalayas (India) and Southeast Asian countries (Appanah and Turnbull, 1998). Several EM fungi from dipterocarps have been reported from wide areas of Asian countries (Indonesia, Philippines, Sri Lanka and Thailand) (Natarajan et al., 2005). Trees belonging to Dipterocarpaceae serve as potential hosts to many ectomycorrhizal fungi in the Western Ghats (Sridhar and Karun, 2019). Up to 77% of 148 species of EM fungi are supported by eight tree species of Dipterocarpaceae. The EM species in decreasing order associated with Dipterocarpaceae: *Inocybe* > *Russula* > *Amanita* > *Boletus*. Similarly, the tree species of Dipteropcarpaceae possess EM fungi in decreasing order: *Vateria indica* > *Hopea ponga* > *H. parviflora* > *Diospyros malabarica* > *Myristica malabarica* > *Dipterocarpus indicus*. According to Natarajan et al. (2005), *Vateria indica* hosts the highest number of EM fungi in the Western Ghats. Interestingly, Dipterocarpaceae members in the Western Ghats of Kerala possess the highest EM fungal genus *Inocybe* (Latha and Manimohan, 2017).

4. Association with Animals

There is a variety of fauna that directly or indirectly influence the growth, development, reproduction and dispersal of macrofungi (Tang et al., 2015) (Fig. 2). Faunal association provides substrates, moisture, aids in reproduction, dispersal and expansion of the ecological niches of macrofungi. Reciprocally, macrofungi provide nutrition as well as protection for the associated fauna. In addition to dispersal, animals facilitate genetic diversity as well as the genetic recombination of several macrofungi. The interaction of various animals with EM fungi in Mexico has been discussed by Ocañas et al. (Chapter 6). Interestingly, rhizomorphs of Marasmiaceae are used by birds to build their nests (Lodge and Cantrell, Chapter 2). Mutualism and parasitism of wild mushrooms with animals are discussed in Chapter 4.

4.1 Termites, Ants, Wasps and Scale Insects

The association of termites (Macrotermitinae) with termitomycete mushrooms (Basidiomycota) is one of the well-known examples of termite-mushroom symbioses. Termites provide a suitable microclimate as well as a substrate to the fungus by selective inhibition of fungal competitors, while fungus offers nitrogen-enriched nutrients to the termites with specific enzymes to fetch additional nutrients (Korb and Aanen, 2003). Fungal secretion of lignocellulolytic enzymes will be mixed with termite gut enzymes along with termite gut bacteria, which leads to the efficient degradation of plant material (Tang et al., 2015). Thus, the monoculture of selected mushrooms provides superior nutrients that will be selected by the termites for more benefits (Nobre and Aanen, 2012). Termite-macrofungal symbiotic associations are prominent in southern Africa, Asia (south and southeast) and islands of the South Pacific by vertical (local) and horizontal (widespread) transmissions (Licht et al., 2006; Nobre et al., 2010, 2011; Wang et al., 2012; Tang et al., 2015). Mushroom gardens are built by termites as well as ants in neotropics and paleotropics, respectively (Lodge and Cantrell, Chapter 2). Thus, fungus-farming termites and ants play a significant role in the dispersion and genetic materials of termitomycetes. Similar to termitomycetes, *Xylaria* (Ascomycota) is often associated with termite nests (Visser et al. 2011; Karun and Sridhar, 2015). However, *Xylaria* will not damage the termite nest or termitomycetes as it flourishes in the abandoned termite colonies.

Amylostereum areolatum, *A. chailletii* and *A. laevigatum* (basidiomycota) have symbiotic associations with different species of siricid woodwasps (Tabata and Abe, 1999). The female woodwasps possess special features such as intersegmental sacs (to transfer fungal spores and to deposit along with eggs in host trees) and phytotoxin secreted by wasps facilitate congenial conditions for fungi to flourish (Thomsen and Harding, 2011). Such association results in providing nutrients, enzymes and environmental conditions required for the survival and metamorphosis of larvae (Slippers et al., 2003).

The family Septobasidiaceae encompasses five genera with about 175 species and most of them are considered under the genus *Septobasidium* (Tang et al., 2015). This genus has an intimate association with scale insects (Diaspididae). Many species live symbiotically or parasitize several scale insects (Coccoidea). A tripartite association of *Auruculoscypha* (Basidiomycota) with trees of Anacardiaceae and the scale insect *Neogreenia zeylinaca* (Margarodidae) has been reported recently by Manimohan (2019).

4.2 Dispersal

Macrofungi will be dispersed widely in spite of their mutualistic or parasitic or pathogenic association with various fauna. Vašutová et al. (2019) have discussed the role of animals in the transport of mycorrhizal fungi. Many animals (vertebrates and invertebrates) serve as transporters of fungal spores or mycelia to broaden the geographic regions (narrow or wide) of epigeous (Phallelses), hypogeous (truffles), coprophilous (psychedelic) and parasitic (*Cordyceps*) macrofungi. Even though many macrofungi are parasites of insects (e.g., *Cordyceps*), they have the advantage of dispersal to wider niches in the ecosystem.

The stinkhorns (Phallales: *Phallus* spp.; Basidiomycota) produce morphologically distinct with varied coloured fruit bodies. The spore-bearing gleba with mucilage emerges unpleasant odors to attract insects (Magnago et al., 2013). On feeding the gelatinous material of gleba, insects transmit basidiospores to different niches through their excrements (Oliveira and Morato, 2000).

The macrofungi produce fruit bodies underneath the soil surface (hypogeous) and produce large fruit bodies underneath the soil with thick layers of humus or leaf litter (e.g., truffles: *Tuber* spp.; Ascomycota) (Hawker, 1954). Such macrofungi provide nutrients to plants such as ectomycorrhizae and also protect from root pathogens. Such hypogeous fruit bodies serve as a source of nutrition to many mammals in wild (including human beings). The aroma of fruit bodies attracts small mammals (e.g., squirrels, rats and bandicoots). Consumption and defecation by animals in different locations leads to the dispersal of spores (Claridge et al., 2000). Therefore, mycophagy by small mammals leads to dissemination and reproduction of truffles. Saprophagus slugs (e.g., *Meghimatium fruhstorferi*) are also known to consume many macrofungi (*Pleurotus*, *Armillaria* and *Gymnopilus*), excrete viable spores in litter or woody debris, which results in higher germination of spores compared to control spores (Kitabayashi et al., 2022). It is also known that slugs transport huge amounts of spores ($1 \times 10^{8)}$, which is 100–200 times larger than those carried by a mycophagous fly (Tuno et al., 1998; Kobayashi et al., 2017). Similarly, *Oniscus asellus* (woodlouse) through mycophagy also known for the dissemination of spores of truffles to different ecological niches (Thomas and Thomas, 2022).

Coprophilous (dung-preferring) macrofungi are a specific group chooses herbivore dung of various animals to perpetuate (e.g., elephant and bison dung). Many fungi grow successionally on the dung (phycomycetes, ascomycetes and basidiomycetes) (Richardson, 2002). Many macrofungal ascomycetes (*Peziza* spp.) and basidiomycetes (*Coprinus*) are dispersed through herbivore dung. In addition, coprophilous insects transmit the spores of dung-preferring macrofungi to new ecological niches.

5. Pathogens and Parasites

Although fungal parasitism with animals and plants is fairly well known, macrofungal parasitism with animals and plants are not well understood. Some examples of macrofungal-animal and macrofungal-plant pathogenesis and parasitism is discussed.

5.1 *Animal Pathogens and Parasites*

Many macrofungi evolved complex mechanisms to kill animals to fetch their nutrients (e.g., amoebas, nematodes, rotifers and insects) (Barron, 2003; Shrestha et al., 2012). The wood-rot fungus *Hohenbuehelia* sp. attacks and kill nematodes (Barron and Dierkes, 1977). Similarly, *Pleurotus ostreatus* paralyze nematodes by hyphal adhesive secretion to trap nematodes (Lee et al., 2020). Interestingly, some of the *Pleurotus* spp. lead carnivorous life to fetch nitrogen in nutrient-limiting habitats.

Cordyceps (Clavicipitaceae, Cordycipitaceae and Ophiocordycipitaceae) have been assigned to five genera by Sung et al. (2007) (*Cordyceps*, *Elaphocordyceps*, *Metacordyceps*, *Ophiocordyceps* and *Tyrannicordyceps*). *Cordyceps* and allied species are distributed at high altitudes (3,000–5,100 m asl) in different regions in and around the Himalayas (Gansu, Nepalm, Qinghai, Sichuan, Tibet and Yunnan) (Li et al., 2011). *Cordyceps militaris* is entomopathogenic and well-known to parasite insects and other arthropod fauna (Shrestha et al., 2012). *Cordyceps* are known to parasitize the larva, pupae and adult ghost moths (*Hepialus* spp.) and other insects. Although conventionally *Cordyceps* is known from high altitude regions of the Himalayas, recently several *Cordyceps* and allied species have been reported from the Western Ghats and west coast of southern India parasitizing many hosts (e.g., ants, coleopteran larvae, grasshoppers, pentatomid bugs and Scarabaeid bugs) (Dattaraj et al., 2018). In spite of asexual reproduction and the mode of infection of insects, the sexual reproduction of *Cordyceps* seems to trigger only after infecting the hosts (Stone, 2008). *Ophiocordyceps* spp. are known to infect many insects (Araujo et al., 2021; Yang et al., 2021; Karun and Sridhar, 2013). Although several macrofungi are parasites of many animals, they also have the advantage of dispersal to wider regions in the ecosystem.

5.2 Plant Pathogens and Parasites

Fungal pathogens are valuable for healthy forest ecosystems as they involve in the elimination of weaker plants, highly abundant species and old or weak trees species, which in turn, create space for other plant species leading to increased biodiversity (Castello et al., 1995; Old et al., 2000; Burgess and Wingfield, 2006). In addition, pathogenic wood-decaying brown-rot and white-rot fungi are capable to degrade lignocellulosic polymers by their enzymes and recycle the nutrients in the forest ecosystems.

Many macrofungi are pathogenic to roots and trunks of tropical tree species (Lodge and Cantrell, Chapter 2). Arun Kumar and Vinjsha emphasized the macrofungal pathogenesis of several economically valuable live tree species in India and other parts of the world (e.g., *Amauroderma*, *Bjerkandera*, *Ganoderma*, *Perenniporia*, *Rigidoporus* and *Trametes*) (Chapter 9).

Several macrofungi are responsible for the dieback disease of the black alder (*Alnus glutinosa*) in Poland (Piętka and Grzywacz, 2018). The most common macrofungal causative agents of macrofungi include *Daedaleopsis confragosa*, *Stereum hirsutum*, *S. subtomentosum* and *Xanthoporia radiata*. Many orchids are parasitized by a wide array of saprotrophic macrofungi (Lodge and Cantrell, Chapter 2).

6. Disturbance and Conservation

Specific conditions prevailing in a specific habitat may be natural or by human interference (e.g., habitat degradation, low nutrients, low organic matter, extreme pH, extreme temperature, extreme moisture, mine habitats, clear-cut forests, intensive agricultural lands and fire-damaged habitats).

6.1 Disturbance

The EM fungi are capable to establish in soils low in organic matter, low in minerals, disturbed, extreme pH, drastic fluctuations in soil temperature, severe fluctuation of moisture and pollution (Münzenberger et al. 2004; Staudenrausch et al. 2005; Huang et al. 2012). Such capability of EM fungi depends on the rehabilitation by specific tree species which serve as a potential tool for phytoremediation. The EM fungi are also adapted to habitats with anthropogenic disturbance (e.g., clear-cut or timber harvest regions, post-agricultural lands, mining areas and fire-damaged habitats) (Kałucka and Jagodziński, 2017).

The EM fungi are very low in severely disturbed soils as well as young mine spoils (Malajczuk et al., 1994; Lunt and Hedger, 2003; Bois et al., 2005). With time-lapse, the quantity of inoculum with spore bank raises mainly by spore dispersal via air as well as animal carriers (Bois et al., 2005). Through a succession of vegetation over a period of time, slowly diverse EM fungi increase (Malajczuk et al., 1994). Similar observations have been made in coal-mine and metal-contaminated tailings (Parraga-Aguado et al., 2014; Kałucka and Jagodziński, 2017). Degradation of habitat has strong damaging effects on the macrofungal diversity in the forests of the Western Ghats of India (Brown et al., 2006). Suzuki and Bougher discussed ammonia fungi and their contributions to the recovery of forests after nitrogen disturbance, including forests in different climatic and vegetation regions (Chapter 3).

Intentional and unintentional forest fires have a severe effect on the vegetation, soil and microbial communities in forests. The growth and perpetuation of macrofungi in fire-influenced forests depend on the intensity of the fire. Although epigeous macrofungi face severe threats, hypogeous macrofungi fairly escape the intensity of fire impacts. However, severe fires lead to the eradication of forest stands, which leads to the loss of EM fungi owing to the destruction of host trees. Forest fires are responsible for ecosystem modification by influencing the physical, chemical and biological quality of forest soils (Agbeshie et al., 2022). The wildfire is detrimental to fungi owing to litter removal and increase in pH (Erkovan et al., 2016). In sporadic fires in scrub jungles of southwest India, a positive correlation was found between macrofungal species richness and soil phosphorus level in the fire-impacted region, whereas the richness of sporocarp was negatively correlated with soil pH in the control region (Greeshma et al., 2016).

6.2 Conservation

Conservation guidelines in forests are mainly based on wood-inhabiting polypores (Edman et al., 2004). Consideration of more wood-inhabiting macrofungal groups is necessary to develop suitable conservation strategies in forests. Such strategies are hampered owing to a lack of knowledge on many unknown macrofungi in forest ecosystems. Macrofungal conservation policies need to concentrate on the status of the forest (e.g., pristine, degraded, monoculture and agroforestry) for rehabilitation (Brown et al., 2006). As habitat degradation has severe impacts on macrofungal diversity and activity, rehabilitation of human interfered habitats needs to maintain the naturalness for the successful conservation of macrofungi (Edman et al., 2004; Brown et al., 2006).

Conclusion

Macrofungi occupied an ecologically important place in the fungal kingdom owing to their multifarious benefits and ecological services. Currently, their global diversity is speculated as 10% of the total fungal diversity. They grow and reproduce on various substrates and developed saprophytic, mutualistic, pathogenic and parasite modes of lifestyles with plants and animals. Macrofungal activities are dependent on various climatic, environmental and substrate conditions. They are capable to degrade lignocellulosic materials, nutrient recycling and trigger biogeochemical cycles in different ecosystems worldwide. Macrofungal growth, reproduction, propagation and dissemination depend on the nature of the ecosystem they inhabit. Macrofungal conservation is necessary to maximize the benefits to plants, animals and ecosystems. Macrofungal capability to occupy and function in several natural, unnatural, extreme and hostile conditions and ecosystems have attracted attention to employ them for bioremediation, environmental rehabilitation and biotechnological endeavours.

Acknowledgements

Authors gratefully acknowledge the contributions of several authors to fulfill the desire to project the macrofungal ecology along with knowledge gaps.

References

Agbeshie, A.A., Abugre, S., Atta-Darkwa, T. and Awuah, R. (2022). A review of the effects of forest fire on soil properties. J. For. Res., 33: 1419–1441. 10.1007/s11676-022-01475-4.

Appanah, S. and Turnbull, J.M. (1998). A review of dipterocarps: Taxonomy, ecology and sylviculture. Centre for International Forestry Research, Indonesia, p. 220.

Arantes, V. and Goodell, B. (2014). Current understanding of brown-rot fungal biodegradation mechanisms: A review. Am. Chem. Soc. Symp. Ser., 3–21. 10.1021/bk-2014-1158.ch001.

Araújo, J.P.M., Moriguchi, M.G., Uchiyama, S., Kinjo, N. and Matsuura, Y. (2021). *Ophiocordyceps salganeicola*, a parasite of social cockroaches in Japan and insights into the evolution of other closely-related Blattodea-associated lineages. IMA Fungus, 12: 3. 10.1186/s43008-020-00053-9.

Barron, G.L. (2003). Predatory fungi, wood decay, and the carbon cycle. Biodiversity, 4: 3–9. 10.1080/14888386.2003.9712621.

Barron, G.L. and Dierkes, Y. (1977). Nematophagous fungi: *Hohenbuehelia*, the perfect state of *Nematoctonus*. Can. J. Bot., 55: 3054–3062. 10.1139/b77-345.

Bois, G., Piché, Y., Fung, M.Y.P. and Khasa, D.P. (2005). Mycorrhizal inoculum potentials of pure reclamation materials and revegetated tailing sands from the Canadian oil sand industry. Mycorrhiza, 15: 149–158. 10.1007/s00572-004-0315-4.

Brown, N., Bhagwat, S. and Watkinson, S. (2006). Macrofungal diversity in fragmented and disturbed forests of the Western Ghats of India. J. Appl. Ecol., 43: 11–17.

Brundrett, M.C. and Tedersoo, L. (2018). Evolutionary history of mycorrhizal symbioses and global host plant diversity. New Phytol., 220: 1108–1115. 10.1111/nph.14976.

Burgess, T. and Wingfield, M.J. (2006). Impact of fungal pathogens in natural forest ecosystems: A focus on eucalypts. pp. 285–306. *In*: Sivasithamparam, K., Dixon, K.W. and Barrett, R.L (eds.). Microorganisms in Plant Conservation and Biodiversity. Kluwer Academic Publishers, Dordrecht. 10.1007/0-306-48099-9_11.

Carteron, A., Cichonski, F. and Laliberte, E. (2022). Ectomycorrhizal stands accelerate decomposition to a greater extent than arbuscular mycorrhizal stands in a northern deciduous forest. Ecosystems, 25: 1234–1248. 10.1007/s10021-021-00712-x.

Castello, J.D., Leopold, D.J. and Smallidge, P.J. (1995). Pathogens, patterns, and processes in forest ecosystems. Bioscience, 45: 16–24. 10.2307/1312531.

Chen, Y., Yuan, Z., Bi, S., Wang, X., Ye, Y. and Svennin, J.-C. (2018). Macrofungal species distributions depend on habitat partitioning of topography, light, and vegetation in a temperate mountain forest. Sci. Rep., 8: 13589. 10.1038/s41598-018-31795-7.

Claridge, A.W., Cork, S.J. and Trappe, J.M. (2000). Diversity and habitat relationships of hypogeous fungi - I. Study design, sampling techniques and general survey results. Biodivers. Conserv., 9: 151–173. 10.1023/A:1008941906441.

Corrales, A., Henkel, T.W. and Smith, M.E. (2018). Ectomycorrhizal associations in the tropics e biogeography, diversity patterns and ecosystem roles. New Phytol., 220: 1076–1091. 10.1111/nph.15151.

Dattaraj, H.R., Jagadish, B.R., Sridhar, K.R. and Ghate, S.D. (2018). Are the scrub jungles of southwest India potential habitats of *Cordyceps*? KAVAKA - Trans. Mycol. Soc. India, 51: 20–22.

Edman, M., Kruys, N. and Jonsson, B.G. (2004). Local dispersal sources strongly afect colonization patterns of wood-decaying fungi on spruce logs. Ecol. Appl., 14: 893–901.

Erkovan, Ş., Koc, A., Güllap, M.K., Erkovan, H.I. and Bilen, S. (2016). The effect of fire on the vegetation and soil properties of ungrazed shortgrass steppe rangeland of the Eastern Anatolia region of Turkey. Turk. J. Agric. For., 40: 290–299.

Geml, J. (2017). Altitudinal Gradients in Mycorrhizal Symbioses. Springer, Cham, pp. 107–123. 10.1007/978-3-319-56363-3_5.

Ghate, S.D. and Sridhar, K.R. (2016a). Contribution to the knowledge on macrofungi in mangroves of the Southwest India. Pl. Biosyst., 150: 977–986.

Ghate, S.D. and Sridhar, K.R. (2016b). Spatiotemporal diversity of macrofungi in the coastal sand dunes of Southwestern India. Mycosphere, 7: 458–472.

Goodell, B., Nicholas, D.D. and Schults, T.P. (2003). Wood Deterioration and Preservation: Advance in our Changing World. American Chemical Society, Washington, DC, p. 197.

Greeshma, A.A., Sridhar, K.R., Pavithra, M. and Ghate, S.D. (2016). Impact of fire on the macrofungal diversity of scrub jungles of Southwest India. Mycology, 7: 15–28.

Hawker, L.E. (1954). British hypogeous fungi. Phil. Trans. Royal Soc. Ser. B., 237: 429–546. 10.1098/rstb.1954.0002.

Hawksworth, D.L. (1991). The fungal dimension of biodiversity: Magnitude, significance, and conservation. Mycol. Res., 95: 641–655.

Hawksworth, D.L. (2019). The macrofungal resource: Extent, current utilization, future prospects and challenges. pp. 1–9. *In*: Sridhar, K.R. and Deshmukh, S.K. (eds.). Advances of Macrofungi: Diversity, Ecology and Biotechnology. CRC Press, Boca Raton.

Hawksworth, D.L. and Lücking, R. (2017). Fungal diversity revisited: 2.2 to 3.8 million species. MicrobiolSpectr 5: FUNK-0052-2016.

He, M.Q., Zhao, R.L., Hyde, K.D., Begerow, D., Kemler, M. et al. (2019). Notes, outline and divergence times of Basidiomycota. Fungal Divers., 99: 105–367. 10.1007/s13225-019-00435-4.

Hu, J.-J., Zhao, G.-P., Tuo, Y.-L., Qi, Z.-X. et al. (2022). Ecological factors influencing the occurrence of macrofungi from eastern mountainous areas to the central plains of Jilin Province, China. J. Fungi, 8: 871. 10.3390/jof8080871.

Huang, J., Nara, K., Lian, C., Zong, K., Peng, K. et al. (2012). Ectomycorrhizal fungal communities associated with Masson pine (*Pinus massoniana* Lamb.) in Pb-Zn mine sites of central south China. Mycorrhiza, 22: 589–602. 10.1007/s00572-012-0436-0.

Kałucka, I.L. and Jagodziński, A.M. (2017). Ectomycorrhizal Fungi: A major player in early succession. pp. 187–229. *In*: Varma, A., Prasad, R. and Tuteja, N. (eds.). Mycorrhiza - Function, Diversity, State of the Art. Springer International Publishing AG, Switzerland.

Karun, N.C. and Sridhar, K.R. (2013). The stink bug fungus *Ophiocordyceps nutans* – A proposal for conservation and flagship status in the Western Ghats of India. Fungal Conserv., 3: 43–49.

Karun, N.C. and Sridhar, K.R. (2015). *Xylaria* complex in the South Western India. Pl. Pathol. Quarant., 5: 83–96.

Karun, N.C. and Sridhar, K.R. (2016). Spatial and temporal diversity of macrofungi in the Western Ghat forests of India. Appl. Ecol. Environ. Res., 14: 1–21.

Kirk, P., Cannon, P., Minter, D. and Stalpers, J. (2008). Dictionary of the Fungi, 10th Edition, CABI, Wallingford, p. 2600.

Kitabayashi, K., Kitamura, S. and Tuno, N. (2022). Fungal spore transport by omnivorous mycophagous slug in temperate forest. Ecol. Evol., 12:e8565. 10.1002/ece3.8565.

Kobayashi, M., Kitabayashi, K. and Tuno, N. (2017). Spore dissemination by mycophagous adult drosophilids. Ecol. Res., 32: 621–626. 10.1007/s11284-017-1477-1479.

Korb, J. and Aanen, D.K. (2003). The evolution of uniparental transmis sion of fungal symbionts in fungus-growing termites (Macrotermitinae). Behav. Ecol. Sociobiol., 53: 65–71.

Langer, G.J., Bußkamp, J., Terhonen, E. and Blumenstein, K. (2021). Fungi inhabiting woody tree tissues. For. Microbiol., 175–205. 10.1016/B978-0-12-822542-4.00012-7.

Latha, K.P.D. and Manimohan, P. (2017). Inocybes of Kerala. SporePrint Books, Calicut, India, p. 181.

Lee, C.-H., Chang, H.-W., Yang, C.-T., Wali, N., Shie, J.-J. and Hsueh, Y.-P. (2020). Sensory cilia as the Achilles heel of nematodes when attacked by carnivorous mushrooms. Proc. Natl. Acad. Sci., 117: 6014–6022. 10.1073/pnas.1918473117.

Li, Y., Wang, X.L., Jiao, L., Jiang, Y., Li, H. et al. (2011). A survey of the geographic distribution of *Ophiocordyceps sinensis*. J. Microsc., 49: 913–919.

Licht, H.H.D.F., Boomsma, J.J. and Aanen, D.K. (2006). Presumptive horizontal symbiont transmission in the fungus-growing termite *Macrotermes natalensis*. Mol. Ecol., 15: 3131–3138. 10.1111/j.1365-294X.2006.03008.x.

Lunt, P.H. and Hedger, J.N. (2003). Effects of organic enrichment of mine spoil on growth and nutrient uptake in oak seedlings inoculated with selected ectomycorrhizal fungi. Restor. Ecol., 11: 125–130. 10.1046/j.1526-100X.2003.09968.x.

Magnago, A.C., Trierveiler-Pereira, L. and Neves, M.A. (2013). *Phallales* (Agaricomycetes, Fungi) from the tropical Atlantic Forest of Brazil. J. Torrey Bot. Soc., 140: 236–244. 10.3159/TORREY-D-12-00054.1.

Malajczuk, N., Reddell, P. and Brundrett, M. (1994). Role of ectomycorrhizal fungi in minesite reclama tion. pp 83–100. *In*: Pflfleger, F.L. and Linderman, R.G. (eds.). Mycorrhizae and Plant Health. The American Phytopathological Society Symposium Series. APS Press, Minnesota.

Manimohan, P. (2019). *Auroculoscypha*: A fascinating instance of fungus-insect-plant interactions. pp. 34–39. *In*: Sridhar, K.R. and Deshmukh, S.K. (eds.). Advances in Macrofungi - Diversity, Ecology and Biotechnology. CRC Press, Boca Raton.

Mueller, G.M., Schmit, J.P., Leacock, P.R., Buyck, B., Cifuentes, J. et al. (2007) Global diversity and distribution of macrofungi. Biodivers. Conserv., 16: 37–48.

Münzenberger, B., Golldack, J., Ullrich, A., Schmincke, B. and Hüttl, R.F. (2004). Abundance, diversity, and vitality of mycorrhizae of Scots pine (*Pinus sylvestris* L.) in lignite recultivation sites. Mycorrhiza, 14: 193–202. 10.1007/s00572-003-0257-2.

Natarajan, K., Senthilarasu, G., Kumaresan, V. and Riviere, T. (2005). Diversity in ectomycorrhizal fungi of a dipterocarp forest in Western Ghats. Curr. Sci., 88: 1893–1895.

Netherway, T., Bengtsson, J., Krab, E.J. and Bahram, M. (2021). Biotic interactions with mycorrhizal systems as extended nutrient acquisition strategies shaping forest soil communities and functions. Basic Appl. Ecol., 50: 25–42. 10.1016/j.baae.2020.10.002.

Niego, A.G.T., Rapior, S., Thongklang, N., Raspe, O., Hyde, K.D. and Mortimer, P. (2023). Reviewing the contributions of macrofungi to forest ecosystem processes and services. Fungal Biol. Rev., 44: 100294. c0.1016/j.fbr.2022.11.002.

Nobre, T. and Aanen, D.K. (2012). Fungiculture or termite husbandry? The ruminant hypothesis. Insects, 3: 307–323. 10.3390/insects3010307.

Nobre, T., Eggleton, P. and Aanent, D.K. (2010). Vertical transmission as the key to the colonization of Madagascar by fungus-growing termites? Proc. Royal Soc. B - Biol. Sci., 277: 359–365.

Nobre, T., Fernandes, C., Boomsma, J.J., Korb, J. and Aanen, D.K. (2011). Farming termites determine the genetic population structure of *Termitomyces* fungal symbionts. Mol. Ecol., 20: 2023–2033. 10.1111/j.1365-294X.2011.05064.x.

Old, K.M., See, L.S., Sharma, J.K. and Yuan, Z.Q. (2000). A manual of diseases of tropical acacias in Australia, South-East Asia and India. Center for International Forestry Research, Bogor, Indonesia, p. 104. 10.17528/cifor/000639.

Oliveira, M.L. and Morato, E.F. (2000). Stingless bees (Hymenoptera, Meliponini) feeding on stinkhorn spores (Fungi, Phallales): Robbery or dispersal? Rev. Bras. Zool., 17: 881–884. 10.1590/S0101-81752000000300025.

Parraga-Aguado, I., Querejeta, J.-I., González-Alcaraz, M.-N., Jiménez-Cárceles, F.J. and Conesa, H.M. (2014). Usefulness of pioneer vegetation for the phytomanagement of metal(loid)s enriched tailings: Grasses vs. shrubs vs. trees. J. Environ. Manage., 133: 51–58. 10.1016/j.jenvman.2013.12.001.

Piętka, J. and Grzywacz, A. (2018). Macrofungi found on black alder *Alnus glutinosa* (L.) Gaertn. in alder stands showing signs of a dieback. Sylwan, 162: 22–31.

Rahi, D.K., Rahi, S. and Chaudhary, E. (2021). Enzymes of white rot fungi and their industrial potential. pp. 198–220. *In*: Sridhar, K.R. and Deshmukh, S.K. (eds.). Advances in Macrofungi—Industrial Avenues and Prospects. CRC Press, Boca Raton.

Read, D.J. and Perez-Moreno, J. (2003). Mycorrhizas and nutrient cycling in ecosystems—A journey towards relevance? New Phytol., 157: 475–492.

Richardson, M.J. (2003). Coprophilous fungi. Field Mycol., 4: 41–43. 10.1016/S1468-1641(10)60185-5.

Rudolf, K., Morschhauser, T. and Pál-Fám, F. (2012). Macrofungal diversity in disturbed vegetation types in North-East Hungary. Cent. Eur. J. Biol., 7: 634–647.

Shrestha, B., Zhang, W., Zhang, Y. and Liu, X. (2012). The medicinal fungus *Cordyceps militaris*: Research and development. Mycol. Prog., 11: 599–614. 10.1007/s11557-012-0825-y.

Slippers, B., Coutinho, T., Wingfield, B. and Wingfield, M. (2003). A review of the genus *Amylostereum* and its association with woodwasps. South Afr. J. Sci., 99: 70–74.

Sridhar, K.R. and Karun, N.C. (2019). Diversity and ecology of ectomycorrhizal fungi in the Western Ghats. pp. 479–507. *In*: Singh, D.P., Gupta, V.K. and Prabha, R. (eds.). Microbial Interventions in Agriculture and Environment, Volume 1, Research Trends, Priorities and Prospects. Springer Nature Singapore Pte Ltd., Singapore.

Staudenrauschm, S., Kaldorfm, M., Renker, C., Luis, P. and Buscot, F. (2005). Diversity of the ectomycorrhiza community at a uranium mining heap. Biol. Fert. Soils, 41: 439–446. 10.1007/s00374-005-0849-4.

Stone, R. (2008). Mycology: Last stand for the body Snatcher of the Himalayas? Science, 322: 1182–1182. 10.1126/science.322.5905.1182.

Sung, G.-H., Hywel-Jones, N.L., Sung, J.-M., Luangsa-ard, J.J., Shrestha, B. and Spatafora, J.W. (2007). Phylogenetic classification of *Cordyceps* and the clavicipitaceous fungi. Stud. Mycol., 57: 5–59. 10.3114/sim.2007.57.01.

Suryadi, H., Judono, J.J., Putri, M.R., Eclessia, A.D., Ulhaq, J.M. et al. (2022). Biodelignification of lignocellulose using ligninolytic enzymes from white-rot fungi. Heliyon, 8e08865. 10.1016/j.heliyon.2022.e08865.

Tabata, M. and Abe, Y. (1999). *Amylostereum laevigatum* associated with a horntail *Urocerus antennatus*. Mycoscience, 40: 535–539. 10.1007/BF02461032.

Tang, X., Mi, F., Zhang, Y., He, X. and Cao, Y. (2015). Diversity, population genetics, and evolution of macrofungi associated with animals. Mycology, 6: 94–109.

Thomas, P.W. and Thomas, H.W. (2022). Mycorrhizal fungi and invertebrates: Impacts on *Tuber melanosporum* ascospore dispersal and lifecycle by isopod mycophagy. Food Webs, 33: e00260. 10.1016/j.fooweb.2022.e00260.

Thomsen, I.M. and Harding, S. (2011). Fungal symbionts of siricid woodwasps: Isolation techniques and identification. For. Pathol., 41: 325–333. 10.1111/j.1439-0329.2010.00677.x.

Trierveiler-pereira, L., Santos, P.J.P. and Baseia, I.G. (2013). Ecological aspects of epigeous gasteromycetes (Agaricomycetes, Basidiomycota) in four remnants of the Brazilian Atlantic Forest. Fungal Ecol., 6: 471–478.

Tuno, N. (1998). Spore dispersal of *Dictyophora* fungi (Phallaceae) by flies. Ecol. Res., 13: 7–15.

Vaario, L. and Matsushita, N. (2021). Conservation of edible ectomycorrhizal mushrooms: Understanding of the ECM fungi mediated carbon and nitrogen movement within forest ecosystems. *In*: Ohyama, T. and Inubushi, K. (eds.). Nitrogen in Agriculture - Physiological, Agricultural and Ecological Aspects IntechOpen. 10.5772/intechopen.95399.

van der Heijden, M.G.A., Martin, F.M., Solosse, M.-A. and Samnders, I.R. (2015). Mycorrhizal ecology and evolution: The past, the present, and the future. New Phytol., 205: 1406–1423. 10.1111/nph.13288.

Vašutová, M., Mleczko, P., López-García, A., Maček, I., Boros, G. et al. (2019). Taxi drivers: The role of animals in transporting mycorrhizal fungi. Mycorrhiza, 29: 413–434.

Visser, A.A., Kooihj, P.W., Debets, A.J.M., Kuyper, T.W. and Aanen, D.K. (2011). *Pseudoxylaria* as stowaway of the fungus-growing termite nest: Interaction asymmetry between *Pseudoxylaria*, *Termitomyces* and free-living relatives. Fungal Ecol., 4: 322–332.

Wang, P.F., Juan, H.E., Zhou, W., Li, B., Wu, P. and Li, Z.J. (2012). A survey on the studies of *Termitomyces*. Microbiol. China, 39: 1487–1498.

Yang, Y., Xiao, Y., Yu, G., Wen, T., Deng, C. et al. (2021). Ophiocor*dyceps aphrophoridarum* sp. nov., a new entomopathogenic species from Guizhou, China. Biodivers. Data J., 9e66115. 10.3897/BDJ.9.e66115.

Ye, L., Li, H., Mortimer, P.E., Xu, J., Gui, H. et al. (2019). Substrate preference determines macrofungal biogeography in the Greater Mekong Sub-Region. Forests, 10. 824. 10.3390/f10100824.

Zotti, M., Persiani, A.M., Ambrosio, E., Mazzini, A. and Venturella, G. (2013). Macrofungi as ecosystem resources: Conservation versus exploitation. Pl. Biosyst., 147: 1–7. 10.1080/11263504.2012.753133.

Forest Macrofungi

2

The Roles of Macrofungi in Humid Tropical Forests and the Effects of Disturbance

D Jean Lodge[1] and *Sharon A Cantrell*[2,*]

1. Introduction

Macrofungi play important roles in humid tropical forests, including their prominence in nutrient cycling and decomposition, acting as beneficial mycorrhizal symbionts by providing nutrients to host plants, and parasitizing or killing plants and invertebrates. Macrofungi are defined as those that produce fruiting bodies visible to the unaided eye, but we include examples of agaric fungi (Agaricales, commonly called mushrooms) that have largely or entirely replaced production of mushroom fruiting bodies with highly visible root-like structures (rhizomorphs and cords) or nonsporulating macroscopic structures used for dispersal by animals. An important characteristic of most terrestrial fungi is the production of linear structures such as hyphal strands, cords and rhizomorphs that can translocate nutrients between sources and sinks. The ability of macrofungi to concentrate nutrients in decaying substrata, nodules and fruit bodies makes them nutritionally attractive to both vertebrate (mostly small mammals but also turtles) and invertebrate consumers if they have the enzymes to break down chitin or dissolve calcium from calcium oxalate (Lodge, 1996; Guevara and Dirzo, 1999; Vasco-Palacios, 2008; Trappe et al., 2013; Trierveiler-Pereira et al., 2016). This includes mushroom-farming ants in the Neotropics and termites in the Paleotropics (Cherrett, 2012; Hanki, 2012; Wood and Thomas, 2012).

[1] University of Georgia, Odum School of Ecology and Dept. of Plant Pathology, Athens, GA 30606, USA.
[2] Department of Biology, Universidad Ana G Méndez – Gurabo, PO Box 3030, Gurabo, Puerto Rico 00778.
* Corresponding author: scantrel@uagm.edu

While our emphasis is on the roles of macrofungal linkages in humid tropical forests, we first compare macrofungal roles in temperate and tropical forests. One of the most frequent questions posed by first-time visitors from temperate zones to humid tropical forests is why there are so few large fleshy fungi (mushrooms). This is partly attributable the rarity of woody ectomycorrhizal host plants that subsidize their fungal root symbionts with sugar (up to 30% of net primary production), and partly to the faster decay of leaf litter and thus lower litter standing stocks on the tropical forest floor.

Mycorrhizal fungi facilitate uptake of growth-limiting nutrients and water to their host plants. Although mycorrhizal associations are generally similar in tropical and higher latitude forest, the proportion of mycorrhizal types differs between low and high latitudes (Dickie and Moyersoen, 2008; Tedersoo et al., 2010, 2014). Bennet et al. (2017) showed in their study of 520 populations of 52 temperate North American tree species that trees in wet environments can have strongly negative plant soil feedbacks on seedlings from conspecific mature trees if they are not protected from root pathogens by their mycorrhizal fungi, as is typical in arbuscular mycorrhizal trees. In contrast, ectomycorrhizal fungi form fungal mantles that sheath the fine root tips of their host, protecting them from root pathogens and producing positive plant soil feedbacks from conspecific trees (Laliberte et al., 2015; Bennett and Klironomos, 2018). These patterns are consistent with the Jansen-Connell hypothesis (reviewed by Laliberte et al., 2015) that the high diversity of tropical rainforest trees may in part be maintained by specialist seed and seedling pathogens and seed predators that obtain their highest densities close to the parent tree. It is therefore not surprising that high-diversity wet tropical forests are typically dominated by tree species that form arbuscular rather than ectomycorrhizal associations.

While forest stands dominated by a single ectotrophic tree species or related species in the same family or order are more common in temperate and boreal forests than in the tropics, monodominant ectotrophic forests do occur in the tropics. Some examples of monodominant ectotrophic tropical forests having high diversity of ectomycorrhizal fungi are Dipterocarpaceae forests of SE Asia (Peay et al., 2010), Eucalyptus forest type in Australia (mostly hypogeous fruiting bodies, Lebel and Castellano, 2002; Trappe et al., 2013), *Gymnopodium floribundum* (Polygonaceae) stands in lowland deciduous forest in Chiapas, Mexico (Bandala and Montoya, 2015), *Coccoloba* (Polygonaceae) stands in Central America and the Caribbean (Séne et al., 2010; Delgat et al., 2020) and leguminous *Dicymbe* stands in Guyana in N South America (Henkel et al., 2002, 2005; Smith et al., 2011) and in Cameroon in Africa (Michaëlla Ebenye et al., 2017). While trees in monodominant tropical forests may facilitate establishment of conspecific seedlings (McGuire, 2007), it is not clear to what extent these effects are due to protection of roots from pathogens, transfer of nutrients from parent trees via fungal networks, or ability of ectomycorrhizal fungi to access nutrients in decomposing litter in ways not available to arbuscular mycorrhizal fungi (see discussion in Newberry and Neba, 2019).

2. Nutrient Transfers by Mycorrhizal Fungi

2.1 *Tropical Ectomycorrhizal Fungi and Direct Nutrient Cycling from Litter*

One hypothesis on maintenance of ectotrophic forests, known as the Gadgil hypothesis (Gadgil and Gadgil, 1971), is that the accumulation of litter is due to extraction of limiting nutrients directly from litter by ectomycorrhizal fungi which inhibits decay by decomposer fungi. The Gadgil and Gadgil (1971) study was conducted in a *Pinus radiata* plantation in New Zealand. In contrast, subsequent studies in northeastern USA in Pitch pine forest (Zhu and Ehrenfeld, 1996) and in hardwood forests with a range of dominance by ectotrophic and arbuscular mycorrhizal trees (Lang et al., 2020), found that mycorrhizae stimulated rather than inhibited litter decomposition. Went and Stark (1968), working in an ectotrophic forest growing on extremely nutrient depauperate white sands in Amazonian Venezuela inferred that slow leaf decomposition there was related to direct nutrient cycling by ectomycorrhizal fungi which outcompeted decomposer fungi for limiting nutrients. Stark and Jordan (1978) subsequently applied radioactive isotopes of Ca and P to litter at the same ectotrophic forest on white sands and found almost none of it was leached into the zero-tension lysimeters below but was instead incorporated into the mycorrhizal root mat. The Stark and Jordan (1978) results showed that nutrient cycling was tight and that ectomycorrhizal fungi in the litter mat were good at scavenging the applied mineral nutrients, but it did not demonstrate direct nutrient cycling in the sense of partial breakdown of litter to obtain mineral nutrients in organic forms. Tight nutrient cycling likely occurs wherever growth-limiting mineral nutrients are in short supply, regardless of mycorrhizal types. For example, dilute P solutions added to intact arbuscular mycorrhizal root-litter mats in wet tropical forest in Puerto Rico showed that the mass of P escaping the mat and captured in the zero-tension lysimeter below after 4 days was inversely proportional to the hyphal lengths of arbuscular mycorrhizal fungi (Pearson correlation -0.57; $R^2 = 0.33$) but not root length (Sanchez and Lodge, 2017). At the same site in Puerto Rico (the El Verde Research Area), Luce (1970) and Stark (1971) showed using applied radioactive tracers that humus was the main source of nutrients for surface roots but there was no evidence for direct nutrient cycling. Virtually all of the applied ^{32}P was retained in the litter root mats (Luce, 1970).

Production of exo-enzymes by some ectomycorrhizal fungi may facilitate access to N and P in organic compounds (Bending and Read, 1995), and some ectomycorrhizal fungi were found to break down organic matter (Trojanowski et al., 1984; Haselwandter et al., 1990; Read et al., 1989). The ability of mycorrhizal fungi to obtain N through breakdown of organic N-compounds, such as amino acids, has been shown for ectomycorrhizal agaric fungi (*Hebeloma* spp. and *Laccaria bicolor*) and an ericoid mycorrhizal cup fungus (*Hymenoscyphus ericae*) that forms mycorrhizae with *Vaccinum* (Bajwa and Read, 1986; Nuutinen and Timonen, 2008). The Gadgil hypothesis is therefore plausible but still controversial. Mayor and Henkel (2005) found that access to litter decomposition bags by *Dicymbe* ectomycorrhizae in Guyana depleted Ca, but there was no effect of mycorrhizae on litter mass loss and thus no effect on decomposers.

2.2 *Ectomycorrhizal Fungal Net Transfer of Carbon from Adults to Seedlings*

Whether seedlings growing below conspecific ectomycorrhizal trees are sustained by net carbon transfers via the fungal net is controversial (Newberry and Neba, 2019). Although seedlings in monodominant ectotrophic forests in Cameroon in Africa (Michaëlla Ebenye et al., 2017) and Guyana in South America (McGuire, 2007) were shown to share mycorrhizae with mature trees, the results of treatments can be difficult to interpret because the bags with three different mesh sizes used to exclude mycorrhizal fungal hyphae and/or roots are confounded by other effects, such as root competition and exclusion of pathogenic fungi (Newberry and Neba, 2019). Newberry and Neba (2019) found in their Cameroon study that seedling stem dry mass was significantly greater in medium-mesh bags and inferred that ectomycorrhizal fungi increased growth by 13.6% but root competition lowered growth by 31.2%, with an overall negative effect of access to mycorrhizae of 17.6%. In contrast, McGuire (2007) found in Guyana that contact by *Dicymbe* seedlings with ectomycorrhizal fungal networks increased their growth, but the mechanism was not determined.

Using radioactive tracers in a north temperate forest in the Pacific NW USA, Simard et al. (1997) demonstrated a net transfer of carbon from *Betula papyrifera* seedlings to shaded *Pseudotsuga menziesii* seedlings in the field via a shared ectomycorrhizal fungus. Subsequent work by Simard and others showed nutrient transfers from adult to seedling trees (Simard and Durall, 2004; Simard et al., 2012), but it remains uncertain as to how significant these transfers are (Selosse et al., 2006; Newberry and Neba, 2019). Hoeksema (2015) reviewed the difficulties of distinguishing among alternative hypotheses regarding testing of direct nutrient transfers between individuals of ectomycorrhizal trees in the field. Wu et al. (2001) found some cases where the transferred nutrients ended up in the fungal mantle of the recipient and not in the plant itself. Wilkinson (1998) argued against significant transfers of carbon to seedlings of another species at the cost of its own fitness. Newberry and Neba (2019) suggested that losses of carbon through transfers to seedlings of other species is probably minor in comparison to the benefits of supplying carbon to conspecific seedlings.

3. Ectomycorrhizal Fungi used as Epiparasites by Achlorophyllous Plants

A curious relationship has been unraveled in both temperate and tropical forests in which plants lacking chlorophyll extract carbon from ectomycorrhizal fungi, which in turn receive their carbon from photosynthetic trees (Bitartondo et al., 2004; Selosse et al., 2004, 2021). The achlorophyllus plants in these three-part associations are known as epiparasitic mycoheterotrophs (Merckx, 2013; Selosse, 2021). Mycoheterotrophy has evolved more than 26 times in the Orchidaceae (Merckx, 2013).

3.1 *Orchids Parasitizing Saprotrphic Macromycetes*

Some mycoheterotrophic orchids parasitize macromycete fungi that are putatively saprotrophic rather than ectomycorrhizal. *Mycena* species have been reported forming orchid mycorrhizae (OM) by Fan et al. (1996), Guo et al. (1997), Selosse et al. (2010) and Zhang et al. (2012). Saprotrophic *Coprinus*, *Mycena* and *Psathyrella* spp. (Selosse et al., 2010, 2020; Lee et al., 2015) and assorted species in the Marasmiaceae (Selosse et al., 2010; Dearnaley and Bougoure, 2010) have been repeatedly reported to form OM.

Selosse et al. (2021) addressed the multiple phylogenetic lineages of fungi that independently formed mycorrhizae with orchids (Martos et al., 2009) and proposed that their ancestors may have been root endophytes. There is support for that hypothesis in the agaric genus, *Mycena*, which is known to decompose leaf litter and wood. Frankland (1998) confirmed that *Mycena galopus*, a temperate forest litter decomposer, was able to degrade all substrates. Later, Grelet et al. (2017) showed that *M. galopus* was endophytic with *Vaccinum* roots and stimulated plant growth, while Thoen et al. (2020) showed that *Mycena pura* and *M. galopus* formed mantles around root tips of *Betula pendula* and stimulated seedling growth. Litter decomposer *Mycena* species were found associated with the mycoheterotrophic orchids *Wullschlaegelia aphylla* in Guadeloupe (Martos et al., 2009) and *W. calocera* on Puerto Rico (DJL, pers. obs.; Fig. 1A, B) in the Caribbean, and fungal clamp connections were observed in the OM of the latter confirming they belonged to a basidiomycete (Fig. 1C).

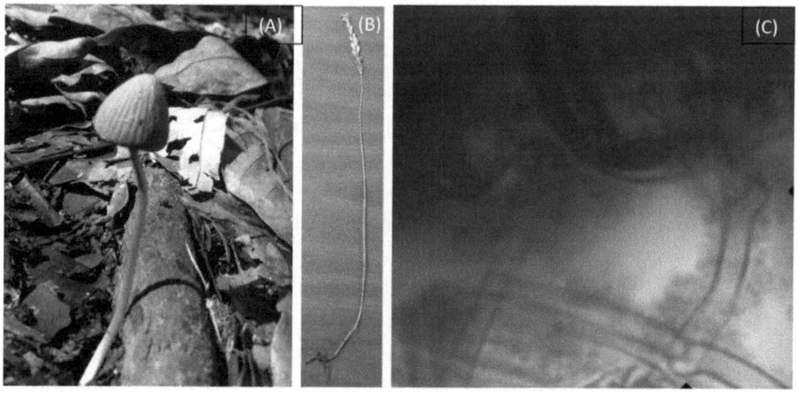

Fig. 1. *Mycena* sp. basidiome (A) associated with achlorophyllous, mycoheterotrophic orchid, *Wulschlegelia calcrata* (B), Microphoto of the orchid mycorrhiza of *W. calcrata* showing two clamp connections of the basidiomycete fungus (C) (Photos by D.J. Lodge).

3.2 *Mixotrophy: Photosynthetic Plants that Parasitize Macrofungi for Carbon*

Photosynthetic (green) orchid plants and some plants in the Ericaceae obtain N, P and C from their mycorrhizal fungal symbionts (Zimmer et al., 2007; Selosse, 2017). Orchids have minute seeds that depend on mycorrhizal fungi for all their nutrition to be able to germinate and grow as seedlings. As the orchids grow and become photosynthetic, they augment C extracted from their mycorrhizal fungi with C from photosynthesis, and are called mixotrophic (Julou et al., 2005; Selosse and Roy, 2009; Dearnaley et al., 2017; Jacquemyn and Merckx, 2019). Mixotrophic orchids are found in both temperate and tropical forests, though there are more epiphytic orchids in the tropics and more terrestrial orchids in the temperate zones. Most OM fungi are asexual states of macromycete lineages, such as the non-fruiting polyphyletic rhizoctonia fungi, but a growing number of mixotrophic orchids such as *Anoctochilus*, *Cymbidium* and *Dendrobium* form mycorrhizae with macromycete fungi that produce fruiting bodies, such as *Mycena* (Fan et al., 1996; Guo et al., 1997; Ogura-Tsujita et al., 2009).

4. Nutrient Transfers by Decomposer Fungi

Decomposer fungi in both temperate and tropical forest use rootlike structures (rhizomorphs, cords and hyphal strands) to colonize new resource bases and translocate nutrients between food bases (Fig. 2A–F). The ability of macrofungi to move mineral nutrients allows them to build biomass rapidly in fresh nutrient-poor substrata and compete with other fungi for resources. Importation of nutrients accelerates both macrofungal growth and the rate of decomposition in forests.

4.1 Wood Decomposer Macrofungi

Researchers in the UK were first to recognize the importance of mineral nutrient translocation by macrofungi to wood decomposition and ecosystem processes (Frankland, 1982; Swift, 1982; Rayner and Todd, 1982). Subsequent work in the UK and Scandinavia confirmed the importance of nutrient movements through fungal cords and rhizomorphs (Fig. 2B, C) for the accumulation of mineral nutrients in nutrient-poor woody substrata, rates of wood decomposition and interactions between competing mycelia (Beare et al., 1992; Boddy, 1993, 1999; Boddy and Watkinson, 1995; Frey et al., 2000; Cairney, 2005; Lindahl et al., 1999, 2001).

4.1.1 Nutrient Translocation by Decomposer Macrofungi

Although wood represents a considerable energy resource in carbon compounds, wood is a patchy resource that is widely distributed on the forest floor and it has concentrations of N and P that are too low to support rapid growth of decomposers, so translocation of nutrients from other parts of the mycelium and substrata is essential (Lindahl et al., 2002). Although fruiting bodies of wood decomposer macrofungi produce airborne spores that can colonize wood at a distance (Frankland, 1998), the dominant means of dispersal, colonization and competition for fresh substrata is via a 'ground war' mediated by fungal conducting elements (Boddy, 1993; Lindahl et al., 1999).

Wood decomposer fungi can capture nearly all the N and 90% of the P in their substrate (Swift 1977; Frankland, 1982). The efficiency of fungal nutrient capture is facilitated by autodigestion of their dead mycelia (Jennings, 1982; Watkinson, 1984).

Fig. 2. Nutrient-conducting structures of macromycete fungi: Hyphal strands and cords of wood-decomposer fungi (A,B), Fungal cords (B) are used primarily to colonize new pieces of wood and move nutrients from old to new food bases, but hyphal strands (A) also scavenge nutrients from the soil: *Phanerochete flava* (A), *Laternea triscapa* (B) in the stinkhorn group (Phallales). Agaric fungi are important in binding leaf litter into mats that resist erosion (C-D), Bleaching of the leaves (C-F) results from delignification by agaric fungi. Diffuse fungal mycelium and hyphal strands (C,D) connecting leaf litter in a leaf litter mat formed by agaric fungi in the genus *Gymnopus*. Agaric fungi also trap litter in the understory of humid tropical forests (E-F) and are often plant pathogens: Abundant shiny rhizomorphs of the horse-hair mushroom (*Marasmius crinis-equi*), that has trapped, killed and decomposed leaves of an understory shrub (E), an undescribed agaric in the genus *Amyloflagellula* in Puerto Rico. Saprotrophic state and early stage of leaf colonization via cords and hyphal strands with developing necrotrophic stage and reversion to saprotophy at white arrows indicate fungal disease fronts in live plant tissue (F) (Photos by D.J. Lodge, except D by S.A. Cantrell).

Wells et al. (1999) showed using P^{32} that P was preferentially translocated to high carbon substrates. Watkinson (1984) showed that *Serpula lacrimans* extracted 75% of the N from one piece of wood and moved it across an inert surface to a fresh piece of wood. Nitrogen is translocated in the form of amino acids, as demonstrated using radioactively labeled N (Watkinson et al., 2006; Tlalka et al., 2007). Video imaging using a scintillation counter showed that some wood decomposer macromycetes have a bi-directional, non-circadian oscillation of N in amino acids in relation to the center of the colony, but not all basidiomycete wood decomposers that translocate nutrients via fungal cords show oscillation (Tlalka et al., 2002, 2003, 2007, 2008; Bebber et al., 2007).

Macrofungi may translocate nutrients to a fresh, energy-rich substratum where competition for these limiting resources may be less (Ettema and Wardle, 2002), or tip the outcome of competition for a new substratum to the mycelium that reaches it first and colonizes the greatest surface area or volume (Boddy, 1993). Wood decomposer macrofungi in wet tropical forest in Puerto Rico were found to have species-specific preferences for a particular range of diameter classes (Lodge, 1996). Similarly, Boddy (1999) and Boddy and Jones (2007) showed that wood decay fungi in the UK had specific preferences for large or small pieces of wood. Cord-forming wood decomposers allocated nutrients based on the sizes of a potential new food bases and whether they were already occupied (Boddy, 1993; Tlalka et al., 2008).

Translocation of N by cord-forming wood decomposer fungi from a source with high N concentration to a sink with low N but high carbon has been found to increase accumulation of fungal biomass in the sink, and increase rates of decomposition (Beare et al., 1992; Frey et al., 2000; Tlalka, 2008). Mycelial fans of cord-forming wood decomposer absorb mineral nutrients from soil when available (Fig. 2B.; Tlalka, 2008). In a wet tropical forest on Puerto Rico, abundant woody debris from hurricane Hugo damage stimulated growth of cord-forming stinkhorn fungi (*Laternea triscapa* and *Dictyophora indusiata*), and several species of *Phanerochaete* (Fig. 2A, B); Lodge, 1996), which coincided with disappearance of nitrate from soil solution (Zimmerman et al., 2005). Plots where woody debris was removed had more available N in soil and more rapid canopy closure than plots where woody debris left in place, suggesting that the wood decomposers outcompeted trees for limiting N when presented with overwhelming quantities carbon in wood with low N:C ratios (Zimmerman et al., 2005).

4.1.2 Wood Decomposer Macrofungi that Kill Nematodes to Obtain N

Several lineages of wood decomposer macrofungi have adapted to the low C:N ratios in wood and have evolved nematode trapping as means to augment their supply of N (Thorn and Barron, 1984). *Hohenbuehelia*, *Pleurotus* and *Resupinatus* in the basidiomycete agaric family Pleurotaceae use toxic exudates on sticky knobs to trap and immobilize nematodes (Thorn and Barron, 1984; Thorn et al., 2000, 2005). A group of ascomycete cup fungi in the Orbiliaceae independently evolved nematode trapping sticky nets. A study of *Orbilia oligospora* indicated that the fungus exudes compounds that mimic sex hormones in nematodes thereby attracting them to their death in the toxic sticky nets (Chen et al., 2013). The fungus goes through a cycle of autolysis prior to producing the sticky nematode trapping nets in response to the presence of nematodes (Chen et al., 2013).

4.2 Leaf Decomposer Macrofungi on the Forest Floor

Agaric fungi are keystone decomposers in humid tropical forests. They preferentially decompose high-lignin, low-nutrient leaf litter that is characteristic of many humid tropical forests growing on highly weathered soils. These agaric macrofungi mimic wood decomposer fungi in their ability to move nutrients between resource bases via hyphal strands and rhizomorphs, which gives them a competitive advantage in nutrient-poor leaf litter over other decomposers such as microfungi and bacteria. Furthermore, leaf decomposer agaric fungi accelerate litter decay by breaking down lignin and related compounds that protect labile C substrates. The loss of lignin-like compounds results in bleaching of the litter and is referred to as white-rot (Fig. 2C-F). In addition, the root-like structures that agaric fungi use to colonize new litter and translocate nutrients also bind litter together to form surface litter mats that reduce soil erosion (Fig. 2C, D).

4.2.1 N and P Availability in Leaf Litter in Relation to Agaric Litter Decomposers

Ratios of carbon to mineral nutrients in leaf litter influence whether agaric fungi colonize litter early, late, or in all stages of decomposition across all forest ecosystems (Osono, 2020; Lodge et al., 2022; Osono et al., 2021). Low N concentrations in litter are related to reduced rates of leaf decomposition in higher latitude coniferous and broadleaved forests (Berg and Staff, 1980; Mellilo et al., 1982). Among broadleaved trees, P but not N concentrations in foliage decrease strongly with increasing temperature and latitude which is attributable to greater soil weathering in the tropics (Reich and Oleksyn, 2004). For example, concentrations of P but not N in freshly fallen *Quercus rubra* leaves in Lithuania were considerably higher (1.5 mg g^{-1} P and 11.2 mg g^{-1} N) than in *Quercus mongolica* leaf litter in Korea (0.23 mg g^{-1} P and 10.0 mg g^{-1} N; Mun, 2009; Junkšienė et al., 2017). There are also regional differences in foliar nutrient concentrations attributable to soil parent material type and age (Reich and Oleksyn, 2004). For example, in the tropical rainforest at El Verde, on Puerto Rico which has relatively younger volcanic soil, P concentrations in fresh leaf litterfall is 3-4 times higher than in Dipterocarp forest in SW China growing on older, highly weathered soil. In Puerto Rico, P concentrations in freshly fallen leaves of the dominant tree on ridges, *Dacryodes excelsa*, were 0.9–1.1 mg g^{-1} (Lodge et al., 2014) and mixed freshly fallen leaves on ridges was 1.3 mg g^{-1} (Zou et al., 2015). In contrast, freshly fallen leaves of *Dipterocarpus retusus* in SW China had 0.3 mg g^{-1} P in freshly fallen leaves (Paudel et al., 2015). Further, Dent and Burslem (2017) found differences based on soil parent material in the same region, with higher N and P concentrations in leaf litter from Dipterocarp saplings growing on fertile alluvial soil than on nutrient-depauperate sandstone soil in SE Asia (16.3 vs 11.7 mg g^{-1} N, and 1.0 vs 0.5 mg g^{-1} P). Low concentrations P in leaf litter are typical of lowland tropical forests while both N and P are typically low in forests growing at high elevations on tropical mountains (Vitousek, 1982; Tanner et al., 1998; Cleveland et al., 2006; Vincent et al., 2010).

4.2.2 Effects of Sclerophylly and High Lignin Content

Leaf litter from trees growing on nutrient-depauperate soils, in addition to having lower concentrations of limiting nutrients, also tend to be sclerophyllous (Medina et al., 1990; Read et al., 2007) with higher quantities of lignin and related phenolic compounds. Lignin-like compounds have been reported to protect more labile C in cellulose and hemicellulose which contributes further to slowing decomposition (Meetemeyer, 1978; Melillo et al., 1982; Devi and Yadava, 2007). As noted in the review by Rahman et al. (2013), however, high lignin content in leaf litter is not always associated with slower decomposition, and the relationship may differ depending on the presence of N and the decomposer organisms involved. For example, a study in forest plantations in Costa Rica found increasing rates of leaf litter turnover with higher lignin concentrations (Raich et al., 2007). Slow decay may occur in leaves that have both high lignin and N concentrations, as N has been shown to inhibit production of ligninolytic agaric fungal enzymes (Carreiro et al., 2000; Waldrop et al., 2004). Raich et al. (2007), Anderson and Swift (1983) and Lavelle et al. (1993) have suggested that some decomposer organisms in tropical rain forests may prefer lignin and override the effects of leaf chemistry. Agaric litter decomposers that transfer limiting mineral nutrients from decomposed litter via rhizomorphs and hyphal strands to build biomass in freshly fallen leaves that are deficient in P fill this role.

4.2.3 Agaric Litter Mat Fungi Concentrate and Conserve Nutrients

The concentration and translocation of limiting mineral elements such as P in humid tropical forests by agaric litter mat fungi helps to conserve them against losses via leaching (Stark, 1972; Lodge et al., 1994). Lodge et al. (2014) showed a significant increase in P content of a layer of freshly fallen *Dacryodes excelsa* leaves placed between screens of litterbaskets in control plots that had undisturbed canopies at El Verde in Puerto Rico. Fungal connectivity between the forest floor and the litter layers above were predictive of P accumulation (Lodge et al., 2014). Disturbances that open the forest canopy reduced fungal litter mat cover (Lodge et al., 2022) and was associated with greater nutrient fluxes from litter to soil (Moreno et al., 2022; see Effects of Disturbances on Tropical Forest Fungi). Nutrients accumulated via recycling by agaric litter mat fungi is often released in pulses following drought and rewetting (Lodge et al., 1994). These pulsed nutrient releases are important for nutrient cycling and forest productivity because nutrient dynamics are non-linear (Lodge et al., 1994). Pulsed releases of P in tropical forests that are P-limited results in a greater proportion becoming available for root uptake by swamping soil P-fixation sites and the capacity of the soil microbial biomass to immobilize P (Lodge et al., 1994).

4.2.4 White-rot from Agaric Litter Decomposer Fungi Increases Decay Rates

In addition to using translocated nutrients to build biomass rapidly in fresh litterfall (Lodge and Asbury, 1988; Lodge et al., 2008, 2014), agaric macrofungi produce white-rot (Fig. 2C-F) which contributes to accelerated rates of leaf decomposition (Fig. 3). White-rot fungi obtain energy from labile carbon (sugars and cellulose) which fuels the break down resistant lignin-like compounds called acid-unhydrolyzable residues (AUR) comprising lignin, lignocellulose, and other aromatic or polyphenolic substances such as cutin, tannin and humic acids (Lodge et al., 2008, 2014, 2022; Osono et al., 2021). Agaric fungi are important secondary decomposers of leaf litter

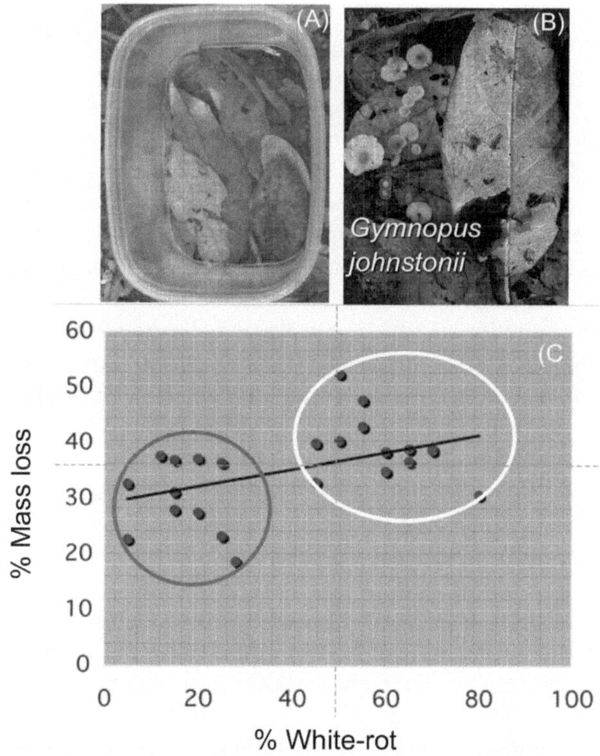

Fig. 3. One-mm mesh litterbags containing 10 g freshly fallen leaves of *Dacryodes excelsa* were placed on litter colonized by microfungi (A) or on white-rot leaf decomposer mushroom litter mats of *Gymnopus johnstonii* (B). White-rot mycelia were deterred from colonizing non-white-rot litter (A) or restricted from migrating out of the white-rot treatment using plastic containers with the bottoms removed. Nevertheless, some white-rot mycelia invaded the non-white-rot treatments over 3 months, so the slope of the regression of % mass loss against % of leaf area with white rot of +16.3% was used as an estimate of the effect of white-rot on mass loss © (Photos and data by D.J. Lodge).

in boreal forests (Hintikka, 1970; Berg and Staff, 1980; Lindahl and Boberg, 2008) and temperate forests (Frankland et al., 1995; Frankland, 1998; Osono, 2006; Osono and Takeda, 2001; Klotzbücher et al., 2011; Schneider et al., 2012) where microfungi utilize labile C first leading to accumulation of lignin-protected C in the humus layer. In humid and semi-arid tropical and subtropical forests, agaric fungi play roles of both primary and secondary decomposers, depending on the fungal genus and species (Hedger, 1985; Lodge and Asbury, 1988; Torres et al., 2005; Osono, 2006, 2020; Osono et al., 2021).

Agaric fungi that produce white-rot have been shown to increase rates of mass loss in leaf litter in boreal, temperate and tropical litter (Hintiika, 1970; Santana et al., 2005; Lodge et al., 2008; Osono and Takeda, 2002, 2006). Experiments by Hintiika (1970) and Santana et al. (2005) used litter placed in microcosms while Osono and Takeda (2002, 2006) incubated sterilized litter inoculated with fungal cultures placed on agar. Santana et al. (2005) found that tropical leaf litter mass loss in microcosms containing a white-rot basidiomycete in addition to a dominant microfungal ascomycete was 22% higher than those with an ascomycete alone. Osono and Takeda (2002, 2006) found rates of leaf litter mass loss from decomposer basidiomycetes was generally higher than loss from ascomycetes and zygomycetes.

Field experiments to determine the effects of white-rot on leaf decomposition rates are more challenging. Lodge et al. (2008) conducted a field experiment at Sabana, Puerto Rico by placing 10 g dry wt of freshly fallen *Dacryodes excelsa* leaf litter enclosed in 1-mm mesh litterbags either on or off of white-rot leaf litter mats with 12 replicates per treatment and found 8.4% faster mass loss from leaves on the white-rot mats despite the invasion of the non-white-rot treatment by white-rot fungi and out-migration of basidiomycete mycelia from the white-rot treatment over the 3-month trial. DJL conducted a modified field experiment at Sabana in Puerto Rico as described here using plastic containers with the bottoms removed to restrict lateral movements of white-rot basidiomycete mycelia in the litter layer (Fig. 3A). White-rot accelerated mass loss by 15.3% over the 3-month trial. Because the amount of white-rot varied, including the presence of some white-rot in the non-white-rot treatment, the slope of percent mass loss was regressed against percent of leaf area bleached by white-rot to estimate the effect of white-rot on mass loss (Fig. 3C).

SAC tested a third experimental design with undergraduate students at El Verde in Puerto Rico using sixteen 16.5 cm diameter PVC rings to restrict lateral movements of white-rot mycelia, eight without an air gap between the fungal litter mat and the screen supporting 15 g of fallen leaf litter above to allow white-rot fungal colonization, and eight with a 5-cm air gap to inhibit fungal colonization (Fig. 4A). The freshly fallen litter was an air-dried equal mix of two species, *Dacryodes excelsa* and *Manilkara bidentata*. The rings were located in a 10 × 10 m plot. The objectives were to determine if the 5 cm air gap was effective in inhibiting colonization by white-rot mycelia, if moisture levels were the same in the two treatments (K.4B-C) and if arthropods colonizing litter were the same with or without an air gap. The summer undergraduate research program was too short to measure the effects of white-rot on mass loss.

The t-tests and box-plot analyses were performed using RStudio Version 1.1.456. Fungal connections between litter layers were counted to determine if the 5 cm air gap was effective in excluding white-rot fungi from the fresh leaf litter. A t-test showed there were significantly more fungal connections in rings without air gaps than in rings with air gaps (Fig. 4C). Forest floor litter had significantly greater moisture than the fresh litter layer (mean 79.5% vs. 66.8% moisture, respectively; Fig. 4B). Moisture in the fresh litter was not significantly different between the gap and no-gap treatments (Fig. 4B). Enumeration of arthropod orders extracted from the litter using Tullgren funnels showed that the same arthropod orders were present in litter in the gap and no-gap treatments.

4.2.5 *White-Rot Litter Mats Reduce Soil Erosion on Steep Slopes*

Some agaric fungal decomposers colonize a single leaf and are described as unit-restricted. However, it is the non-unit-restricted agaric leaf decomposers that bind

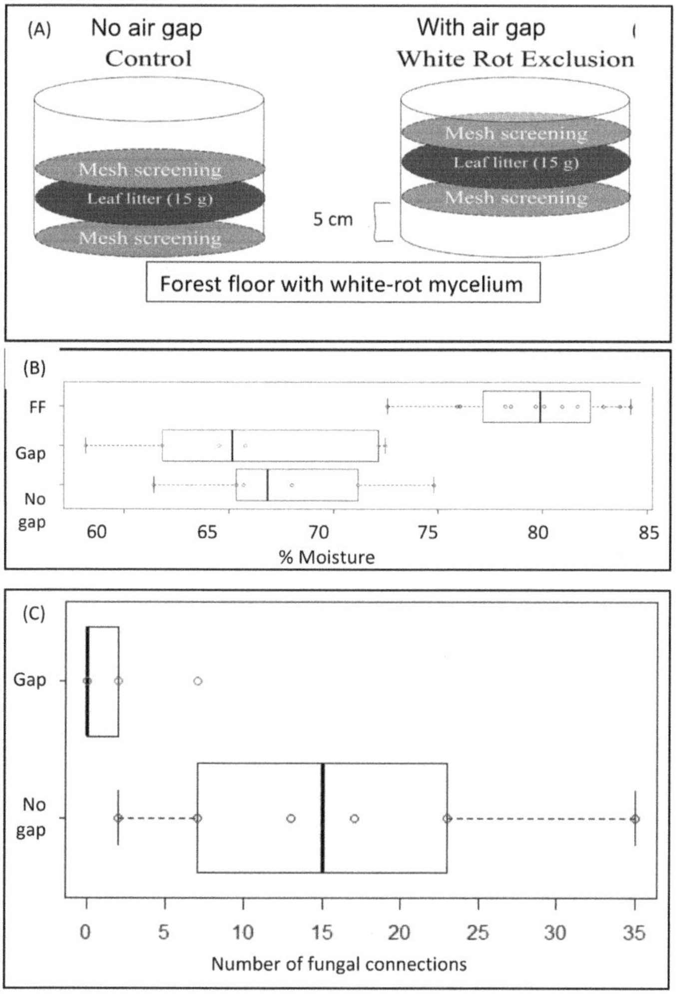

Fig. 4. Experimental design test for comparing litter decomposition with and without white-rot litter decomposer fungi. Experiment rings with and without a 5 cm air gap between the forest floor with white-rot mycelium and the freshly fallen litter layer between 1 mm mesh screens (A). Percent moisture in forest floor litter (FF) and fresh leaf litter with or without an air gap (B). Number of fungal connections in the gap vs. the no-gap treatment (C) (S.A. Cantrell).

leaves together into surface litter mats using hyphal strands, cords and rhizomorphs (Fig. 2C, D) that have the greatest impact on nutrient cycling (Osono, 2006; Lodge et al., 2008, 2014, 2022). During the rainier season in wet tropical forest in Puerto Rico, half of the freshly fallen leaves were bound to litter mats by rhizomorphs, cords and hyphal strands of primary decomposer agaric fungi within the first two days, and the remaining leaves were bound to the litter at the same rate over subsequent days (Lodge and Asbury, 1998). Steep slopes in wet tropical forests that have loose leaf litter lose their litter cover during heavy rains. In addition to loss of nutrients in litter that is exported from the ecosystem in streams, exposed surface soil is also exposed to surface erosion and stream export. (Lodge and Asbury, 1998; Lodge et al., 2008). Experiments conducted at El Verde in Puerto Rico showed variation among agaric litter mat formers in their ability to trap freshly fallen leaves when exposed to partial shade. *Marasmius crinis-equi* incorporated fresh leaf litter rapidly into litter mats in partial shade and greatly reduced soil erosion on a steep road embankment exposed to partial shade (Lodge et al., 2008).

The extent of white-rot litter mats in forests varies among sites depending on climate, the mineral nutrient and lignin concentrations in leaf litter, as noted in Sections 3.2.1 and 3.2.2. Bleached humus produced by unit-unrestricted basidiomycetes covered 0.4–11.3% of the forest floor of a boreal forest in Finland (Hintikka, 1970), and 9.8% of the forest floor in temperate broadleaved forest of Japan (Osono, 2006). White-rot litter mat cover is somewhat higher in tropical than in temperate and boreal forests (Lodge, 2022). Lodge et al. (2022) found in undisturbed wet tropical forest in Puerto Rico that fungal litter mat cover varied from 10% to 53%, similar to the 17.4% cover found in tropical Japan by Osono (2006).

4.2.6 Tropical Leaf Decomposer Macrofungi in Tropical Forest Understory

A unique feature common to humid tropical forests is the presence of leaf litter trapped in the understory by rhizomorph nets of decomposer agaric fungi (Fig. 2E, F); Hedger, 1990; Hedger et al., 1993; Koch et al., 2018). Although there are wood decomposer polypore fungi and a few agaric fungi that use rhizomorphs to colonize dead branches in live tree canopies in the temperate zone, understory leaf litter trapping is unknown outside the tropics (Hedger et al., 1993). Significant amounts of leaf litter have been found decomposing while suspended 1–2 m above the forest floor in Ecuador (Hedger et al., 1993). The understory leaf-trapping fungi that produce rhizomorphs are agaric fungi in the Marasmiaceae (*Marasmius*, *Marasmiellus*, and *Crinipellis*; Hedger et al., 1993; Koch et al., 2018). Some of these fungi are not strictly saprotrophic as they can convert to a pathogenic phase and kill the green leaves that support them (see Section 4.1; Fig. 2E, F).

5. Macrofungal Plant Pathogens in Tropical Forest

5.1 Pathogenic Macrofungi on Leaves in Tropical Forest Understory

Reports of plant pathogenic mushroom species in tropical forest understory have a long history dating back to a report by Petch (1915) of a thread blight in an Indian tea plantation. All the agaric leaf and twig pathogens are necrotrophic, meaning they kill plant cells rapidly and then obtain their nutrition saprotrophically from the dead tissue. The species in Marasmiaceae that form understory litter traps can persist for long times in a saprotrophic phase and then suddenly turn pathogenic and produce defoliation-inducing compounds (Su et al., 2011). *Marasmius crinis-equi* (Fig. 2E) was reported as the causal agent of horsehair blight on cacao (*Theobroma*) by Bunting and Dale (1924) and Dennis (1951), and on tea (Hu, 1984). Another arboreal plant pathogen in the Marasmiaceae, *Brunneocorticium corynecarpon* was described as a pathogen of citrus from Surinam in the enigmatic, polyphyletic genus *Rhizomorpha* because it consisted entirely of rhizomorphs (Fig. 5) and lacked fruiting bodies

Fig. 5. *Brunneocorticium corynecarpon* is a non-fruiting agaric mushroom pathogen on leaves and twigs in wet neotropical forests (Photo in Surinam by J.M. Wunderle).

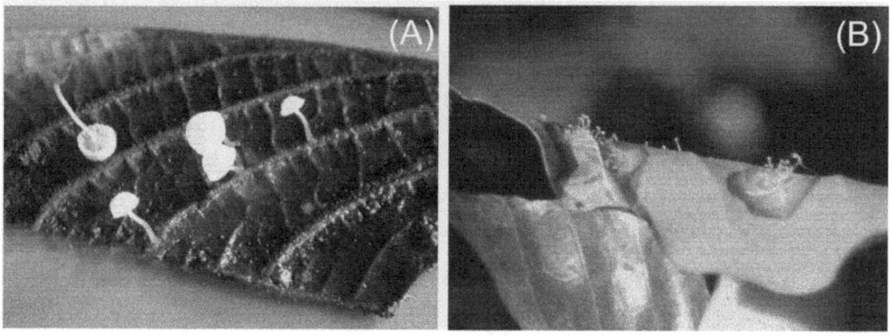

Fig. 6. *Mycena citricolor*, an agaric leaf pathogen on coffee and other plants in wet tropical areas. Mushroom fruiting bodies of *M. citricolor* (A) and *M. citricolor* gemmifers, which are asexual structures dispersed by raindrops and possibly insects (B) (Photos by D.J. Lodge).

(Weigelt and Kunze, 1928). It was subsequently reported as causing crown dieback of trees in Brazil (Berkeley, 1856; Hennings, 1904) and Guyana (Wakefield, 1934), and various native trees in Surinam and Belize (Koch et al., 2018). An unidentified, non-fruiting cord-forming species was found killing shrubs at El Verde and Sabana in Puerto Rico and was photographed in the necrotrophic phase and cultured by DJL (Fig. 2F). DNA sequencing of the ribosomal DNA ITS and LSU regions by M.C. Aime and R. Koch (pers. comm., 9 Oct. 2017 and 30 2021, respectively) were used to identify it as an undetermined species of *Amyloflagellula* in the Marasmiaceae.

Mycena citricolor (Fig. 6A, B) is a well-known agaric pathogen causing American Leafspot on coffee in Central America and the Caribbean beginning with the first report by Buller (1934). Unlike the thread blights caused by *Marasmius* species, *M. citricolor* does not produce rhizomorphs. *M. citricolor* produces macroscopic asexual stalked structures with sticky heads called gemmifers (Fig. 6B) that are dispersed by raindrops in addition to producing sexual spores on mushroom fruiting bodies (Fig. 6A) (Hedger, 1985).

5.2 *Pathogenic Macrofungi on Roots and Trunks of Tropical Trees*

A few basidiomycete macrofungi are known to be pathogens of tropical trees. Most are in the Polyporales but a few are in the Agaricales, Russulales and Gomphales. The macromycete tree pathogens are mostly necrotrophic, shifting from a saprotrophic mode to a pathogenic phase in which they produce toxins, plant growth-regulators and compounds that induce tissue death or apical branching. *Moniliophthora perniciosa* alters plant growth regulators causing a loss of apical dominance which stimulates branching, thus forming the witches' broom (Meinhardt et al., 2008).

Various species in the agaric genus *Armillaria* are pathogens that kill trees in both temperate and tropical forests (Kile et al., 1991) as well as in tea plantations in Africa (Otieno, 2002). *Armillaria* species are commonly called the shoestring fungus because of the black shiny rhizomorphs they produce. All pathogenic *Armillaria* produce rhizomorphs. *Armillaria mellea* (the honey mushroom) is a pathogenic mushroom of the N temperate zone but has been spread to tropical and subtropical areas such as in Africa. Using mating tests, Mohammed and Guillaumin (1993) showed that *A. mellea* from high elevation in tropical Africa was partly compatible with *A. mellea* sensu stricto. Currently 76 species of *Armillaria* are recognized in Index Fungorum and they are distributed around the world. Some *Armillaria* species are necrotrophic, using energy from dead wood to fuel production of toxins emitted through rhizomorphs that kill nearby trees in an expanding ring pattern (Shaw and Kile, 1991). Pathogenic *Armillaria* can cause extensive damage to native forests if there is stress from climate change, an abundance of tree stumps left by logging operations, or in monoculture plantations (Kile et al., 1991).

Mushrooms in the genus *Moniliophthora* are in the Marasmiaceae and are found in tropical and subtropical forests. Many species of *Moniliophthora* appear to be pathogens of trees shrubs and woody vines (Fig. 7H-J), but one is known as a non-pathogenic tree endophyte (Lisboa et al., 2020) and another is known only as an endophyte of grass (Niveiro et al., 2020). The most widely known member of this genus is *M. perniciosa* (Fig. 7I) which is the causal agent of witches' broom in cacao (*Theobroma cacao*), a native understory tree in the Neotropics (Aime and Phillips-Mora, 2005; Teixeira et al., 2015; Lisboa et al., 2020). There are additional biotypes on other tree species and a vine (Griffith and Hedger, 1994; Meinhardt et al., 2008; Lisboa et al., 2020). *M. perniciosa* has an odd life cycle, changing from a biotrophic (parasitic) stage to a necrotrophic stage and then fruiting on the dead branches of the witches' broom (Teixeira et al., 2015). Conversion to the saprotrophic/necrotrophic stage involves fusion of opposite mating type hyphae in the cocoa biotype, but a vine biotype appears able to self-dikaryotize (secondary homothallism) (Griffith and Hedger, 1994).

Fig. 7. Root and trunk and branch pathogens of tropical trees. Undetermined genus in the Russulales, a pathogen on Sierra palm roots in the Caribbean (A). *Beenakia informis* (B-D): *B. informis* on dying mahogany (B), soft hairy upper surface of fruit body (C), soft white teeth that stain yellow on lower surface of fruit body (D), *Cristataspora flavipora* (=*Ganoderma flaviporum*) on tree roots in Caribbean (E), *Cristataspora coffeata* (=*Humphreya coffeatum*) on dying tree roots in moist forest in Caribbean (F), *Rigidoporus microporus*, a pathogen of rubber tree roots in SE Asia and Africa but not the genetic variant in the Neotropics (G), *Moniniophthora ticoi* in Argentina (Photo by N. Niveiro) (H), *Moniliophthora perniciosa*, a pathogen of Cacao causing witch's broom (Photo by Scott Bauer was released by USDA ARS, ID k8626-1 and is in the public domain) (I) and *Moniniophthora mayarum* dying *Ceiba pentandra* roots in Belize (J) (Photos except H and I by D.J. Lodge).

Rigidoporus microporus causes white root disease in rubber trees (*Hevea brasiliensis*) in SE Asia and Africa (Go et al., 2021). Oddly, *R. microporus* was described from Jamaica in the Caribbean and is widespread in the neotropics (Fig. 7G), but it is not reported as a pathogen of rubber trees where they are native in Brazil (Oghenekome, 2004). Examination of the molecular phylogeny in Go et al. (2021) indicates that the tropical American clade is distinct from the African and SE Asian clades.

Beenakia informis was found by DJL to be an aggressive tree pathogen in tropical forests in Puerto Rico and Belize. The fungus was found fruiting on dying 50-yrold large leaved mahogany (*Swietenia macrophylla*) at Guajataca and El Verde in Puerto Rico, and at Douglas da Silva in Belize. Additional infections by *B. informis* were found on *Guarea guidonia* and *Buchenavia capitata* at El Verde in Puerto Rico (Lodge, 2002). All of the trees had crown dieback, and the cambium of large roots and the trunks up to one meter above ground was dead and had white mycelium together with golden brown, warty chlamydospores that were also present in fruiting bodies of *B. informis* (Lodge, 2002). White mycelium with yellow chlamydospores can be seen under the bark associated with the fruiting body in the upper right of Fig. 7C. Infection fronts were traced using the diagnostic chlamydospores through root grafts to neighboring susceptible trees.

Both *Cristataspora coffeata* (=*Humphreya coffeatum*) and *C. flavipora* (=*Ganoderma flaviporum*) in the Polyporaceae were found fruiting from roots of dying trees exhibiting dieback in tropical dry forest on Guana Island in the British Virgin Islands by DJL (Fig. 7E, F). These species have not previously been reported as tree pathogens, and most species of *Ganoderma* are decomposers rather than strict pathogens. *Ganoderma* and *Amauroderma* species in the Polyporaceae, however, have been reported as pathogens of trees, especially of tropical palms (Flood et al., 2000; Kües et al., 2015) and *Acacia mangia* in plantations in Indonesia and Malaysia (Glen et al., 2009). Tropical species of *Cristataspora* and *Ganoderma* that fruit directly from roots are potentially root pathogens.

6. Macrofungal Interactions with Animals

Some macrofungi use invertebrates as dispersal agents as accidental carriers (phoresy) while in other cases there is a mutualistic association between the fungus and the dispersal agent. The ability of macrofungi to concentrate nutrients makes them nutritionally attractive to both invertebrate and vertebrate consumers. On the other end of the association spectrum some macro-ascomycetes in the Cordycipitaceae and Ophiocordycipitaceae parasitize invertebrates.

6.1 Dissemination of Macrofungi by Animals

In temperate and some tropical forests, truffle fungi (ascomycetes) and truffle-like fungi (basidiomycetes) fruit at or below the soil surface and use rodents or marsupials to disperse their spores within the same forest habitat (Trappe and Claridge, 2005; Trappe et al., 2005; Połatyńska, 2014). Squirrels and other rodents in the Americas, Eurasia and SE Asia are attracted to truffles by their odor, and the truffle spores pass unharmed through the gut and are deposited in feces (Trappe et al., 2005; Połatyńska, 2014). True truffles and most truffle-like fungi are ectomycorrhizal symbionts of trees, so it is most efficient if the spores are only distributed near potential hosts (Trappe and Claridge, 2005; Trappe et al., 2005). In Australia, marsupials, primarily wallabies and bettongs disperse truffles (Połatyńska, 2014; Trappe et al., 2005).

6.1.1 Flies Dispersing Spores of Phallales

In the stinkhorn fungi (Phallales), flies consume the slimy spore mass and transmit the spores in their feces (Tuno, 1998). Most of the commonly known stinkhorn fungi

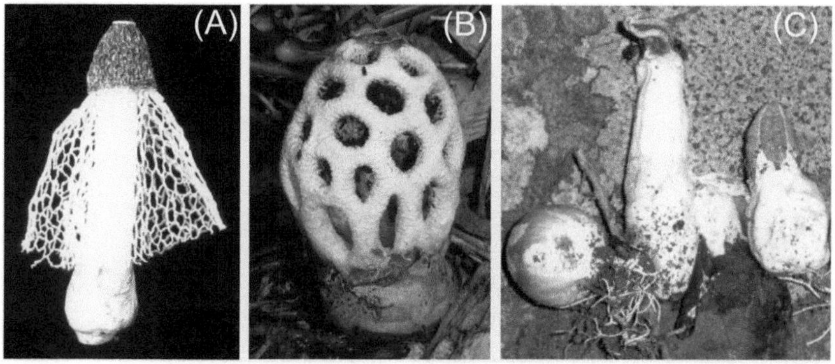

Fig. 8. Dispersal of macrofungi by animals: *Dictyophora indusiata* with a flesh fly on the slimy green spore mass (A), *Clathrus roseovolvatus* with an odor of cheese, shown with a fungus gnat (B), *Lysurus* sp. with flies, from Guana Island, BVI with an odor of rotting figs (C) (Photos by D.J. Lodge).

have odors of rotting meat and typically red and green colors to attract flesh flies (Fig. 2B; 8A). Some stinkhorn fungi, however, lack red colors and have odors of cheese or rotting fruit and attract different species of flies for their spore dispersal (Fig. 8B, C).

6.1.2 *Rhizomorphs of Marasmiaceae Transported by Birds for Nests*

Some of the agaric species that trap litter in the understory of humid tropical forests, including pathogenic species, are dispersed by birds (Aubrecht et al., 2013; Koch et al., 2018). The fungi produce rhizomorphs that are collected by birds for nesting material (Fig. 9; Aubrecht et al., 2013; Koch et al., 2018). This is the primary means of medium-range dispersal by these marasmioid fungi, and the only means of dispersal for *Brunneocorticium corynecarpon* as it has never been found to produce sexual or asexual spores (Fig. 5; Koch et al., 2018). Aubrecht et al. (2013) hypothesized that birds and rhizomorph-forming fungi such as *Marasmius crinis-equi* (Fig. 2E) had evolved a mutualistic association, and the fungal rhizomorphs produce antibiotics that may reduce mortality of nestlings in wet tropical forests. Freyman (2008) found that fungal rhizomorphs were longer and retained less moisture than plant fibers, which are beneficial attributes of nest material. Rhizomorphs of *B. corynecarpon* (Fig. 5) may be especially attractive to birds for nesting material as they can be up to one meter long, it has knobs that may help anchor it in the nest, and the bright white outer surface is hydrophobic and stands out in low understory light (Koch et al., 2018).

6.1.3 *Phoretic Dispersal of Mycena Veil Elements by Invertebrates*

Section Sacchariferae of the agaric genus *Mycena* produce a sticky, loose veil covering on their mushrooms (Fig. 10A) that are reported to be dispersed by invertebrates, especially collembola and mites (Desjardin, 1995; Bougher, 1995). The etymology of the section name is 'sugared', for the powdery covering comprising a universal veil on the mushrooms' surface (Desjardin, 1995). Both Singer (1983) and Desjardin (1995) have germinated cherocytes of *Mycena* in sect. Sacchariferae, which may be

Fig. 9. Black rhizomorphs of an agaric mushroom incorporated into the nest of a bird in tropical wet forest of Puerto Rico. The rhizomorphs most likely belong to a marasmioid fungus, *Marasmiellus brevipes*, and the nest my belong to a Stripe-Headed Tananger (Spindalis) (Photo by D.J. Lodge).

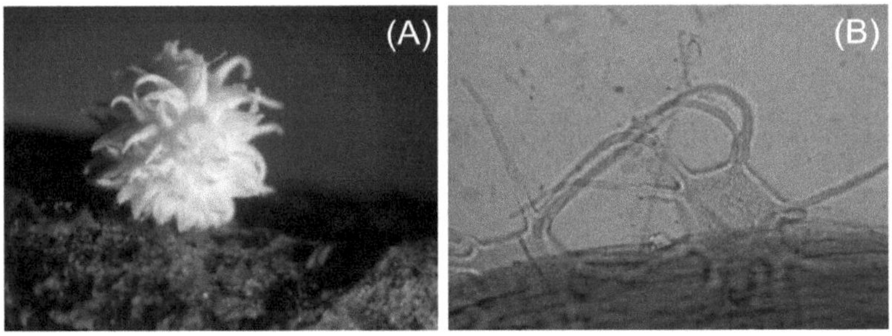

Fig. 10. Veil elements disseminated by invertebrates in *Mycena* sect. Sacchariferae: *Mycena spinocissima* (=*Amparoina spinosissima*) mushroom primordium covered in sticky, soft, detersible spines of the universal veil (A) (Photo by T. Læssøe). Microphoto of loose veil elements (cherocytes) of *Mycena biornata* var. *manausensis* (sect. Sacchariferae) from Doyle's Delight, Belize (B) (Photo by D.J. Lodge).

thin- or thick-walled and have long spines (Fig. 10B). The dispersal of veil elements by invertebrates is likely incidental on the part of the invertebrate (i.e., phoresy).

6.2 *Macrofungi as Food for Invertebrate and Vertebrates*

Because macrofungi concentrate carbon and mineral nutrients obtained through decomposition of recalcitrant plant material or from a mycorrhizal host tree and then translocate these into their fruiting bodies for reproduction, mushrooms and other types of fruitifications are consumed by a variety of invertebrate (Fig. 11) and vertebrate animals. N and P concentrations in agaric macromycetes fruiting on wood at El Verde in Puerto Rico were found to be 12–24 times higher than in their wood substrate (Lodge, 1996). The leaf decomposer mushroom, *Gymnopus johnstonii* that was analyzed in Puerto Rico had 2.4–2.8 times more N, Ca and K than their leaf litter

Fig. 11. Examples of invertebrates that feed on macromycete fruiting bodies: Snails, such as *Caracolus caracolla* consuming *Marasmiellus coilobasis* fruit bodies in Puerto Rico (A). A cricket and a chrysomelid beetle eating a fruit body of *Mycena* sp. in Puerto Rico (B). Pleasing fungus beetles, *Megalodacne fasciata* feeding on a shelf fungus, *Earliella scabrosa* in Belize (C). Land Planaria foraging on *Xeromphalina tenuipes* in Puerto Rico (D). The larger insect is a homopteran 'plant' hopper in the family Derbidae that sucks cell sap from agaric fruit bodies in Puerto Rico (E). The smaller insects are fruit flies that use large fruitings of brightly colored mushrooms for mating leks. Missing pieces of the caps are from snail feeding (Photos by D.J. Lodge).

substrate (Lodge, 1987, 1996). P concentrations were 12 times higher than those in freshly fallen leaves but concentrations were found to increase by an order of magnitude prior to fruiting (Lodge, 1987, 1996). The increase in P concentrations in pre-fruiting mycelium represents a change from 22% to 85% of the total P pool in white-rot litter mats at El Verde (Lodge, 1987, 1996). Calcium concentrations in agaric fruit bodies at El Verde varied greatly among species (Lodge, 1987, 1996), and most was likely in the form of calcium oxalate crystals (Frankland et al., 1978; Cromack et al., 1997). Fungi are thought to be an important source of calcium for invertebrates, and earthworms and collembolans have gut microflora that can break down calcium oxalate (Cromack et al., 1977).

6.2.1 Fungivory by Invertebrate and Vertebrate Animals

Given the high carbon and mineral nutrient content of macrofungal fruit bodies, the many reports of mycophogy by animals is not surprising. Trierveiler-Pereira (2016) observed black-capped squirrel monkey (*Saimiri boliviensis*) eating ascomata of two species of *Ascopolyporus* (*Cordycipitaceae, Hypocreales*) in Amazonian forest in NW Brazil. *Ascopolyporus* species resemble the closely related *Hyperdermium bertonii* (Fig. 13C) but the fruiting bodies are larger and filled with gelatinous material (Bischoff et al., 2005). Trierveiler-Pereira (2016) also observed Azara's agouti in Southern Brazil eating primordia ('eggs', resembling the one in Fig. 8C) of *Itajahya galericulata* in the Phallales. Vasco-Palacios et al. (2008) reported that the Andoke and Muinane indigenous tribes in Amazonian Colombia observed the morrocoy tortoise (*Geochelonea denticulada*) consuming fungi, and that squirrels consumed *Lentinula raphanica*, *Trogia* aff. *buccinulis* and *Hydropus* cf. cavipes var. *muinialbus*. As noted previously in Section 5.1, ectomycorrhizal fungi that produce truffles and underground truffle-like fungi have employed olfactory and dietary attractiveness to vertebrates such as squirrels, wallabies and bettongs in order to disperse their spores in suitable forest habitat (Połatyńska, 2014; Trappe and Claridge, 2005; Trappe et al., 2005). Macias et al. (2019) found a wide range of fungi were utilized as food by *Brachycybe lecontii* in the southeastern USA but that wood decay fungi were toxic to them in feeding trials. Surprisingly, fungi in the Hypocreales were more commonly ingested despite many being parasites of invertebrates (Macias et al., 2019). A related fluid-feeding millipede genus, *Colobognatha,* that is found in the tropics and has an ancient lineage was thought to have fungi as their main food source based on gut contents (Moritz et al., 2022).

6.2.2 Mushroom-Farming Ants in the Neotropics and Termites in the Paleotropics

The fungus-gardening Neotropical leaf-cutter ants and Paleotropical termites have a highly co-evolved, mutualistic relationship between the insects and the fungi they grow in their gardens. *Termitomyces* belonging to the family Lyophyllaceae has species with ecologies ranging from saprotrophic to symbiotic mutualists and parasites. The genus *Termitomyces* is a mutualistic symbiont of Paleotropical termites and present an enigma as to how this symbiosis arose. Van de Peppel et al. (2021) showed that traits scattered among related fungal lineages in the Lyophyllaceae, including the Neotropical genus, *Arthromyces*, likely predisposed these fungi to symbiotic relationships with insects. Despite evidence of ancestral lineages of Lyophyllaceae in the Neotropics associated with termites, and that Paleotropical termites are most abundant in savannas, the main lineage leading to full symbiosis (Fig. 12A) apparently evolved in African rainforest (Durr and Eggleton, 2005). The termites via their associated microbial symbionts are among the most successful

Fig. 12. Fungus-farming insects. J.M. Wunderle in front of 3 m-tall termite mound in Southern Africa (A) (Photo by D.J. Lodge). Leaf-cutter ants (*Atta* sp.) bringing a leaf back to their fungus garden in Belize (B) (Photo by S. Schmieding, USDA Forest Service, in public domain).

animal group in the world via enzyme activity that can convert nearly all the energy from any type of biomass into food for their gardeners (da Costa et al. 2019). The Neotropical analog of Paleotropical fungus-gardening termites are ants in the tribe Attiani (Fig. 12B). Switching of cultivars among Attine colonies of different *Cyphomyremex* species was shown to be detrimental to the ants whereas switching of fungal cultivars between ant colonies of the same species was not detrimental (Mehdiabadi et al., 2005). The ants co-cultivate an antibiotic producing bacterium to protect its fungus garden from infection (Currie et al 1999).

6.3 Fungal Parasites of Invertebrates

Typically brightly colored ascomycetous fungal parasites of invertebrates are well-known in the N temperate zone, especially *Cordyceps militaris* that produces red-orange fruit bodies from insect pupae. Another commonly encountered *Cordyceps* in both temperate and tropical areas is a variety that parasitizes wasps and emerges from their heads, *C. militaris* var. *sphecocephala*. It is in the tropics, however, where the diversity of macro-ascomycete parasites of invertebrates reaches its zenith (e.g., examples in Fig. 13; Sourell and Lodge, 2015; Sourell et al., 2018, 2020; Cárdenas Medina et al., 2019). Kepler et al. (2017), revised the nomenclature of the families comprising ascomycete parasites of invertebrates (Clavicipitaceae, Cordycipitaceae and Ophiocordycipitaceae), including reassignment of asexual states to genera of sexual states. These fungi are known to alter the behavior of insects in a way that increases the likelihood of fungal transmission to other individuals of the same species or group (Hughes et al., 2011; Lovett et al., 2020). Lovett et al. (2020)

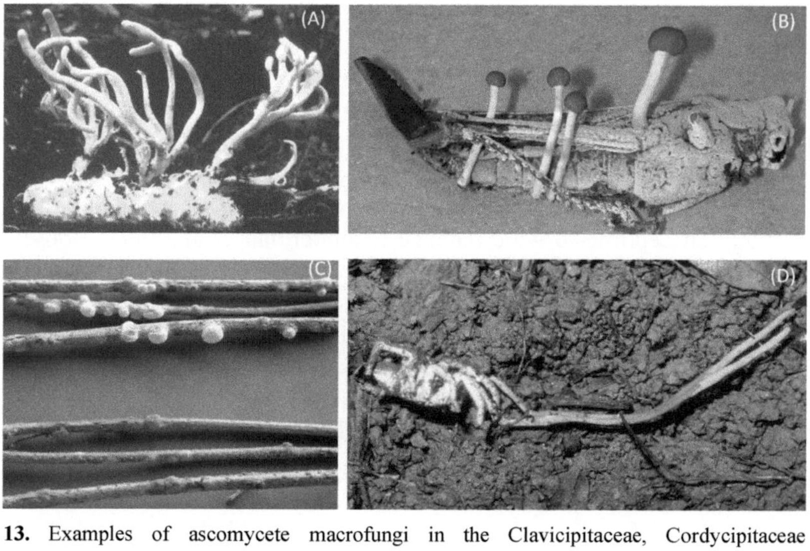

Fig. 13. Examples of ascomycete macrofungi in the Clavicipitaceae, Cordycipitaceae and Ophiocordycipitaceae that parasitize invertebrates in humid tropical forests: *Cordyceps* sp. on beetle or lepidopteran larva or pupa (A), *Ophiocordyceps amazonica* on grasshopper in Belize (B), *Hyperdermium bertonii*, a parasite of scale insects, which are parasitic on vines in Belize (C) and *Cordyceps* cf., *ignota* on tarantula in Belize (D) (Photos by D.J. Lodge).

showed that a hallucinogen was involved in fungal manipulation of host behavior. For example, ants in tropical forests that are parasitized by *Ophiocordyceps* become attracted to light and then clamp their mandibles down on vegetation before dying. Insects whose behavior has been altered by a fungal parasite to favor fungal dispersal are sometimes referred to as 'zombies' (Hughes et al., 2011; Lovett et al., 2020).

7. Effects of Disturbance and Stress on Macrofungi

Forest disturbances are acute events that are of short duration and high impact, such as direct strikes from high intensity tropical cyclones that remove the canopy and topple trees, whereas stresses occur of longer periods of time, such as droughts. Forest disturbances include anthropogenic activities such as logging (Miller and Lodge, 2007). Macrofungi on the forest floor and in the understory of tropical forest respond to both disturbance and drought stress, especially species that grow on small debris such as leaf litter and twigs (Hedger, 1983; Hedger et al., 1993; Lodge et al., 2008, 2014, 2022). Disturbances that open the forest canopy and drought stress are often related.

7.1 *Effects Canopy Opening on Macrofungi*

Both natural disturbances from tropical cyclones and anthopogenic disturbances such as building roads, logging and conversion of forest to agriculture can open the canopy exposing macrofungi on the forest floor and in the understory to direct sunlight and greater evaporation potential (Miller and Lodge, 2008; Lodge et al., 2008, 2014, 2022). Long-term ecological studies in the Luquillo Mountains in NE Puerto Rico have provided extensive data on fungal responses to disturbances. NE Puerto Rico was struck by three major hurricanes over 30 years, beginning with class 3 hurricane Hugo on 18 Sept. 1989, then 9 years later by class 3 hurricane Georges on 21 Sept. 1998 and class 4 hurricane Maria on 20 September 2017. Hurricane Hugo was a relatively dry storm and the pattern of tree damage reflected topographic exposure to winds, proximity to the center of the storm track, past storm history, and previous human disturbance which promoted recruitment of tree species that were differentially susceptible to wind damage (Zimmerman et al., 1995; Boose et al., 2004). It took 5–6 years for the canopy to close at El Verde after Hurricane Hugo (Brokaw, 2012). Hurricane Georges did not cause as much structural damage to trees since the previous hurricane had removed many large branches. Hurricane Maria was the most intense hurricane to recently strike Puerto Rico and caused extensive mortality and damage to trees across the island, including in the Luquillo Mts., with greater tree mortality attributable to rain and flooding than to wind (Uriarte et al., 2019).

Opening of the forest canopy at El Verde in Puerto Rico by hurricane Hugo exposed litter mat macrofungi to drying which caused local extinctions of the previously dominant agaric species on ridges, *Gymnopus johnstonii* (Fig. 3B; Lodge and Cantrell, 1995). The ability of leaf litter mat fungi to colonize freshly fallen leaves and translocate P from their mycelia in previously decomposed litter was studied by Lodge et al. (2014) in a large-scale canopy trimming experiment (CTE) at El Verde to partially simulate the effects of hurricane damage (Shiels et al., 2015). Significant decreases in both fungal connectivity between litter cohorts and upward translocation of P into new fungal biomass in freshly fallen leaves occurred in the open canopy treatments compared to the closed canopy control which affected rates of decomposition (Lodge et al., 2014). Mass loss in the senesced leaf cohort differed significantly among treatments and generally decomposed faster under closed canopy (Lodge et al., 2014). Mass loss in senesced leaves at 7 weeks was predicted by P concentration while mass loss at 14 weeks was predicted by abundance of fungal connections between the senesced litter cohort and forest floor at 7 weeks (Lodge et al., 2014). Cantrell et al. (2014) found that microbial succession in senesced leaf litter stopped when both the canopy and forest floor debris were removed, and a suggestive trimming by time interaction indicated susceptibility of the leaf litter microbial community to canopy opening. Such changes are important as Liu et al. (2022) found in a global analysis that the richness of fungal decomposers was consistently and positively related to ecosystem stability.

The CTE was designed to determine the effects of increased frequency of hurricanes, and the first canopy trim occurred in 2004–2005. In the second iteration of the CTE in 2014, percent fungal litter mat cover decreased significantly in response to canopy opening (Lodge et al., 2022). Modeling of factors potentially causing decreases in fungal litter mat cover in 2014–2015 indicated the main factor was exposure to solar radiation. Damage to the forest canopy from hurricane Maria in 2017 opened the canopy again and percent fungal litter mat cover decreased to near zero, but high rainfall may have contributed to mortality of agaric fungi in litter mats (Lodge et al., 2022). Canopy openness was associated with lower litter moisture in the first iteration of the CTE (Richardson et al., 2010) which suggests that increased evaporative loss may be involved (Lodge et al., 2014, 2022).

Construction of access roads in forests exposes the forest floor edge to greater solar irradiation with the potential to affect litter decomposer fungi. Experiments testing the tolerance of litter-binding mushrooms to partial canopy opening and increased evaporative potential along an access road at El Verde showed that one fungus, *Marasmius crinis-equi*, attached to freshly fallen leaves faster and controlled soil erosion on the steep road bank more in partial shade than in full shade. Interestingly, *M. crinis-equi*, which is a necrotrophic plant pathogen more frequently found in the understory than on the forest floor (Sections 4.1 and 5.1.2), trapped more litter in the understory of disturbed areas than in undisturbed cocoa plantation in Ecuador (Hedger et al., 1993). Hedger et al. (1993) found the differences resulted from the disturbed site having more contact zones (fungal holdfasts) than in the undisturbed site but not greater rhizomorph length in the litter trapping nets. Hedger et al. (1993) also found that *M. crinis-equi* tolerated moisture stress (see Section 6.2 below).

7.2 *Effect of Moisture Stress on Macrofungi in Humid Tropical Forests*

Hedger et al. (1993) found that litter-trapping fungal isolates from Ecuador and Papua New Guinea grew on agar with water potentials adjusted using salts to between

–4 and –8 MPa, produced rhizomorphs down to –2 MPa, and tolerated 33% relative humidity for up to three days. Dr. R.E. Petersen (pers. comm. to DJL) found he could grow *M. crinis-equi* from rhizomorphs in dried herbarium specimens that had been stored for longer periods. A 2015 drought in Puerto Rico caused significant reductions in fungal litter mat cover at El Verde, except in one untrimmed plot where the litter mats were dominated by *M. crinis-equi* (Lodge et al., 2022). However, most agaric litter decomposer fungi that are adapted to wet tropical forests are highly susceptible to drying (Lodge and Cantrell, 1995; Lodge et al., 2022). Fungal biovolume in the litter layer at El Verde in Puerto Rico, which is dominated by agaric litter decomposers, decrease by half following three days without sufficient rain to saturate the canopy and reach the forest floor (Lodge, 1993). Drying and wetting cycles lead to non-linear responses to pulsed nutrient releases in wet tropical forests (Lodge et al., 1994). Rewetting of rainforest litter and soil induces a large flux of mineral nutrients from litter to soil when soil microbial biomass is low thus lowering competition for limiting nutrients and leading to greater root uptake (Lodge et al., 1994). Cleveland et al. (2010) found in Panama that experimental exclusion of 25% and 50% of rainfall did not change the amount of dissolved organic carbon (DOC) flux from litter to soil but it did increase loss of C from soil through respiration owing to pulse dynamics and the non-linear response of the soil microbial community to concentrated fluxes of DOC. Buscardo et al. (2021) found that both seasonal drought in an Amazonian rainforest as well as long-term drought induced through a 50% rainfall exclusion experiment over 14 years reduced soil saprotrophic fungal relative abundance and dominance but increased species richness in the mineral soil. The increase in soil fungal richness under drought was from dark septate microfungi whereas basidiomycete fungi declined (Buscardo et al., 2021). Further, they found that soil C, N and P cycling were also altered by drought, but their sampling design did not include the litter layer on the forest floor (Buscardo et al., 2021).

Current and future increases in drought frequency caused by climate change such as those in Puerto Rico and the Virgin Islands (Bhardwag et al., 2018; Bowden et al., 2020) will undoubtedly affect macrofungal community composition, abundance and function in humid tropical forests. There will likely be some adaptation to drier conditions resulting from environmental filtering that favors species adapted to drought. Studies by Hedger et al. in Ecuador and Papua New Guinae and (1993), Lodge et al. in Puerto Rico (2008, 2022) indicate some resilience to drought among marasmoid macrofungal species in humid tropical forests.

Conclusions and Future Outlook

In humid tropical forests, macrofungi that decompose litter, of which some are also orchid symbionts, are especially sensitive to canopy opening caused by intense tropical storms, anthropogenic disturbances such as deforestation and drought. Both the frequency of intense tropical cyclones and drought are increasing in the humid tropics because of global climate change. Such climatic events together with removal of native forests for agriculture and tree plantations will disrupt nutrient cycling and reduce populations of plant and fungal species that are intricately dependent on each other. There are some indications of resiliency among litter decomposers that can tolerate or thrive following canopy opening, but some of these are also pathogenic to plants.

References

Aime, M.C. and Phillips-Mora, W. (2005). The causal agents of witches' broom and frosty pod rot of cacao (chocolate, *Theobroma cacao*) form a new lineage of Marasmiaceae. Mycologia, 97: 1012–1022. https://doi.org/10.3852/mycologia.97.5.1012.

Anderson, J.M. and Swift, M.J. (1983). Decomposition in tropical forests. pp. 287–308. *In*: Sutton, S.L., Whitmore, T.C. and Chadwick, A.C. (eds.). Tropical Rain Forest: Ecology and Management, Oxford: Special Publication No. 2 of the British Ecological Society, Blackwell Scientific Publications, Oxford, UK.

Aubrecht, G., Huber, W. and Weissenhofer, A. (2013). Coincidence or benefit? The use of *Marasmius* (horse-hair fungus) filaments in bird nests. Avian Biol. Res., 6: 26–30.

Bandala, V.M. and Montoya, L. (2015). *Gymnopodium floribundum* trees (*Polygonaceae*) harbour a diverse ectomycorrhizal fungal community in the tropical deciduous forest of Southeastern Mexico. Res. Rev. J. Bot. Sci., 4: 73–75.

Beare, M.H., Parmalee, R.W., Hendrix, P.F., Cheng, W., Coleman, D. and Crossley, D.A. (1992). Microbial and faunal interactions and effects on litter nitrogen and decomposition in ecosystems. Ecol. Monogr., 62: 569–591.

Bebber, D., Tlalka, Hynes, J.M., Darrah, P.R., Ashford, A.E., Watkinson, S.C., Boddy, L. and Fricker, M.D. (2007). Imaging complex nutrient dynamics in mycelial networks. pp. 3–21. *In*: Gadd, G.M., Dyer, P. and Watkinson, S.C. (eds.). Fungi in the Environment. Cambridge University Press, Cambridge, UK.

Bell, T., Freckleton, R.P. and Lewis, O.T. (2006). Plant pathogens drive density-dependent seedling mortality in a tropical tree. Ecol. Lett., 9: 569–574. doi:10.1111/j.1461-0248.2006.00905.x

Bending, G.D. and Read, D.J. (1995). The structure and function of the vegetative mycelium of ectomycorrhizal plants. VI. Activities of nutrient mobilising enzymes in birch litter colonised by *Paxillus involutus* (Fr.) Fr. New Phytologist, 130: 411–417.

Berg, B. and Staff, H. (1980). Decomposition rates and chemical changes of Scots Pine litter. II. Influence of chemical composition. Ecol. Bull., 32: 373–90.

Bhardwaj, A., Vasubandhu, M., Misra, V., Misra, A., Wooten, A., Boyles, R., Bowden, J.H. and Terando, A.J. (2018). Downscaling future climate change projections over Puerto Rico using a non-hydrostatic atmospheric model. Climatic Change, 147: 133–147. https://doi.org/10.1007/s10584-017-2130-x.

Bitartondo, M.I., Burghardt, B., Gebauer, G., Bruns, T.D. and Read, D.J. (2004). Changing partners in the dark: Isotopic and molecular evidence of ectomycorrhizal liaisons between forest orchids and trees. Proc. R. Soc. B, 271: 1799–1806.

Boddy, L. (1993). Saprotrophic cord-forming fungi: Warfare strategies and other ecological aspects. Mycological Research, 97: 641–655. https://doi.org/10.1016/S0953-7562(09)80141-X.

Boddy, L. (1999). Saprotrophic cord-forming fungi: Meeting the challenge of heterogeneous environments. Mycologia, 91: 13–32.

Boddy, L. and Jones, H.T. (2007). Mycelial responses to heterogeneous environments: Parallels with macroorganisms. pp. 112–140. *In*: Gadd, G.M., Dyer, P. and Watkinson, S.C. (eds.). Fungi in the Environment. Cambridge University Press, Cambridge, UK.

Boddy, L. and Watkinson, S.C. (1995). Wood decomposition, higher fungi, and their role in nutrient redistribution. Can. J. Bot., 73 (suppl. 1): S1377–S1383.

Boose, E.R., Serrano, M.I. and Foster, D.R. (2004). Landscape and regional impacts of hurricanes in Puerto Rico. Ecol. Monogr., 74: 335–352.

Bougher, N.L. (2009). Two intimately co-occurring species of *Mycena* section Sacchariferae in south-west Australia. Mycotaxon, 108: 59–174.

Bowden, J.H., Terando, A.J., Misra, V., Wooten, A., Bhardwaj, A., Boyles, R., Gould, W., Collazo, J.A. and Spero, T.L. (2021). High-resolution dynamically downscaled rainfall and temperature projections for ecological life zones within Puerto Rico and for the U.S. Virgin Islands. RMetS, 41: 1305–1327. https://doi.org/10.1002/joc.6810.

Bunting, R.H. and Dale, H.A. (1924). Gold Coast Plant Diseases. Waterloo & Sons, Ltd., London, UK.

Buscardo, E., Souza, R.C., Meir, P., Geml, J., Schmidt, S.K., da Costa, A.C.L. and Nagy, L. (2021). Effects of natural and experimental drought on soil fungi and biogeochemistry in an Amazon rain forest. Comm. Earth & Envir., 2: 55. https://doi.org/10.1038/s43247-021-00124-8.

Cairney, J.W.G. (2005). Basidiomycete mycelia in forest soils: Dimensions, dynamics and roles in nutrient distribution. Mycol. Res., 109: 7–20.

Cantrell, S.A., Molina, M., Lodge, D.J., Rivera-Figueroa, F.J., Ortiz, M., Marchetti, A.A., Cyterski, M.J. and Pérez-Jiménez, J.R. 2014. Effects of a simulated hurricane disturbance on forest floor microbial communities. Forest Ecology & Manage., 332: 22–31. doi: 10.1016/j.foreco.2014.07.010.

Cárdenas Medina, A., Garcia Roca, M., Shrestha, B. and Pavlich, M. (2019). Tambopata – Macrofungi of Tambopata. Field Museum of Natural History, Rapid Field Guides #1183. Chicago, IL, USA. https://fieldguides.fieldmuseum.org/sites/default/files/rapid-color-guides-pdfs/1183_peru_macrofungi_of_tambopata.pdf.

Carreiro, M.M., Sinsabaugh, R.L., Repert, D.A. and Parkhurst, D.F. (2000). Microbial enzyme shifts explain litter decay responses to simulated nitrogen deposition. Ecology, 81: 2359–2365.

Chen, Y.L., Gao, Y., Zhang, K.Q. and Zou, C.G. (2013). Autophagy is required for trap formation in the nematode-trapping fungus *Arthrobotrys oligospora*. Environ. Microbiol. Rep., 5: 511–7. doi: 10.1111/1758-2229.12054. Epub 2013 Apr 19. PMID: 23864564.

Cherrett, J.M., Powell, R.J. and Stradling, D.J. (2012). The mutualism between leaf-cutting ants and their fungus. *In*: Wilding, N., Collins, N.M., Hammond, P.M. and Webber, J.F. (eds.). Insect-Fungus Interactions, 14th Symposium of the Royal Entomological Society of London in Collaboration with the British Mycological Society, 16–17 Sept. 1987, Imperial College, London, UK. Harcort Brace Javonovich Publishers, Academic Press, London.

Cleveland, C.C., Reed, S.C. and Townsend, A.R. (2006). Nutrient regulation of organic matter decomposition in tropical rain forest. Ecology, 87: 492–503.

Cleveland, C.C., Wieder, W.R., Reed, S.C. and Townsend, A.R. (2010). Experimental drought in a tropical rain forest increases soil carbon dioxide losses to the atmosphere. Ecology, 91: 2313–2323. https://doi.org/10.1890/09-1582.1.

Currie, C.R., Scott, J.A., Summerball, R.C. and Malloch, D. (1999). Fungus growing ants use antibiotic-producing bacteria to control garden parasites. Nature, 398: 701–704.

da Costa, R.R., Hu, H., Li, H. and Poulsen, M. (2019). Symbiotic plant biomass decomposition in fungus-growing termites insects, 10: 87. doi:10.3390/insects10040087.

Dearnaley, J., Perotto, S. and Selosse, Selosse, M.-A. (2017). Structure and development of orchid mycorrhizas. *In*: Martin, F. (ed.). Molecular Mycorrhizal Symbiosis, 1st Edn. John Wiley & Sons, Inc., NJ, USA.

Dearnaley, J.D.W. and Bougoure, J.J. (2010). Isotopic and molecular evidence for saprotrophic Marasmiaceae mycobionts in rhizomes of *Gastrodia sesamoides*. Fungal Ecology, 3: 288–294.

Delgat, L., Courtecuisse, R., De Crop, E., Hampe, F., Hofmann, T.A., Manz, C., Piepenbring, M., Roy, M. and Verbeken, A. (2020). *Lactifluus* (Russulaceae) diversity in Central America and the Caribbean: Melting pot between realms. Persoonia, 44: 278–300.

Dennis, R.W.G. (1951). Some tropical American Agaricaceae referred by Berkeley and Montagne to *Marasmius*, *Collybia* and *Heliomyces*. Kew Bull., 3: 386–410.

Dent, D.H. and Burslem, D.F.R.P. (2016). Leaf traits of dipterocarp species with contrasting distributions across a gradient of nutrient and light availability. Plant Ecol. Divers. 9: 521–533, DOI: 10.1080/17550874.2016.1265018.

Desjardin, D.E. (1995). A preliminary account of the worldwide members of *Mycena* sect. *Sacchariferae*. Bibl. Mycol., 159: 1–89.

Devi, A.S. and Yadava, P.S. (2007). Wood and leaf litter decomposition of *Dipterocarpus tuberculatus* Roxb. in a tropical deciduous forest of Manipur, Northeast India. Current Sci., 93: 243–246.

Dickie, I.A. and Moyersoen, B. (2008). Towards a global view of ectomycorrhizal ecology. New Phytol., 180: 263–265.

Ettema, C.H. and Wardle, D. (2002). Spatial soil ecology. Trends Ecol. Evol., 17: 177–183.

Fan, L., Guo, S.X., Cao, W.Q., Xiao, P.G. and Xu, J.T. (1996). Isolation, culture, identification and biological activity of *Mycena orchidicola* sp. nov. in *Cymbidium sinense* (Orchidaceae). Acta Mycol. Sinica, 15: 251–255.

Flood, J., Bridge, P.D. and Holderness, M. (eds.). (2000). *Ganoderma* Diseases of Perennial Crops. CABI Publishing, Wallingford, UK.

Frankland, J.C. (1982). Biomass and nutrient cycling by decomposer basidiomycetes. pp. 241–261. *In*: Frankland, J.C., Hedger, J.N. and Swift, M.J. (eds.). Decomposer Basidiomycetes: Their Biology and Ecology. Cambridge University Press, Cambridge, U.K.

Frankland, J.C. (1998). Fungal succession: Unravelling the unpredictable. Mycol. Res., 102: 1–15.

Frankland, J.C., Poskitt, J.M. and Howard, D.M. (1995). Spatial development of populations of a decomposer fungus, *Mycena galopus*. Can. J. Bot., 73: S1399–S1406.

Frey, S.D., Elliott, E.T., Paustian, K. and Peterson, G.A. (2000). Fungal translocation as a mechanism for soil nitrogen inputs to surface residue decomposition in no-tillage agroecosystems. Soil Biol. Biochem., 32: 689–698.

Freymann, B.P. (2008). Physical properties of fungal rhizomorphs of marasmioid basidiomycetes used as nesting material by birds. Ibis, 150: 395–399.

Gadgil, R.A. and Gadgil, P.D. (1971). Mycorrhizae and litter decomposition. Nature, 233: 133.

Go, W.Z., Chin, K.L., H'ng, P.S., Wong, M.Y., Luqman, C.A., Surendran, A., Tan, G.H., Lee, C.L., Khoo, P.S. and Kong, W.J. (2021). Virulence of *Rigidoporus microporus* isolates causing white root rot disease on rubber trees (*Hevea brasiliensis*) in Malaysia. Plants (Basel, Switzerland), 10(10): 2123. https://doi.org/10.3390/plants10102123.

Grelet, G.A., Ba, R., Goeke, D.F., Houliston, G.J., Taylor, A.F. and Durall, D.M. (2017). A plant growth-promoting symbiosis between *Mycena galopus* and *Vaccinium corymbosum* plantlets. Mycorrhiza, 27: 831–839.

Griffith, G.W. and Hedger, J.N. (1994). The breeding biology of biotypes of the witches' broom pathogen of cocoa, *Crinipellis perniciosa*. Heredity, 72: 278–289.

Guevara, R. and Dirzo, R. (1999). Consumption of macrofungi by invertebrates in a Mexican tropical cloud forest: Do fruit body characteristics matter? J. Trop. Ecol., 15: 603–617.

Guo, S.X., Fan, L., Cao, W.Q., Xu, J.T. and Xiao, P.G. (1997). *Mycena anoectochila* sp. nov. isolated from mycorrhizal roots of *Anoctochilus roxburghii* from Xishuangbanna, China. Mycologia, 89: 952–954.

Hanski, I. (2012). Fungivory: Fungi, insects and ecology. pp. 25–68. *In*: Wilding, N., Collins, N.M., Hammond, P.M. and Webber, J.F. (eds.). Insect-Fungus Interactions, 14th Symposium of the Royal Entomological Society of London in Collaboration with the British Mycological Society, 16–17 Sept. 1987, Imperial College, London, UK. Harcort Brace Javonovich Publishers, Academic Press, London.

Hedger, J., Lewis, P. and Gitay, H. (1993). Litter-trapping by fungi in moist tropical forest. pp 15–36. *In*: Isaac, S., Frankland, J.C., Watling, R. and Whalley, A.J.S. (eds.). Aspects of Tropical Mycology. Cambridge University Press, Cambridge, U.K.

Hedger, J.N. (1985). Tropical agarics: resource relations and fruiting periodicity. pp. 41–86. *In*: Moore, D., Casselton, L.A., Wood, D.A. and Frankland, J.C. (eds.). Symposium on Developmental Biology of Higher Fungi, April 1984, Manchester, England, British Mycological Society Symposium Series, Academic Press, Cambridge University Press, Cambridge, U.K.

Henessy, C. and Daly, A. (2007). Ganoderma Diseases. Darwin: Northern Territory Government. Plant Pathology, Diagnostic Services. Darwin, Australia.

Henkel, T.W., Mayor, J.R. and Woolley, L.P. (2005). Mast fruiting and seedling survival of the ectomycorrhizal, monodominant *Dicymbe corymbose* (Caesalpiniaceae) in Guyana. New Phytol., 167: 543–556.

Henkel, T.W., Terborgh, J. and Vilgalys, R.J. (2002). Ectomycorrhizal fungi and their leguminous hosts in the Pakaraima Mountains of Guyana. Mycol. Res., 106: 515–531.

Hintikka, V. (1970). Studies on white-rot humus formed by higher fungi in forest soils. Communicationes Inst. For. Fenniae, 69: 1–68.

Hu, C.C. (1984). Horse-hair blight, new disease of tea bush caused by *Marasmius equicrinis* Mull in Taiwan. Taiwan Tea Res. Bull., 3: 1–4.

Hughes, D.P., Andersen, S.B., Hywel-Jones, N.L., Himaman, W., Billen, J. and Boomsma, J.J. (2011). Behavioral mechanisms and morphological symptoms of zombie ants dying from fungal infection. BMC Ecol., 11: 13. pmid:21554670.

Jacquemyn, H. and Merckx, V.S.F.T. (2019). Mycorrhizal symbioses and the evolution of trophic modes in plants. J. Ecol., 107: 1567–1581.

Jennings, D.H. (1982). The movement of *Serpula lacrimans* from substrate to substrate over nutritionally inert surfaces. pp. 91–108. *In*: Frankland, J.C., Hedger, J.N. and Swift, M.J. (eds.). Decomposer Basidiomycetes: Their Biology and Ecology. Cambridge University Press, Cambridge, U.K.

Julou, T., Burghardt, B., Gebauer, G., Berveiller, D., Damesin, C. and Selosse, M.-A. (2005). Mixotrophy in orchids: Insights from a comparative study of green individuals and nonphotosynthetic individuals of *Cephalanthera damasonium*. New Phyto., 166: 639653.

Junkšienė, G., Janušauskaitė, D., Armolaitis, K. and Baliuckas, V. (2017). Leaf litterfall decomposition of pedunculate (*Quercus robur* L.) and sessile (*Q. petraea* [Matt.] Liebl.) oaks and their hybrids and its impact on soil microbiota. Dendrobiology, 78: 51–62.

Kepler, R.M., Luangsa-ard, J.J., Hywel-Jones, N.L., Quandt, C.A., Sung, G.-H., Rehner, S.A., Aime, M.C., Henkel, T.A., Sanjuan, T., Zare, R., Chen, M., Li, Z., Rossman, A.Y., Spatafora, J.W. and Shrestha, B. (2017). A phylogenetically-based nomenclature for Cordycipitaceae (Hypocreales). IMA Fungus, 8: 335–353. https://doi.org/10.5598/imafungus.2017.08.02.08.

Kile, G.A., McDonald, G.I. and Byler, J.W. (1991). Ecology and disease in natural forests pp. 101–121. *In*: Shaw III, C.G. and Kile, G.A. (eds.). *Armillaria* Root Disease. Agriculture Handbook No. 691, US Department of Agriculure, Forest Service, Washington, DC, USA.

Klotzbücher, T., Kaiser, K., Guggenberger, G., Gatzek, C. and Kalbitz, K. (2011). A new conceptual model for the fate of lignin in decomposing plant litter. Ecology, 92: 1052–1062.

Koch, R.A., Lodge, D.J., Sourell, S., Nakasone, K.K., McCoy, A.G. and Aime, M.C. (2018). Tying up loose threads: Revised taxonomy and phylogeny of an avian-dispersed Neotropical rhizomorph-forming fungus. Mycol. Progr., 17: 989–998. doi: 10.1007/s11557-018-1411-8.

Kües, U., Nelson, D.R., Liu, C., Yu, G.-J., Zhang, J., Li, J., Wang, X.-C. and Sun, H. (2015). Genome analysis of medicinal Ganoderma spp. with plant-pathogenic and saprotrophic life-styles. Phytochemistry, 114: 18–37. doi: 10.1016/j.phytochem.2014.11.019.

Laliberte, E., Lambers, H., Burgess, T.I. and Joseph Wright, S. (2015). Phosphorus limitation, soil-borne pathogens and the coexistence of plant species in hyperdiverse forests and shrublands. New Phytol., 206: 507–521.

Lavelle, P., Blanchart, E., Martin, A., Martin, S., Spain, A., Toutain, F., Barois, I. and Schaefer, R. (1993). A hierarchical model for decomposition in terrestrial ecosystems: Application to soils of the humid tropics. Biotropica, 25: 130–150.

Lebel, T. and Castellano, M.A. (2002). Type studies of sequestrate Russulales II. Australian and New Zealand species related to *Russula*, Mycologia, 94: 327–354. DOI: 10.1080/15572536.2003.11833240.

Lee, Y.I., Yang, C.K. and Gebauer, G. (2015). The importance of associations with saprotrophic non-Rhizoctonia fungi among fully mycoheterotrophic orchids is currently under-estimated: Novel evidence from subtropical Asia. Annals of Botany, 116: 423–435.

Lindahl, B. and Boberg, J. (2008). Distribution and function of litter basidiomycetes in coniferous forests. pp. 183–196. *In*: Boddy, L. and Frankland, J.C. (eds.). Ecology of Saprobic Basidiomycetes. Academic Press, Elsevier LTD., Amsterdam, the Netherlands.

Lindahl, B., Finlay, R. and Olsson, S. (2001). Simultaneous, bidirectional translocation of ^{32}P and ^{33}P between wood blocks connected by mycelial cords of *Hypholoma fasciculare*. New Phytol., 150: 189–194.

Lindahl, B., Stenlid, J., Olsson, S. and Finlay, R. (1999). Translocation of ^{32}P between interacting mycelia of a wood decomposing fungus and ectomycorrhizal fungi in microcosm systems. New Phytol., 144: 183–193.

Lindahl, B., Taylor, A.F.S. and Finlay, R.D. (2002). Defining nutritional constraints on carbon cycling in boreal forests – Towards a less 'phytocentric' perspective. Plant Soil, 242: 123–135.

Lisboa, D.O., Evans, H.C., Araújo, J.P.M., Elias, S.G. and Barreto, R.W. (2020). *Moniliophthora perniciosa*, the mushroom causing witches' broom disease of cacao: Insights into its taxonomy, ecology and host range in Brazil. Fungal Biol., 124: 983–1003.

Liu, S., García-Palacios, P., Tedersoo, L., Guirado, E., van der Heijden, M.G.A., Wagg, C., Chen, D., Wang, Q., Wang, J., Sing, B.K. and Delgado-Baquerizo, M. (2022). Phylotype diversity within soil fungal functional groups drives ecosystem stability. Nat. Ecol. Evol. https://doi.org/10.1038/s41559-022-01756-5.

Lodge, D.J. (1993). Nutrient cycling by fungi in wet tropical forests. pp. 37–57. *In*: Isaac, S., Frankland, J.C., Watling, R. and Whalley, A.J.S. (eds.). Aspects of Tropical Mycology. BMS Symposium Series 19. Cambridge Univ. Press., Cambridge, U.K.

Lodge, D.J. (1996). Microorganisms. pp. 53–108. *In*: Reagan, D.P. and Waide, R.B. (eds.). The Food Web of a Tropical Forest. Chicago: University of Chicago Press, Chicago, IL, USA.

Lodge, D.J. (2002). Recognizing and treating disease and insect problems of Caribbean trees. *In*: Proceedings of the 5th Annual Caribbean Urban Forestry Conference, St. Croix, Virgin Islands, May 22–25, 2000. Univ. of the Virgin Islands (Cooperative Extension Service), St. Croix, VI.

Lodge, D.J. and Asbury, C.E. (1988). Basidiomycetes reduce export of organic matter from forest slopes. Mycologia, 80: 888–890.

Lodge, D.J. and Cantrell, S. (1995). Fungal communities in wet tropical forests: Variation in time and space. Can. J. Bot. (suppl. 1): S1391–S1398.

Lodge, D.J., Cantrell, S.A. and González, G. (2014). Effects of canopy opening and debris deposition on fungal connectivity, phosphorus movement between litter cohorts and mass loss. For. Ecol. Manag., 332: 11–21. doi: 10.1016/j.foreco.2014.03.002.

Lodge, D.J., McDowell, W.H. and McSwiney, C.P. (1994). The importance of nutrient pulses in tropical forests. Trends in Ecology and Evolution, 9: 384–387.

Lodge, D.J., McDowell, W.H., Macy, J., Ward, S.K., Leisso, R., Claudio Campos, K. and Kuhnert, K. (2008). Distribution and role of mat-forming saprobic basidiomycetes in a tropical forest. pp. 197–209. *In*: Boddy, L. and Frankland, J.C. (eds.). Ecology of Saprobic Basidiomycetes. Academic Press, Elsevier LTD., Amsterdam, the Netherlands.

Lodge, D.J., Van Beusekom, A.E., González, G., Sánchez-Julia, M. and Stankavich, S. (2022). Disturbance reduces fungal white-rot litter mat cover in a wet subtropical forest. Ecosphere, 13: e3936. https://doi.org/10.1002/ecs2.3936.

Luce, R.A. (1970). The phosphorus cycle in a tropical rain forest. pp. H161–H166. *In*: Reagan, D.P. and Waide, R.B. (eds.). The Food Web of a Tropical Forest. Chicago: University of Chicago Press, Chicago, IL, USA.

Macias, A.M., Marek, P.E., Morrissey, E.M., Brewer, M.S., Short, D.P.G., Stauder, C.M., Wickert, K.L., Berger, M.C., Metheny, A.M., Stajich, J.E., Boyce, G., Rio, R.V.M., Panaccione, D.J., Wong, V., Jones, T.H. and Kasson, M.T. (2019). Diversity and function of fungi associated with the fungivorous millipede, *Brachycybe lecontii*. Fungal Ecology, 41: 187–197.

Martos, F., Dulormne, M., Pailler, T., Bonfante, P., Faccio, A., Fournel, J., Dubois, M.-P. and Selosse, M.-A. (2009). Independent recuitment of saprotrophic fungi as mycorrhizal partners by tropical achlorophyllous orchids. New Phytologist, 184: 668–681.

Mayor, J.R. and Henkel, T.W. (2005). Do ectomycorrhizas alter leaf-litter decomposition in monodominant tropical forests of Guyana? New Phytol., 169(3): 579–88 doi: 10.1111/j.1469-8137.2005.01607.x.

McGuire, K.L. (2007). Common ectomycorrhizal networks may maintain monodominance in a tropical rain forest. Ecology, 88: 567–574.

Medina, E., Garcia, V. and Cuevas, E. (1990). Sclerophylly and oligotrophic environments: Relationships between leaf structure, mineral nutrient content, and drought resistance in tropical rainforests of the Upper Río Negro region. Biotropica, 22: 51–64.

Meentemeyer, V. (1978). Macroclimate and lignin control of litter decomposition rates. Ecology, 59: 465–472.

Mehdiabadi, N.J., Hughes, B. and Mueller, M.G. (2005). Cooperation, conflict, and coevolution in the attine ant-fungus symbiosis. Behav. Ecol., 2005: 291–296. doi:10.1093/beheco/arj028.

Meinhardt, L.W., Rincones, J., Bailey, B.A., Aime, M.C., Griffith, G.W., Zhang, D. and Pereira, G.A. (2008). *Moniliophthora perniciosa*, the causal agent of witches' broom disease of cacao: What's new from this old foe?. Molec. Plant Path., 9: 577–588. https://doi.org/10.1111/j.1364-3703.2008.00496.x.

Melillo, J.M., Aber, J.D. and Muratore, J.F. (1982). Nitrogen and lignin control of hardwood leaf litter decomposition dynamics. Ecology, 63: 621–626.

Merckx, V.S.F.T. ed. (2013). Mycoheterotrophy: The Biology of Plants Living on Fungi. Springer. Berlin.

Michaëlla Ebenye, H.C., Taudière, A., Niang, N., Ndiaye, C., Sauve, M., Onguene Awana, N., Verbekens, M., De Kesel, A., Sène, S., Dièdhiou, A.G., Sarda, V., Sadio, O., Cissoko, M., Ndoye, I., Sellose, M.-A. and Ba, M. (2017). Ectomycorrhizal fungi are shared between seedlings and adults in a monodominant *Gilbertiodendron dewevrei* rain forest in Cameroon. Biotropica, 49: 256–267.

Miller, R.M. and Lodge, D.J. (2007). Fungal responses to disturbance—Agriculture and forestry. pp. 44–67. *In*: Esser, K., Kubicek, P. and Druzhinina, I.S. (eds.). The Mycota, 2nd Ed., IV, Environmental and Microbial Relationships. Springer-Verlag, Berlin.

Mohammed, C. and Guillaumin, J.J. (1993). pp. 206–217. *In*: Frankland, J.C., Watling, R. and Whalley, A.J.S. (eds.). Aspects of Tropical Mycology. Isaac., Cambridge University Press, Cambridge, U.K.

Moritz, L., Borisova, E., Hammel, J.U., Blanke, A. and Wesener, T. (2021). A previously unknown feeding mode in millipedes and the convergence of fluid feeding across arthropods. Sci. Adv., 8: 1–9, eabm0577.

Mun, H.-T. (2009). Weight loss and nutrient dynamics during decomposition of *Quercus mongolica* in Mt. Worak National Park. J. Ecol. Field Biol., 32: 123–127.

Nelson, A., Vandergrift, R., Carroll, G.C. and Roy, B.A. (2020). Double lives: Transfer of fungal endophytes from leaves to woody substrates. PeerJ., 8: e9341. Published online 2020 Aug 28. doi: 10.7717/peerj.9341.

Newberry, D.M. and Neba, G.A. (2019). Micronutrients may influence the efficacy of ectomycorrhizas to support tree seedlings in a lowland African rain forest. Ecosphere, 10(4): e02686. 10.1002/ecs2.2686.

Niveiro, N., Ramírez, N.A., Michlig, A., Lodge, D.J. and Aime, M.C. (2020). Studies of Neotropical tree pathogens in *Moniliophthora*: A new species, *M. mayarum*, and new combinations for *Crinipellis ticoi* and *C. brasiliensis*. Mycokeys, 66: 39–54. doi: 10.3897/mycokeys.66.48711.

Oghenekome, U.O. (2004). Natural rubber, *Hevea brasiliensis* (Willd. ex A. Juss.) Müll. Arg., germplasm collection in the Amazon Basin, Brazil: A retrospective. Econ. Bot., 58: 544–555.

Ogura-Tsujita, Y., Gebauer, G., Hashimoto, T., Umata, H. and Yukawa, T. (2009). Evidence for novel and specialized mycorrhizal parasitism: The orchid *Gastrodia confusa* gains carbon from saprotrophic *Mycena*. Proc. R. Soc. B, 276: 761–767. doi:10.1098/rspb.2008.1225.

Osono, T. (2006). Fungal decomposition of lignin in leaf litter: Comparison between tropical and temperate forests. pp. 111–117. *In*: Meyer, W. and Pearce, C. (eds.). Proceedings of the 8th International Mycological Congress, Cairns, Australia, August 20–25. International Mycological Association.

Osono, T. (2020). Functional diversity of ligninolytic fungi associated with leaf litter decomposition. Ecol. Res., 35: 30–43.

Osono, T. and Takeda, H. (2001). Organic chemical and nutrient dynamics in decomposing beech leaf litter in relation to fungal ingrowth and succession during 3-year decomposition processes in a cool temperate deciduous forest in Japan. Ecol. Res., 16: 649–70.

Osono, T. and Takeda, H. (2002). Comparison of litter decomposing ability among diverse fungi in a cool temperate deciduous forest in Japan. Mycologia, 94: 421–427.

Osono, T. and Takeda, H. (2006). Fungal decomposition of Abies needle and Betula leaf litter. Mycologia, 98: 172–179.

Osono, T., Hirose, D. and Matsuoka, S. (2021). Variability of decomposing ability among fungi associated with the bleaching of subtropical leaf litter. Mycologia, 113: 703–714.

Paudel, E., Dossa, G.G.O., de Blécourt, M., Beckschäffer, P., Xu, J. and Harrison, R.D. (2015). Quantifying the factors affecting leaf litter decomposition across a tropical forest disturbance gradient. Ecosphere, 6(12): 267. http://dx.doi.org/10.1890/ES15-00112.1.

Peay, K.G., Kennedy, P.G., Davies, S.J., Tan, S. and Burns, T.D. (2010) Potential link between plant and fungal distributions in a dipterocarp rainforest: Community and phylogenetic structure of tropical ectomycorrhizal fungi across a plant and soil ecotone. New Phytol., 185: 529–542.

Petch, T. (1915). Horse-hair blights. Ann. Royal Bot. Gard. Peradeniya, 6: 1–26.

Połatyńska, M. (2014). Small mammals feeding on hypogeous fungi. Folia Biologica et Oecologica, 10: 89–95.

Rahman, M.M., Tsukamoto, J., Rahman, Md.M., Yoneyama, A. and Mostafa, K.M. (2013). Lignin and its effects on litter decomposition in forest ecosystems. Chem. Ecol., 29: 540–553.

Raich, J.W., Russell, A.E. and Bedoya Arrieta, R. (2007). Lignin and enhanced litter turnover in tree plantations of lowland Costa Rica. For. Ecol. Manag., 239: 128–135.

Rayner, A.D.M. and Todd, N.K. (1982). Population structure in wood-decomposing basidiomycetes. pp. 109–128. *In*: Frankland, J.C., Hedger, J.N. and Swift, M.J. (eds.). Decomposer Basidiomycetes: Their Biology and Ecology. Cambridge University Press, Cambridge, U.K.

Read, J., Sanson, G.D. and Lamont, B.B. (2005). Leaf mechanical properties in sclerophyll woodland and shrubland on contrasting soils. Plant Soil, 276: 95–113. https://doi.org/10.1007/s11104-005-3343-8.

Reich, P.B. and Oleksyn, J. (2004). Global patterns of plant leaf N and P in relation to temperature and latitude. PNAS, 101: 11001–11006.

RStudio Team. (2020). RStudio: Integrated Development for R. RStudio, Version 1.1.456, PBC, Boston, MA URL http://www.rstudio.com/.

Sanchez, M. and Lodge, D.J. (2017). Field trial – Estimating phosphorus uptake and translocation by arbuscular mycorrhizal fungi in surface root mats. Mycological Society of America Meeting, 16–19 July 2017, Athens, GA, USA. https://www.researchgate.net/publication/360947228_Field_Trial-Estimating_phosphorus_uptake_and_translocation_by_arbuscular_mycorrhizal_fungi_in_surface_root_mats.

Santana, M., Lodge, D.J. and Lebow, P. (2005). Relationship of host recurrence in fungi to rates of tropical leaf decomposition. Pedobiologia, 49: 549–564.

Schneider, T., Keiblinger, K.M., Schmid, E., Sternflinger-Gleixner, K., Ellensdorfer, G., Roschitzki, G., Richter, A., Eberl, L., Zechmeister-Boltenstern, S. and Riedel, K. (2012). Who is who in litter decomposition? Metaproteomics reveals major microbial players and their biogeochemical functions. ISME J., 6: 1749–62.

Selosse, M.-A., Scappaticci, G., Faccio, A. and Bonfante, P. (2004). Chlorophyllous and achlorophyllous specimens of Epipactis microphylla (Neottiieae, Orchidaceae) are associated with ectomycorrhizal septomycetes, including truffles. Microbial Ecology, 47: 416–426.

Selosse, M.-A., Schneider-Maunoury, L. and Martos, F. (2018). Time to re-think ecology? Fungal ecological niches are often prejudged. New Phytol., 217: 968–972.

Selosse, M.-A., Martos, F., Perry, B.A., Padamsee, M., Roy, M. and Pailler, T. (2010). Saprotrophic fungal symbionts in tropical achlorophyllous orchids: Finding treasures among the 'molecular scraps? Plant Signaling and Behaviour, 5: 1–5.

Selosse, M.-A., Petrolli, R., María Isabel Mujica, M.I., Laurent, L., Perez-Lamarque, B., Tomáš Figura, T., Bourceret, A., Jacquemyn, H., Li, T., Gao, J., Minasiewicz, J. and Martos, F. (2021). The waiting room hypothesis revisited by orchids: Were orchid mycorrhizal fungi recruited among root endophytes? Ann. Bot., 20: 1–12. https://doi.org/10.1093/aob/mcab134.

Selosse, M.A., Richard, F., He, X.H. and Simard, S.W. (2006). Mycorrhizal networks: Des liaisons dangereuses? Trends in Ecology & Evolution, 21: 621–628.

Selosse, M.-A. and Roy, M. (2009). Green plants that feed on fungi: Facts and questions about mixotrophy. Trends Plant Sci., 14: 64–70.

Séne, S., Avril, R., Chaintreuil, C., Geoffroy, A., Ndiaye, C., Gamby Diédhiou, A., Sadio, O., Courtecuisse, R., Ndao Sylla, S., Selosse M.-A. and Bâ, A. (2015) Ectomycorrhizal fungal communities of *Coccoloba uvifera* (L.) L. mature trees and seedlings in the neotropical coastal forests of Guadeloupe (Lesser Antilles). Mycorrhiza, 25(7). DOI: 10.1007/s00572-015-0633-8.

Shaw III, C.G. and Kile, G.A. (eds.). (1991). Agriculture Handbook No. 691, US Department of Agriculure, Forest Service, Washington, DC, USA. pp. 101–121.

Shaw, M.W. and Vandenbon, A.E. (2007). A qualitative host-pathogen interaction in the *Theobroma cacao-Moniliophthora perniciosa* pathosystem. *Plant Pathol.* doi: 10.1111/j.1365-3059.2006.01549.x.

Shiels, A., González, G., Lodge, D.J., Willig, M.R. and Zimmerman, J.K. (2015). Cascading effects of canopy opening and debris deposition from a large-scale hurricane experiment in a tropical rainforest. Bioscience, 65: 871–881.

Simard, S., Perry, D., Jones, M., Myrold, D.D., Durall, D.M. and Molina, R. (1997). Net transfer of carbon between ectomycorrhizal tree species in the field. Nature, 388: 579–582. https://doi.org/10.1038/41557.

Simard, S.W. and Durall, D.M. (2004). Mycorrhizal networks: A review of their extent, function, and importance. Can. J. Bot., 82: 1140–1165.

Simard, S.W., Beiler, K.J., Bingham, M.A., Deslippe, J.R., Philip, L.J. and Teste, F.P. (2012). Mycorrhizal networks: Mechanisms, ecology and modelling. Fungal Biology Reviews, 26: 39–60.

Singer, R. (1983). Acanthocytes in *Amparoina* and *Mycena*. Cryptogam. Mycol., 4: 111–115.

Smith, M.E., Henkel, T., Aime, M.C., Fremier, A.K. and Vilgalys, R. (2011) Ectomycorrhizal fungal diversity and community structure on three co-occurring leguminous canopy tree species in a Neotropical rainforest. New Phytol., 192: 699–712.

Sourell, S. and Lodge, D.J. (2015). Cristalino Lodge, RPPN Cristalino, Alta Floresta, Mata Grosso, Brazil. Vol. 1. Fungi of Reserva Particular do Patrimonio Natural do Cristalino. Field Museum of Natural History, Rapid Field Guide #719, Chicago, IL, USA. https://fieldguides.fieldmuseum.org/sites/default/files/rapid-color-guides-pdfs/719_brasil_fungi_of_rppn_cristalino_2.pdf.

Sourell, S., Araújo, J.P.M., Sanjua, T., Chaverri, P. and Hoyer, R.C. (2020). Madre de Dios – Entomophathogenic fungi of Tambopata National Reserve. Field Museum of Natural History, Rapid Field Guide #1242, Chicago, IL, USA. https://fieldguides.fieldmuseum.org/sites/default/files/rapid-color-guides-pdfs/1242_peru_entomopathogenic_fungi_of_tambopata.pdf.

Sourell, S., Lodge, D.J., Araújo, J.P.M., Baroni, T., Chaverri, P., Furtado, A., Gilbertoni, T., Karstedt, F., Oliveira, J.J.S., Trierveiler Pereira, L. and Simon Cardoso, J. (2018). Cristalino Lodge, RPPN Cristalino, Alta Floresta, Mata Grosso, Brazil. Vol. 2. Fungi of Reserva Particular do Patrimonio Natural do Cristalino. Field Museum of Natural History, Rapid Field Guide #1047, Chicago, IL, USA. https://fieldguides.fieldmuseum.org/sites/default/files/rapid-color-guides-pdfs/1047_brazil_fungi_of_cristalino_reserve_0.pdf.

Stark, N. (1971). Radiotracer studies of nutrient cycling. Proc. Third Nat. Symposium on Radioecology, Oak Ridge, Tennessee, pp. 130–142.

Stark, N. (1972). Nutrient cycling pathways in litter fungi. Bioscience, 22: 355–360.

Stark, N. and Jordan, C.F. (1978). Nutrient retention by the root mat of an Amazonian rain forest. Ecology, 59: 434–437.

Su, H.J., Thseng, F.M., Chen, J.S. and Ko, W.-H. (2011). Production of volatile substances by rhizomorphs of *Marasmius crinisequi* and its significance in nature. Fungal Divers, 49: 199–202.

Swift, M.J. (1977). The ecology of wood decomposition. Sci. Progr. Oxford, 64: 175–199.

Swift, M.J. (1982). Basidiomycetes as components of forest ecosystems. *In*: Frankland, J.C., Hedger, J.N. and Swift, M.J. (eds.). Decomposer Basidiomycetes: Their Biology and Ecology. Cambridge University Press, Cambridge, U.K.

Tanner, E.V.J., Vitousek, P.M. and Cuevas, E. (1998). Experimental investigations of nutrient limitation of forest growth on tropical mountains. Ecology, 79: 10–22.

Tedersoo, L., May, T.W. and Smith, M.E. (2010). Ectomycorrhizal lifestyle in fungi: Global diversity, distribution, and evolution of phylogenetic lineages. Mycorrhiza, 20: 217–263.

Tedersoo, L., Bahram, M.P., lme, S.K., ljalg, U., Yorou, N.S., Wijesundera, R., Villarreal Ruiz, L., Vasco-Palacios, A.M., Quang Thu, P., Juija, A., Smith, M.E., Sharp, C., Saluveer, E., Saitta, A., Rosas, M., Riit, T., Ratkowsky, D., Pritsch, K.P., ldmaa, K., Piepenbring, M., Phosri, C., Peterson, M., Parts, K., Pärtel, K., Otsing, E., Nouhra, E., Njouonkou, A.L., Nilsson, R.H., Morgado, L.N., Mayor, J., May, T.W., Majuakim, L., Lodge, D.J., Lee, S.S., Larsson, K.-H., Kohout, P., Hosaka, K., Hiiesalu, I., Henkel, T.W., Harend, H., Guo, L.-D., Greslebin, A., Grelet, G., Geml, J., Gates, G., Dunstan, W., Dunk, C., Drenkhan, R., Dearnaley, J., Kesel, A.D., Dang, T., Chen, X., Buegger, F., Brearley, F.Q., Bonito, G., Anslan, S., Abell, S. and Abarenkov, K. (2014). Global diversity and geography of soil fungi. Science Mag., 346: 1078. doi: 10.1126/ science.1256688.

Teixeira, P.J.P.L., Thomazella, D.P.d.T. and Pereira, G.A.G. (2015) Time for Chocolate: Current understanding and new perspectives on Cacao Witches' Broom disease research. PLoS Pathog., 11(10): e1005130. https://doi.org/10.1371/journal.ppat.1005130.

Thorn, R.G. and Barron, G.L. (1984). Carnivorous mushrooms. Science, 224: 76–78.

Thorn, R.G., Moncalvo, J.-M., Reddy, C.A. and Vilgalys, R. (2000). Phylogenetic analyses and the distribution of nematophagy support a monophyletic Pleurotaceae within the polyphyletic pleurotoid-lentinoid fungi. Mycologia, 92: 241–252.

Thorn, R.G., Moncalvo, J.-M., Redhead, S.A., Lodge, D.J. and Martin, M.P. (2005). A new poroid species of *Resupinatus* from Puerto Rico, with a reassessment of the cyphelloid genus *Stigmatolemma*. Mycologia, 97: 1140–1151.

Tlalka, M., Bebber, D., Darrah, P.R. and Watkinson, S.C. (2008). Mycelial networks: Nutrient uptake, translocation, and roles in ecosystems. pp. 43–62. *In*: Boddy, L., Frankand, J.C. and Van West, P. (eds.). Ecology of Saprotrophic Basidiomycetes. British Mycological Society Symposium Series, Academic Press, Elsevier, LTD.

Tlalka, M., Darrah, P.R., Hensman, D., Watkinson, S.C. and Fricker, M.D. (2003). Non-circadian oscillations in amino-acid transport have complementary profiles in assimilatory and foraging hyphae of *Phanerochete velutina*. New Phytol., 158: 325–335.

Tlalka, M., Watkinson, S.C. and Darrah, P.R. (2002). Continuous imaging of amino-acid translocation in intact mycelia of *Phanerochete velutina* reveals rapid, pulsatile fluxes. New Phytol., 153: 173–184. https://doi.org/10.1046/j.0028-646X.2001.00288.x.

Tlalka, M., Watkinson, S.C., Darrah, P.R., Bebber, D. and Fricker, M.D. (2007). Emergence of self-organized oscillatory domains in fungal mycelia. Fungal Genet. Biol., 44: 1085–1095. DOI: 10.1016/j.fgb.2007.02.013.

Torres, P.A., Abril, A.B. and Bucher, E.H. (2005). Microbial succession in litter decomposition in the semi-arid Chaco woodland. Soil Biol. Biochem., 37: 49–54.

Trappe, J.M. and Claridge, A.W. (2005). Hypogeous fungi: Evolution or reproductive and dispersal strategies through interactions with animals and ectomycorrhizal plants. *In*: Dighton, J., Oudemans, P. and White, J. (eds.). The Fungal Community. 3rd edn., CRC Press, Boca Raton, Florida, USA.

Trappe, J.M., Castellano, M.A., Halling, R.E., Osmundson, T.W., Binder, M., Fechner, N. and Malajczuk, N. (2013). Australasian sequestrate fungi 18: *Solioccasus polychromus* gen. & sp. nov., a richly colored, tropical to subtropical, hypogeous fungus, Mycologia, 105:4, 888–895. DOI: 10.3852/12-046.

Trappe, J.M., Claridge, A.W. and Jumpponen, A. (2005). Fire, hypogeous fungi and marsupials. Mycol. Res., 109: 516–518.

Tuno, N. (1998). Spore dispersal of *Dictyophora* fungi (Phallaceae) by flies. Ecol. Res., 17: 7–15.

Uriarte, M., Thompson, J. and Zimmerman, J.K. (2019). Hurricane María tripled stem breaks and doubled tree mortality relative to other major storms. Nat. Commun., 10: 1362. https://doi.org/10.1038/s41467-019-09319-2.

van de Peppel, L.J.J., Nieuwenhuis, M., Auxier, B., Grum-Grzhimaylo, A.A., Cárdenas, M.E., de Beer, Z.W., Lodge, D.J., Smith, M.E., Kuyper, T.W., Franco-Molano, A.E., Baroni, T.J. and Aanen, D.K. (2021). Ancestral predisposition toward a domesticated lifestyle in the termite-cultivated fungus Termitomyces. Curr. Biol., 11: 4413–4421.e5. https://10.1016/j.cub.2021.07.070.

Vasco-Palacios, A.M., Suaza, S.C., Castaño-Betancur, M. and Franco-Molano, A.E. (2008). Conocimiento etnoecólogico de los hongos entre los indígenas Uitoto, Muinane y Andoke de la Amazonía Colombiana. Acta Amazônica, 38: 17–30.

Vincent, A.G. Turner, B.L. and Tanner, E.V.J. (2010). Soil organic phosphorus dynamics following perturbation of tropical moist forest. Europ. J. Soil Sci., 61: 48–57.

Vitousek, P. (1982). Nutrient cycling and nutrient use efficiency. Am. Nat., 119: 553–72.

Waldrop, M.P., Zak, D.R. and Sinsabaugh, R.L. (2004). Microbial community response to nitrogen deposition in northern forest ecosystems. Soil Biol. Biochem., 36: 1443–1451.

Watkinson, S.C. (1984). Morphogenesis of the *Serpula lacrimans* colony in relation to its functions in nature. pp. 165–184. *In*: Jennings, D.H. and Rayner, A.D.M. (eds.). The Ecology and Physiology of the Fungal Mycelium. Cambridge University Press, Cambridge, UK.

Watkinson, S.C., Bebber, D., Darrah, P.R., Fricker, M.D., Tlalka, M. and Boddy, L. (2006). The role of wood decay fungi in carbon and nitrogen dynamics of the forest floor. pp. 151–181. *In*: Gadd, G.M. (ed.). Fungi in Biogeochemical Cycles. Cambridge University Press, Cambridge, UK.

Weigelt, C. and Kunze, G. (1828). Surinam Exsiccati.

Wells, J.M., Harris, M.J. and Boddy, L. (1999). Dynamics of mycelial growth and phosphorus partitioning in developing cord systems of *Phanerochaete velutina*: dependence on carbon availability. New Phytol., 89: 325–334.

Went, F.W. and Stark, N. (1968). Mycorrhiza. Bio-Sci., 18: 1035–1039.

Wilkinson, D.M. (1998). The evolutionary ecology of mycorrhizal networks. Oikos, 82: 407–410.

Wood, T.G. and Thomas, R.J. (2012). Mutualistic association between Macrotermitinae and Termitomyces. pp. 69–88. *In*: Wilding, N., Collins, N.M., Hammond, P.M. and Webber, J.F. (eds.). Insect-Fungus Interactions, 14th Symposium of the Royal Entomological Society of London in Collaboration with the British Mycological Society, 16–17 Sept. 1987, Imperial College, London, UK. Harcort Brace Javonovich Publishers, Academic Press, London.

Wu, B.Y., Nara, K. and Hogetsu, T. (2001). Can C-14-labeled photosynthetic products move between *Pinus densiflora* seedlings linked by ectomycorrhizal mycelia? New Phytol., 149: 137–146.

Zhang, L., Chen, J., Lv, Y., Gao, C. and Guo, S. (2012). *Mycena* sp., a mycorrhizal fungus of the orchid *Dendrobium officinale*. Mycol. Progress, 11: 395–401.

Zimmer, K., Hynson, N.A., Gebauer, G., Allen, E.B., Allen, M.F. and Read, D.J. (2007). Wide geographical distribution of nitrogen and carbon gains from fungi in pyroloids and monotropoids (Ericaceae) and in orchids. New Phytologist, 175: 166–175.

Zimmerman, J.K., Pulliam, W.M., Lodge, D.J., Quiñones-Orfila, V., Fetcher, N., Guzman-Grajales, S., Parrotta, J.A., Asbury, C.E., Walker, L.R. and Waide, R.B. 1995. Nitrogen immobilization by decomposing woody debris and the recovery of tropical wet forest from Hurricane damage. Oikos, 72: 314–322.

Zou, X., Zucca, C.P., Waide, R.B. and McDowell, W.H. (2015). Long-term influence of deforestation on tree species composition and litter dynamics of a tropical rain forest in Puerto Rico. For. Ecol. Manag., 78: 147–157.

3

Ecological Aspects of Ammonia Fungi in Various Vegetation Sites

Akira Suzuki[1,*] and *Neale L Bougher*[2]

1. Introduction

Fungi occurring in nitrogen-rich situations have been recognized in many studies (e.g., Hora, 1959, 1972; Laiho, 1970; Lehmann, 1976; Lehman and Hudson, 1977; Ohenoja, 1978; Egli, 2011). The first report of a specific assemblage of fungi after artificial urea treatment was by Sagara and Hamada (1965) in Japan. Thereafter, Sagara (1973, 1975) treated soils of various vegetation sites in Japan with a range of nitrogen-rich and other compounds. He found similar effects on the occurrence of fungi to urea or aqua ammonia, basic nitrogenous compounds such as ammonium acetate, amines, and L-arginine, and also some polymer compounds (peptone and albumin). However, these effects did not occur with non-basic ammonium salts, nitrate salts, $NaNO_2$, and nitrogen-free compounds. Some fungi observed after such chemical treatments also respond to the application of alkalis such as NaOH and KOH (Sagara, 1975).

The stimulation of fungal fruiting by alkali enrichment may be due to the generation of ammonia from litter by chemical and microbial activities. Based on these results, Sagara (1975) concluded that the combination of a high ammonia

[1] Chiba University, Japan.
[2] Western Australian Herbarium, Biodiversity and Conservation Science, Department of Biodiversity, Conservation and Attractions, Western Australia, Australia.
* Corresponding author: asmush@faculty.chiba-u.jp

level and an alkaline condition is required to induce a succession of assemblages of fungi, thus he coined the term 'ammonia fungi'. The ammonia fungi belong to chemoecological group with exclusive or enhanced reproductive structure formation in the presence of ammonia nitrogen, which reacts as a base (Sagara, 1975). In the early phase soon after the presence of ammonia - in neutral to weakly alkaline soil conditions, certain saprotrophic species of anamorphic fungi rapidly occur. They are usually soon followed by various saprotrophic Ascomycetes and Basidiomycetes. The late phase - in acidic soil conditions, occurs during one or more of ensuing fungal fruiting seasons. Most of the fungi in the late phase are ectomycorrhizal species (Sagara, 1995; Sagara et al., 2008). Under natural conditions, individual or successional assemblages of ammonia fungi are restricted to ecological niches such as near animal corpses, urine and dung depositions, or other ammonia-rich post-putrefaction sites (Sagara, 1975, 1992, 1995; Sagara et al., 2008).

Based on the species assemblage observed in these conditions, this ecological group of fungi are referred to as 'post-putrefaction fungi', a category that includes the ammonia fungi (Sagara, 1995; Sagara et al., 2008). Research on ammonia fungi has been undertaken mostly to describe the particular characteristics of fungal species assemblages specific to nitrogen-enriched sites of various types of vegetation in different biogeographical areas in the world (Suzuki, 2017). Changes in the fungal species assemblage as well as litter decomposition rate and soil properties such as pH, inorganic nitrogen, and moisture content, after input of urea, have been examined in several warm temperate plant communities with moist hot summer (Cfa) region in Japan (Table 1). However, such studies had not been undertaken in vegetation sites in any other climatic conditions. In this article, we highlight the results of the studies undertaken in Japan by comparing the changes in mycobiota, soil properties, and litter decomposition rates after artificial input of urea in several forests: Cfa region and in a warm temperate region with dry hot summers (Csa) (Mediterranean climate).

2. Succession of Ammonia Fungi

The assemblage of fungal species after a large amount of urea application has been documented in a wide range of habitats in Cf and Df regions in Japan (coniferous forests, evergreen broad-leaved forests, and deciduous broad-leaved forests) (Sagara, 1975; Yamanaka, 1995a, b; Fukiharu and Hongo, 1995; Fukiharu et al., 1997; Sato and Suzuki, 1997; Suzuki et al., 2002; He and Suzuki, 2004; Imamura and Yumoto, 2004), a *Pinus* dominated forest in the Dwa region in China (Fukiharu et al., 2013), boreal forests in some Dfb regions, i.e., a lodgepole pine (*Pinus contorta*) forest and an aspen (*Populus tremuloides*) forest in Canada (Raut et al., 2011, 2015). Also, studies have been undertaken in *Nothofagus* forests in a Cfb region in New Zealand (Suzuki et al., 2003; Fukiharu et al., 2011), a mixed deciduous/bamboo forest in an Aw region, Thailand (Manusweeraporn et al., 2013), *Pinus* forests, and a *Castanopsis* spp. and a *Lithocarpus* mixed forest in an Aw region in Vietnam (Ho et al., 2013; Nguyen et al., 2019; Pham et al., 2019).

Table 1. Occurrence of ammonia fungi in various vegetation site in the Csa region of Japan, where changes in soil properties have been examined.

Fungal species	**Effective season**	**Vegetation**						
	for the urea treatment*	**I**	**II**	**III**	**IV**	**V**	**VI**	**VII**
Reference	**Sagara et al. (2008)**	**Sagara (1975)**	**Yamanaka (1993)**	**Sato and Suzuki (1997)**	**Fukiharu et al. (1997)**	**Suzuki (2000)**	**Suzuki et al. (2002)**	**He and Suzuki (2004)**
Early phase of ammonia fungi								
Amblypsporium botrytis	Winter and Summer		+	+	+	+	+	+
Ascobolus denudatus			+	+		+	+	+
Peziza moravecii	Winter	+	+	+			+	+
Peziza urinophila	Summer		+					
Psudombrophila petrakii	Winter	+	+	+		+	+	
Lyophyllum ambustum	Winter							
Lyophyllum tylicolor	Winter	+	+			+	+	
Coprinopsis cinerea					+			
Coprinopsis echinospora	Winter	+	+		+			
Coprinopsis phlyctidospora	Summer			+		+	+	+
Crucispora rhombisperma	Summer?				+			
Late phase of ammonia fungi								
Collybia cookei			+					
Hebeloma radicosoides	Winter and Summer		+					
Hebeloma spoliatum				+		+	+	
Hebeloma vinosophyllum	Summer		+					

Table 1 contd. ...

...Table 1 contd.

Fungal species	**Effective season for the urea treatment***	**Vegetation**						
		I	**II**	**III**	**IV**	**V**	**VI**	**VII**
Reference	**Sagara et al. (2008)**	**Sagara (1975)**	**Yamanaka (1993)**	**Sato and Suzuki (1997)**	**Fukiharu et al. (1997)**	**Suzuki (2000)**	**Suzuki et al. (2002)**	**He and Suzuki (2004)**
Laccaria bicolor	Winter and Summer	+	+					
Lepista sordidata								+
Lepista sp.				+				

I: *Pinus densiflora - Chamaecyparis obtuta* forests. Urea (85.8 g/m^2) is applied in December.
II: *Pinus densiflora* forest. Urea (700 g/m^2) is applied in April.
III: *Quercus serrata dominated mixed* forest. Urea (800 g/m^2) is applied in May.
IV: *Phyllostachys reticulata* grove. Urea (700 g/m^2) is applied in July.
V: *Abies firma* and *Quercus seerata* dominated mixed forest. Urea (800 g/m2) is applied in May.
VI: *Abies firma* and *Quercus* spp. dominated mixed forest.
VII: *Lithocarpus edulis* forest in Japan. Urea (1600 g/m^2) is applied in June
*: Determined by suitable season for the urea application to induce occurrence of each species.

Analyses of the changes in the fungal species assemblage coupled with those in soil properties caused by a sudden input of a large quantity of urea have been conducted in several forests including a grove in the Cfa region of Honshu Island, Japan (Table 1), but not in vegetation sites in any other climatic areas. In the ecological studies, the dose of urea applied ranged between 85.8 g/m^2 and 1600 g/m^2 in each plot (Table 1). All those research sites in Cfa regions are characterized by having brown soil with the parent rocks composing sandstone, slate, and or mudstone, and the pH of their forest floors is 3.6–7.1 throughout the year (see Table 3). Nitrogen fertilization for forests, mostly for management, has been carried out in various vegetation sites around the world (Ohenoja, 1978, 1988; Salo, 1979; O'Connell, 1994; Wallenstein et al., 2006; Flinct et al., 2008; Wright et al., 2009; Faustino et al., 2015; Sullivan and Sullivan, 2018), and in some cases, the mycobiota of the fertilized forests has been documented, with a particular focus on the ectomycorrhizal fungi (Ohenoja, 1978, 1988). However, in such studies, the dose of fertilizer used for nitrogen fertilization is 1/3–1/100 folds lower than that of the urea enrichment experiments conducted for the assessment of ammonia fungi.

Building upon such previous research, we conducted a nitrogen enrichment study with different amounts of urea in a Jarrah Forest (Table 2), rich in fire-adapted communities and organisms, in a Csa Mediterranean climatic region in order to document there the 'urea effect' as conceptualized by nitrogen enrichment studies in other areas such as forests in the Cfa region (Sagara, 1975, 1976a). In Cfa forests, vegetation and or the time of urea treatment within a year influences the occurrence of ammonia fungi (Sagara, 1975, 1976a; Sagara et al., 2008). Saprotrophic ammonia fungi are more influenced by the timing of treatment, whereas ectomycorrhizal ammonia fungi are more influenced by vegetation factors (Sagara, 1975; Imamura and Yumoto, 2004). In the Jarrah Forest, autumn rains typically induce annual episodes of fungal fruiting predominantly between May and August. Our studies show that the application of urea in the Jarrah Forest produces a succession of 7 fungi: *Ascobolus denudatus*, *Peziza moravecii*, *Thecotheus urinamans* (Nagao et al., 2003), unidentified *Peziza* species, *Lyophyllum tylicolor* (Suzuki, 2017), *Coprinopsis austrophlyctidospora* (Fukiharu et al., 2011), and *Hebeloma aminophilum* (Fig. 1A, B, Table 2). Three of the fungi, i.e., the ubiquitous saprotrophic species, *As. denudatus*, *P. moravecii* and *L. tylicolor* (Suzuki et al., 2002) occurred on all urea-treated plots (Table 2). However, their fruit bodies were generally more abundant on plots with the higher urea treatments (400 and 800 g urea/m^2) than on plots with the lowest urea treatment (100 g urea/m^2). All of the other fungi were restricted to plots with particular levels of urea treatment: *T. urinamans*, possibly an endemic saprotrophic species, occurred on the 100 g, 400 g, and 800 g urea/m^2 treated plots, an endemic saprotrophic fungus *C. austrophlyctidospora* on the 200–800 g urea/m^2 treated plots. *As. denudatus*, *P. moravecii*, and *Peziza* sp. occurred ca. 4 weeks after the urea application. *T. urinamans*, and *L. tylicolor* occurred 1 week and 2 weeks later than the former three fungi, respectively. Thereafter, *C. austrophlyctidospora* occurred 5 weeks later than the appearance of the initial three fungi, i.e., during the final stage of the occurrence period of *As. denudatus* and *P. moravecii*

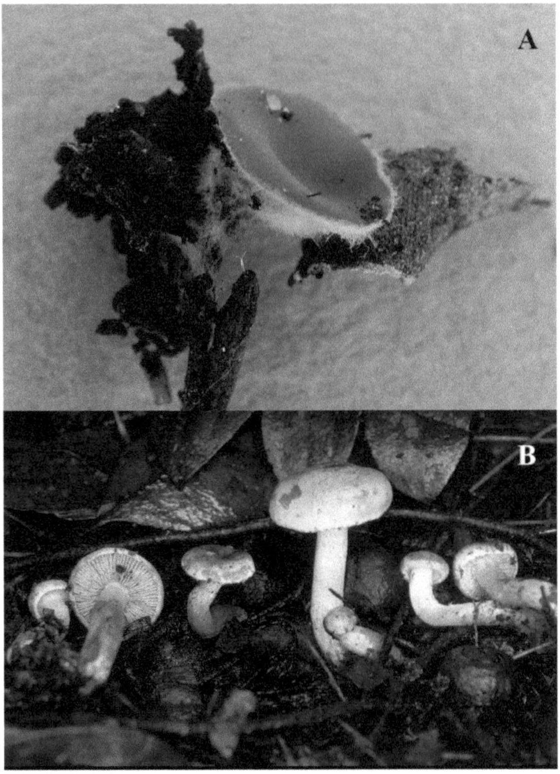

Fig. 1. An ascoma of *Thecotheus urinamans* (PERTH 07598513) appeared 33 days after 400 g/m² of urea application (A); Basidiomata of *Hebeloma aminophilum* (PERTH 7696892) appeared 406 days after 800 g/m² of urea application (B).

(Table 2). Extremely abundant fruit bodies were produced by *As. denudatus*, *P. moravecii* and *L. tylicolor* on most of the urea-treated plots and continued to occur for more than 1 month (Table 2). Fruit bodies of *P. moravecii* and *C. austrophlyctidospora* were generally more abundant on plots with the higher urea treatments (400, 800 g urea/m²) than on plots with the lowest (100, 200 g urea/m²) (Table 2). A putatively mycorrhizal fungus *H. aminophilum* is known as a 'sarcophilous fungus' (Miller and Hilton, 1986/1987; Young, 2002) occurred only during 800 and 1600 g urea/m² treatments of the second year of urea application (Table 2). The fruit bodies collected from the urea plots were relatively small compared to specimens previously observed in other areas near carcasses (Hilton, 1978; Miller and Hilton, 1986/1987; Young, 2002). The occurrence of *H. aminophilum* on the plots reconfirmed its status as an ammonia fungus distributed in Australia as well as New Zealand (Suzuki et al., 2003). In general, most fruiting activity emanated from partially decomposed litter hidden below the surface coarse litter.

Several or more fungi were observed both inside and outside the plots after 1 year of 100, 200 and 400 g/m² of urea application and 2 years after 800 g/m² of urea application (Table 2). Each of these produced only one fruit body inside the

Table 2. Successive occurrence of ammonia fungi after different amounts of urea application in a Jarrah Forest, Western Australia. Experimental plots were established in a Jarrah Forest located north of Dwellingup, Western Australia (32°34'00"S, E116°05'00"E, 330 m alt). The site has a lateritic soil and supports a forest structure typical of the northern jarrah forest - *Eucalyptus marginata* and *Corymbia calophylla* are the dominant overstorey trees; smaller trees include *Banksia grandis*, and the understorey has a large number of shrub and herbaceous plant species. Fertilizer urea was applied as granules to two plots of 4 m x 4 m (100 g urea/m2) and three plots each of: 4 m x 4 m (100 g urea/m^2), 2 m x 4 m (200 g urea/m^2), 2 m x 2m (400 g urea/m^2), and 1 m x 2 m (800 g urea/m^2), on 18 July 1996. Three control plots (1 x 2 m) did not receive urea.

Fungal species	Observation period												
	1996									1997		1998	
	8.6	8.13	8.20	8.26	9.3	9.10	9.19	9.24	11.19	7.7	7.14	7.27	7.27
	[19]	[26]	[33]	[39]	[47]	[54]	[63]	[68]	[124]	[354]	[361]	[374]	[740]
Ascobolus denudatus		○ △	○ △	○ △	△								
		▲	● ▲	● ▲	● ▲	▲	▲	▲					
Peziza moravecii		○ △	○ △	○ △	○ △	○ △	○ △	△					
			● ▲	● ▲	● ▲	● ▲	● ▲	● ▲					
Peziza sp.		△	○ △				△						
		●	●		▲			▲					
Thecotheus urinamans			○										
			●	● ▲									
Lyophyllu tylicolor				△	○ △	○ △	○ △	○ △					
					● ▲	● ▲	● ▲	● ▲					
Coprinopsis austrophlyctidospora							△	△					
							● ▲	● ▲					
Hebeloma aminophilum													
											▲	▲	

Table 2 contd. ...

...Table 2 contd.

Numerical numbers in the bracket indicate days after urea application.

Appearance of fruiting bodies (occurrence) of ammonia fungi in each urea plot, i.e., ○: 100 g urea/m^2, Δ: 200 g urea/m^2, ●:400 g urea/m^2, ▲:800 g urea/m^2.
In the Jarrah Forest, *H. aminophilum* was also collected 1 year after 800 g/m^2 and 1600 g/m^2 of urea treatments on May 8, 1997 and June 30, 1997, respectively.
The occurrence of each fungal species under different urea level is shown by an integrated collection record in all plots of each amount of the urea application.
A. denudatus: PERTH 07598971, PERTH 07598556, PERTH 07599048, PERTH 07598459. PERTH 07598629, *P. moravecii*: PERTH 07598181, PERTH 07598203, PERTH 07599099, PERTH 07598203, *Peziza* sp.: PERTH 07665547, PERTH 07599064, PERTH 07598513; *T. urinamans*: PERTH 07599056, PERTH 07599021, PERTH 07599153, PERTH 07598513; *L. tylicolor*: PERTH 07598246, PERTH 07598254, PERTH 07598262, ERTH 07598270, PERTH 07598335, PERTH 07598343, PERTH 07598378, PERTH 07598386; *C. austrophlyctidospora*: PERTH 07599137, PERTH 07598440; *H. aminophilum*: PERTH 07697295, PERTH 07697090, PERTH 07696833, PERTH 07696841.

Non-ammonia fungi occurred in each urea plot were as follows:

100 g urea/m^2: In 1997, *Cortinarius radicatus, Russula clelandii, Ru. flockonae* and *Mycena* sp.; In 1998, an unidentified Aphyllophorales, *Coltricia oblectans*, *C. radicatus*, *Cortinarius* sp., *Daldinia* sp., *Dermocybe* sp., *Galerina* sp., *Hysterangium* sp., *Mycena* sp., *Hydnoplicata convoluta*, *Ramaria* sp., *Ru. flocktonae*, *Ru. neerimea*, *Stephanospora flava*, *Stephanospora* sp.
200 g urea/m^2: In 1997, *De. austroveneta*; In 1998, *Cortinarius* sp., *Galerina* sp., *Hysterangium* sp., *Nothocastoreum cretaceum*, an unidentified Pezizales, *S. flava, Stephanospora* sp.
400 g urea/m^2: In 1997, dark truffle; In 1998, *Coltricia oblectans*, *Endogone* sp., *N. cretaceum.*
800 g urea/m^2; In 1997, none; In 1998, *Cortinarius* sp., *Daldinia* sp., *Hysterangium* sp., *Inocybe* sp., *Mycena* sp., *Stephanospora* sp. (Unpublished data).

urea-treated plots and they were much more abundant in the surrounding forest. The occurrence frequency of such non-ammonia fungi in the Jarrah Forest site became larger, roughly proportional to decreased levels of applied urea amounts (Table 2). In this site, it suggests that recovering of fungal communities occurring prior to the nitrogen disturbance had already initiated around 1 year after the urea application. Probably these non-ammonia fungi are able to survive in niches where NH_4-N levels are not artificially elevated. This may be partly explained by the different characteristics of the fungal community established in the forests, such as *Eucalyptus* forests in Australia and *Pinus* forests in Canada grown in high fire disturbance areas since fire disturbance induces an increase in pH associated with the increment of NH_4-N nitrogen (Wang et al., 2014). Recovering time for non-ammonia fungi after high nitrogen disturbance in the Jarrah Forest appears to be shorter than that observed in the forests in Csa regions, in spite of the somewhat slower recovering time of pH and inorganic nitrogen concentrations in the Jarrah Forest (see Table 4, Fig. 2).

To confirm the presence or absence of nitrogen-responsive colonization by anamorphic fungi in the Jarrah Forest such as *Amblyosporium botrytis* which has been recorded from Cfa and Df regions (see Table 1) (Suzuki, 2006), a mixture of coarse and fine litters from the Jarrah Forest was placed in plastic pots, and then treated separately with 21.4 or 85.7 mg of urea/g dry litter. Fruit bodies of *L. tylicolor* occurred with the former dose treatment after 28 days of incubation at room temperature, but no colonization of any anamorphic fungi was observed (by naked eye observation) in both doses of treatment. Colonization of anamorphic fungi, which has been reported in all studies in most vegetation sites in the Cfa and Df regions, remains unconfirmed in the Jarrah Forest and will require microscopic examinations to be more accurately assessed.

The assemblage of ammonia fungi in the Jarrah Forest comprised both ubiquitous saprotrophic species such as *As. denudatus*, *P. moravecii* and *L. tylicolor*, and endemic saprotrophic species such as *T. urinamans* and *C. austrophlyctidospora* as well as endemic ectomycorrhizal species such as *H. aminophilum* (Suzuki et al., 2003; Suzuki, 2009, 2017).

The findings in the Jarrah Forest in south Western Australia indicate that there are parallelisms between the successional occurrence patterns and fungal species assemblage of forests in the Cfa region in Japan and that of at least one type of forest in the Csa region in Australia (Suzuki et al., 1998). In both climatic regions, the fungal species assemblage of the early phase and the late phase of ammonia fungi is composed of species with different distribution ranges, such as species distributed in both hemispheres: worldwide distribution species such as *A. denudatus*, *P. moravecii* and *L. tylicolor*; Palearctic ecozone species such as *C. phlyctidospora* and *C. echinospora*; and those restricted to one or two bioregion(s) - endemic species such as *C. austrophlyctidospora*, *T. urinamans*, *H. aminophilum*, and *Hebeloma vinosophyllum* (Suzuki et al., 2003; Suzuki, 2009, 2017). Unlike the saprotrophic early phase and late phase of ammonia fungi is mostly ectomycorrhizal (Sagara, 1995) and are more specific to vegetation type and therefore generally less widely distributed.

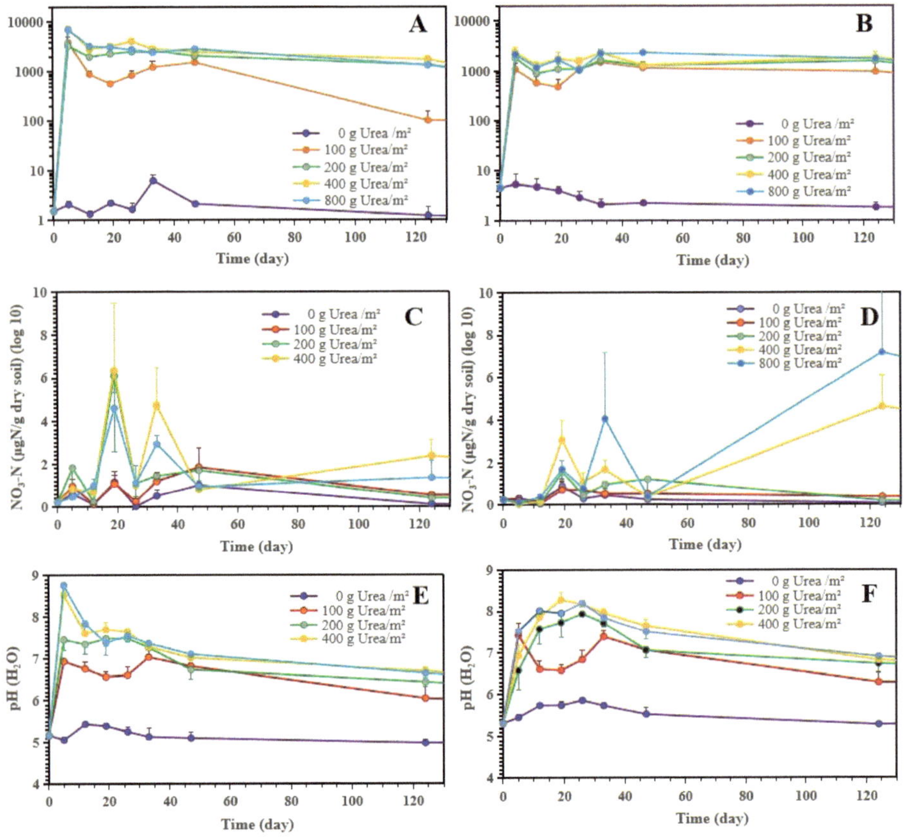

Fig. 2. Changes in soil properties in the Jarrah Forest. Twenty grams of the fresh sample described below was suspended in 80 mL of 1 M KCl (A-D) and 80 mL of distilled water (E, F), respectively. The suspensions kept at 20°C were shaken at 32 rpm for 60 min, and then filtrated by filter paper (Whatman, No. 42). Inorganic nitrogen concentration of the filtrate was analysed by Lachat Instrument Injection Auto analyser Quick Chem. pH of the filtrate was examined by a glass electrode. [NH_4-N content (A, B), NO_3-N content (C, D), and pH (E, F) of the litter]. Coarse litter sample: Litter collected from upper part of O horizon (A, C, E); Fine litter sample: A mixture of litter collected from lower part of O horizon and soil collected from surface of A horizon (B, D, F)] (Suzuki et al., unpublished data (A-F).

3. Changes in Soil Properties after High Ammonia-Nitrogen Impact

The impact of high nitrogen on soil properties has been examined in several plant communities, such as a *Pinus densiflora* forest, an *Abies firma* and *Quercus serrata* dominated mixed forest, an *A. firma* and *Quercus* spp. dominated mixed forest, and a *Lithocarpus edulis* (syn.: *Pasania edulis*) forest, grown on weak acid soils in the Cfa region, Japan (Yamanaka, 1993, 1995a; Suzuki, 2000; Suzuki et al., 2002; He and Suzuki, 2004). The soil properties, especially the inorganic nitrogen content of the above forest floors vary pH 3.4–7.1, NH_4-N 2.3–220 μg N/g dry soil, NO_3-N 0.2–810 μg N/g dry soil (Table 3). In the Jarrah Forest in a Csa region, Australia, pHs, NH_4-N contents, and NO_3-N contents of forest floors are 5.0–5.9, 0.6–6.4 μg N/g dry soil, and 0.1–1.0 μg N/g dry soil, respectively (Table 3). The soil properties indicate that the Jarrah Forest occurring on lateritic soils is a typical nitrogen-poor and weak acidic site (as shown by O'Connell and Grove, 1996). In the *P. densiflora* forest, *A. firma* and *Q. serrata* dominated mixed forest, the *A. firma* and *Quercus* spp. dominated the mixed forest, and the *L. edulis* forest, NH_4-N contents of the forest organic soil rapidly increase 12–5236 folds higher than those of control soils within 2 weeks by 700–1600 g/m^2 of urea treatment, and gradually declines to control level 1.5 or more years after the urea application (Tables 3, 4).

In the Jarrah Forest, NH_4-N contents of coarse and fine litters increased to maximum, i.e., 142–1758 folds higher than the control for the former and 229–978 folds higher than the control for the latter, proportional to the applied amounts of urea (100–800 g/m^2), 1 week after the urea application (Fig. 2A, B). Thereafter, NH_4-N contents of coarse and fine litters maintained a high level for another 6 weeks followed by declining to control level within 1–2 years after the urea application (Tables 3, 4). These results indicate that the pattern of change in NH_4-N contents after a large amount of urea application is similar among the examined forests, irrespective of vegetation, although declining rates in NH_4-N of the Jarrah Forest are slower than those of the examined forests in Cfa areas.

The changing patterns of NO_3-N contents in the examined forests vary depending on vegetation type, i.e., those of L-F as well as H-A horizons in the *A. firma* and *Q. serrata* dominated mixed forest, the *A. firma* and *Quercus* spp. dominated mixed forest, the *P. densiflora* forest, and the *L. edulis* forest show three, three, two, and two peaks, respectively (Suzuki, 2000; Suzuki et al., 2002; Yamanaka, 1995a; He and Suzuki, 2004). The NO_3-N contents of L-F and H-A horizons in the former three forests reach to maxima within 2 weeks after the urea application, namely, at the same reaching time for maximal values of NH_4-N contents of both horizons (Table 4) (Suzuki, 2000, Suzuki et al., 2002, He and Suzuki, 2004). The maximal values of the L-F and H-A horizons show considerable variations, i.e., separately 120–200 and 24–481 folds higher than those in the control level (Tables 3, 4). The maximum NO_3-N contents of the F and H horizons of the urea-treated soils in the *P. densiflora* forest reach separately 20900 and 36000 folds higher than those in the control level about 5 months later than the occurrence of the peak time in NH_4-N content (Table 4) (Yamanaka, 1995a). NO_3-N contents of the urea-treated soils in the four forests return to those control levels within 1.5–2 years (Table 4). In the Jarrah Forest, NO_3-N contents of coarse and fine litters after 100–800 g/m^2 of urea application showed 3 peaks and reached the highest values two or more weeks later than the time elapsed for the maximum of NH_4-N contents in each urea treatment (Fig. 2 C, D). Maximal NO_3-N contents of urea-treated litters in the Jarrah Forest were only 0.03%–0.05% of those typically recorded with urea-treated soils of forests in the Cfa region. Finally, NO_3-N contents of coarse and fine litters of all urea-treated plots declined to control level within 1–2 years after the urea application (Tables 3, 4).

Table 3. Differences in soil properties among different vegetation sites.

Vegetation	Soil horizon	Soil properties					Reference
		pH	NH_4-N (µg N/g dry soil)	NO_3-N (µg N/g dry soil)	NO_2-N (µg N/g dry soil)	Water content (%)	
Pinus densiflora and *Chamaecyparis obtusa* forest	O	3.4–4.1*[3]				63–74*[3]	Sagara (1975)
Pinus densiflora forest	F	3.6–4.0	8.3–46.8	1.5–5.5		64–77	Yamanaka (1993)
	H	3.6–4.1	19.1–46.8	1.1–5.3		62–77	
Quercus serrata dominated mixed forest	L-F	5.6–7.1*[3]				59–72*[3]	Sato and Suzuki (1997)
	H-A	4.9–6.1*[3]				40 - 82*[3]	
Phyllostachys reticulata grove	L-H	5.9–6.7*[3]				41–66*[3]	Fukiharu et al. (1997)
	A	5.7–6.9*[3]				46–79*[3]	
Abies firma and *Quercus serrata* dominated mixed forest	L-F	5.7–6.3*[3]	4.7–155.6*[3]	1.6–45.2*[3]		29–70*[3]	Suzuki (2000)
	H-A	4.9–6.1*[3]	2.3–116.7*[3]	0.2–15.8*[3]		12–64*[3]	
Abies firma and *Quercus* spp. mixed forest	L-F	4.6–5.7*[3]	50–220*[3]	10–810*[3]	0.2–0.8*[3]	38–71*[3]	Suzuki et al. (2002)
	H-A	4.0–5.4*[3]	10–40*[3]	20–460*[3]	0.1–0.5*[3]	38–77*[3]	
Lithocarpus edulis forest	L-F	5.9–6.9*[3]	5.7–25.8*[3]	7.5–91.1*[3]		11–53*[3]	He and Suzuki (2004)
	H-A	6.6–6.9*[3]	4.7–25.8*[3]	4.5–20.3*[3]		27–57*[3]	
Jarrah Forest: *Eucalyptus marginata* and *Corymbia calophylla* dominated mixed forest	O*[1]	5.0–5.4	0.6–6.2	0.1–1.0		30–70	
	O*[2]	5.3–5.9	1.5–6.4	0.1–0.9		27–73	

*[1, 2], See legend for Table 3.
*[2], See legend for Table 4.
*[3], Data estimated from graphs in the references.
Blank: not measured.

Table 4. Differences in soil properties after urea application in different vegetation sites.

Vegetation	Soil horizon	Applied	Soil properties							Reference
		Urea (g/m^2)	Maximum pH	Return time (month)	Max. NH_4-N (µg N/g dry soil)	Return time (month)	Max. NO_3-N (µg N/g dry soil)	Return time (month)	Max. NO_2-N (µg N/g dry soil)	
Pinus densiflora and *Chamaecyparis obtusa* forest	O	85.8	8.9*[3]	ca. 24						Sagara (1975)
Pinus densiflora forest	F	700	9.0*3	Ca. 7, 18	39811*[3]	Ca. 18	6270*[3]	Ca. 18		Yamanaka (1995a)
	H		8.6*3	Ca. 7, 18	10000*3	Ca. 18	10800*[3]	Ca. 18		
Quercus serrata dominated mixed forest	L-F	800	9.0*[3]	Ca. 1, 6						Sato and Suzuki (1997)
	H-A		9.1*[3]	Ca. 6						
Phyllostachys reticulata grove	L-H	800	8.9*[3]	Ca. 5						Fukiharu et al. (1997)
	A		8.9*[3]	Ca. 7						
Abies firma and *Quercus serrata* dominated mixed forest	L-F	800	8.9*[3]	Ca. 4, 17	6222*[3]		2032*[3]			Suzuki (2000)
	H-A		9.0*[3]	Ca. 5, 17	1400*[3]		677*[3]			
Abies firma and *Quercus* spp. mixed forest	L-F	800	9.5*[3]	Ca. 4, 9 <	40000*[3]	9 <	9620*[3]	9 <	15.6*[3]	Suzuki et al. (2002)
	H-A		9.4*[3]	Ca. 4, 9 <	30400*[3]	9 <	24000*3	9 <	38.6*[3]	
Lithocarpus edulis forest	L-F	1600	9.0*[3]	Ca. 6	2832*[3]	18 <	2235*[3]	18 <		He and Suzuki (2004)
	H-A		8.9*[3]	Ca. 10	2832*[3]	18 <	2235*[3]	18 <		

Table 4 contd. ...

...Table 4 contd.

Vegetation	Soil horizon	Applied	Soil properties							Reference
		Urea (g/m²)	Maximum pH	Return time (month)	Max. NH_4-N (µg N/g dry soil)	Return time (month)	Max. NO_3-N (µg N/g dry soil)	Return time (month)	Max. NO_2-N (µg N/g dry soil)	
Jarrah Forest: *Eucalyptus marginata* and *Corymbia calophylla* dominated mixed forest	O*[1]	100	7.0	Ca. 11 - 25	879	Ca. 11 - 25	1.9	Ca. 11–25		
	O*[2]		7.4	Ca. 11 - 25	1467	25 <	0.7	Ca. 4–11		
	O*[1]	800	8.8	Ca. 11 - 25	6980	25 <	4.6	25 <		
	O*[2]		8.2	Ca. 11 - 25	2301	25 <	7.2	Ca. 11–25		

*[1], See legend for Fig. 3.
*[2], See legend for Fig. 4.
*[3], Data estimated from graphs in the references.
Blank: not measured.

Changes in NO_2-N concentration after a large amount of urea application has only been reported in the *A. firma* and *Quercus* spp. mixed forest. The NO_2-N contents of L-F and H-A horizons increases somewhat slowly and reach two weeks later (16–39 µg NO_2-N/g dry soil, 16–77 folds higher than the control level) at the period of maximum contents in NH_4-N and NO_3-N although the maximum contents of NO_2-N are 0.04–0.16% or less of those in NH_4-N and NO_3-N. Thereafter in the *A. firma* and *Quercus* spp. mixed forest, NO_2-N contents of both horizons slowly decline to the control level (Suzuki et al., 2002). Generally, nitrite is toxic to many fungi, particularly under acidic conditions (Cochrane, 1958; Garraway and Evans, 1984), but saprotrophic ammonia fungi, e.g., *Amblyosporium botrytis*, *Ascobolus denudatus*, and *Coprinopsis phlyctidospora* grow well when NO_3-N is the sole nitrogen source under weak acidic to weak alkaline conditions (Barua et al., 2012a).

The soil pH of the vegetation sites in the Cfa region increases up to maximum value, 8.6–9.5 within 2 weeks and drops to marginally higher than the control level about 2–7 months after the urea application, although another 2–18 months elapses until completely returning to the control level by 86–1600 g/m^2 of urea application (Tables 3, 4). In the Jarrah Forest in a Csa region, the pH of the coarse litter increased up to maximum value, 6.9–8.8 within 2 weeks by 100–800 g/m^2 of urea application (Fig. 2E, F) whereas those of the fine litter increased up to maximum value, 7.4–8.2 after 4–5 weeks after 100–800 g/m^2 of urea application. Thereafter, the pH of fine and coarse litters slowly declines to the control level about 1–2 years after the urea application (Tables 3, 4). The increment in pH of the forest soils is tightly linked with the increment of NH_4-N contents in all examined vegetation sites. Declining rates of pH in the vegetation sites in the Cfa region are significantly higher compared to those of inorganic nitrogen contents, whereas declining rates of pH and inorganic nitrogen contents in the Jarrah Forest in the Csa region are similar (Yamanaka, 1995a; Suzuki, 2000; Suzuki et al., 2002; He and Suzuki, 2004) (Fig. 2A–F). Slower declining rates of soil pH and lower level of NO_3-N enhancement subsequent to the urea application appear to be characteristic of the Jarrah Forest.

The water content of the O horizon of the *Pinus*-Chamaecyparis Forest in a Cfa region in Japan was found to increase for 8 months longer than that of the control following an application of 86 g urea/m^2 (Sagara, 1975). Sagara (1975) contended that the increment is possibly due to the increase in the water-holding capacity of organic matter by enhanced decomposition. A similar increase in water content of litter and soils following an application of 700–1,600 urea/m^2 is found in several vegetation sites of Cfa region in Japan such as the *P. densiflora* forest, the *Q. serrata* dominated mixed forest, the *A. firma* and *Q. serrata* dominated mixed forest, the *A. firma* and *Quercus* spp. dominated forest, and the *L. edulis* forest (Yamanaka, 1995a; Sato and Suzuki, 1997; Suzuki, 2000; Suzuki et al., 2002; He and Suzuki, 2004). While, the water content of the A horizon does not change in a *Phyllostachys reticulata* (syn.: *Phyllostachys bambusoides*) grove in a Cfa region following the application of 800 g urea/m^2 (Fukiharu et al., 1997). In the Jarrah Forest, water contents and water holding capacities of fine and coarse litters were not affected by

100–800 g/m^2 of urea application. These findings indicate that the increase of water content observed in several urea-enriched forest floors is not a universal phenomenon after urea treatment.

The rapid increase in NH_4-N concentration associated with rising pH after urea application on the floors of plant communities is likely to be caused by the urease activity of microbes, e.g., ureolytic bacteria which are common in soil (Lloid and Sheaffe, 1973), and fungi which have urease activities (Barua et al., 2012b). This assumption is supported by the following two studies: Detection of high urease activity from urea-treated O and A horizons of a *Castanopsis sieboldii* (syn: *Castanopsis cuspidata*) forest, a *Q.* forest, and a *C. obtusa* forest in Cfa region, Japan (Imamura et al., 2006). Ammonia fungi, especially saprotrophic ammonia fungi have strong urease activity and grow well in a liquid medium containing a high concentration of urea as the sole nitrogen source (Barua et al., 2012b). While, high concentrations of urea are toxic to some non-ammonia fungi (Veterka et al., 2007).

Early phase ammonia fungi, i.e., saprotrophic ammonia fungi, occur during high NH_4-N and NO_3-N concentrations associated with weak alkaline to neutral conditions whereas ectomycorrhizal fungal species of the late phase of ammonia fungi occur after NH_4-N content and pH declining to a somewhat higher than the control level (Yamanaka, 1995a; Suzuki, 2000; Suzuki et al., 2002; He and Suzuki, 2004) (Tables 2–4) (Fig. 2A–F). Generally, saprotrophic ammonia fungi grow well under pH 6 or 7 to pH 8 or pH 9, whereas most ectomycorrhizal ammonia fungi grow well under pH 5 or 6 to pH 7 (Yamanaka, 2003; Suzuki, 2006; Licyayo et al., 2007). Some saprotrophic ammonia fungi such as *Am. botrytis* and *Peziza moravecii* are tolerant (1.1–1.3 M NH_4-N), and grow well under a high concentration of NH_4-N (Licyayo and Suzuki, 2006). Most ammonia fungi, especially, some saprotrophic species and ectomycorrhizal species such as *C. phlyctidospora* and *Hebeloma vinosophyllum* grow well in media containing ammonium salts or nitrate salts as the sole nitrogen source (Yamanaka, 1999; Suzuki, 2006; Barua et al., 2012a). Marsh et al. (2005) based on incubation experiments in the laboratory found that soil fertilization by urea stimulated the nitrate accumulation in soils by the activity of autotrophic-oxidizing bacteria.

Wang et al. (2014) suggested that the elevation of NO_3-N levels in burned soils could be explained by the reduction of uptake, not by increasing in nitrification activities. Certini (2005) mentioned that "nitrate is leached easily downwards while ammonium is held by the soil by its negatively charged surface". In contrast, Attiwill et al. (1996) indicated that "demand for nitrogen by the heterotrophic organisms may be so great that little ammonium nitrogen escapes out of the 'internal cycle' of mineralization followed by immobilization of the NH_4-N in microbial biomass". The dynamics in soil properties after the application of a large amount of urea in several vegetation sites suggest that loss of nitrogen through leaching out of the soil as nitrate following nitrification may be lessened by the nitrogen immobilization activities of ammonia fungi.

4. Litter Decomposition after High Ammonia-Nitrogen Impact

Decomposition of plant organic matter after a large amount of urea application has been examined at several vegetation sites in the Cfa region in Japan. In the *Phyllostachys reticulata* grove, the decomposition rate of the culm pieces buried between L-F and H-A horizons is not affected by an application of 800 g/m^2 of urea (Fukiharu et al., 1997). An increase in the decomposition rate of embedded balsa (*Ochoroma pyramidale*) wood blocks by an application of 800 g/m^2 of urea has been reported in the *Quercus serrata*–dominated mixed forest when the forest floor maintains weak alkaline to a neutral condition (Sato and Suzuki, 1997). In a *Castanopsis sieboldii* and *Quercus* spp. dominated mixed forest, decomposition of embedded balsa wood blocks and embedded leaf litter of broad-leaved forest trees are accelerated subsequent to 800 g/m^2 of urea application, and the decomposition rate of the former is higher than that of the latter. The acceleration observed in both organic matters by the urea treatment continues at least during the first flush of ectomycorrhizal ammonia fungi (Suzuki, 2006). In the *Lithocarpus edulis* forest, the decomposition of buried branch sticks also become greater during the occurrence of the early phase of ammonia fungi by 1600 g/m^2 of urea application. While, decomposition of the buried leaves is not affected during the first year of the urea application, but is strongly suppressed from the second year of the urea application (He and Suzuki, 2004). The higher stimulated decomposition of wood blocks and twigs by nitrogen enrichment is likely to be due to the lower nitrogen content of twigs, bark, and wood compared with that of leaves (Williams and Gray, 1974). For example, the C/N ratio of most wood is between 350 and 500 (Käärik, 1974) whereas that of non-leguminous tree leaves ranges from 25–90 (Cooke and Rayner, 1984).

In the Jarrah Forest, weight loss of the leaf litter (C/N ratio: ca. 50) in the litterbag for 10 months was 35.6% and then became 7.9% for another 13 months (Table 5). The litter decomposition was suppressed by 800 g/m^2 of urea application for 10 months of the exposure to the forest floor, but not by less than 400 g/m^2 of urea application (Table 5). This may be related to the period for a more suitable C/N ratio (31–41: during the occurrence of the early phase of ammonia fungi) associated with non-inhibitive concentration ranges of NH_4-N, for the colonization of saprotrophic ammonia fungi, whereas totally inhibitive for the colonization of non-ammonia fungi, subsequent to different amounts of urea application, Thereafter, the decomposition of the leaf litter once exposed to 800 g/m^2 of urea was stimulated for at least another 13 months or more (Table 5). A similar kind of suppression of leaf litter decomposition by high nitrogen enrichment also has been reported in the *L. edulis* forest, but it occurs after the ceasing of flushes of saprotrophic ammonia fungi (He and Suzuki, 2004).

The period showing a higher effect of urea application on litter decomposing ability coincides with the occurrence period of the early phase of ammonia fungi, *Am. botrytis*, *Ascobolus denudatus*, *Peziza moravecii* and *Coprinopsis phlyctidospora*, etc. (Tables 2, 5). Saprotrophic ammonia fungi have cellulolytic,

Table 5. Decomposition of *Eucalyptus marginata* leaf litter after different amounts of urea application (Leaf litter of *Eucalyptus marginata* packed in litterbags were placed between coarse and fine litters of a mixed forest of *E. marginata* and *Corymbia calophylla* and harvested after 10- and 23-months burial).

Embedded period (month)	Decomposition rate (%) at each urea application (g/m^2)				
	0	100	200	400	800
10	35.6 ± 1.8^a	37.0 ± 0.7^a	36.8 ± 0.5^a	36.7 ± 0.8^a	23.6 ± 4.7^b
23	43.5 ± 1.5^a	42.6 ± 0.8^a	45.3 ± 0.7^a	46.2 ± 1.3^{a,b}	50.8 ± 1.3^b

Values in the table indicate average with standard error (n = 5 for 10 months; n = 8 for 23 months embedded samples).
Superscripts in the same row are significantly different at $p < 0.05$ based on based on Tukey-Kramer method.

hemicellulolytic, and ligninolytic enzyme functions (Soponsathien, 1998a, b; Yamanaka, 1995b). The cellulase optima for the early phase of ammonia fungi is neutral to weakly alkaline conditions (Enokibara et al., 1993). These results indicate that saprotrophic ammonia fungi would more or less contribute to the decomposition of plant organic matter in various types of vegetation after high nitrogen disturbance, although the actual percentage contribution made by the early phase of ammonia fungi to the decomposition is not known. The slow leaf litter decomposition after urea application in the Jarrah Forest could be partly explained by the apparent lack of colonization of anamorphic fungi in the Jarrah Forest. For example, *A. botrytis* is a ubiquitous species (Suzuki et al., 2003) that has high cellulolytic activity and adapts to colonization under high NH_4-N concentrations associated with a neutral to weak alkaline condition (Yamanaka, 1995b, 1999, 2003; Suzuki, 2006; Licyayo and Suzuki, 2006; Licyayo et al., 2007; Barua et al., 2012a).

Most of the late phase of ammonia fungi are ectomycorrhizal (Sagara, 2005) and have weak cellulolytic, hemicellulolytic, and ligninolytic activities under acidic conditions (Enokibara et al., 1993; Yamanaka, 1995b; Soponsathien, 1998a, b). Therefore, the majority of fungi participating in the decomposition of the urea-treated plant organic matters during the period when the late phase of ammonia fungi is present would be not only the late phase of ammonia fungi but also the early phase of ammonia fungi maintaining certain amounts of mycelia under weak acidic condition and or re-colonizing non-ammonia fungi. However, the actual percentage contribution made by ammonia fungi to the decomposition is not known since their colonizing abilities on the urea-treated plant organic matters are determined as a result of interactions among ammonia fungi and non-ammonia fungi under gradually changing soil properties, especially returning pH to the original level (Licyayo et al., 2007; Barua et al., 2012a).

The contents of nutrient elements in the leaf litter of *Eucalyptus marginata* stored in a room were in the order Ca > N, Mg > Na > K > P (Table 6). Total N content in the leaf litter increased by exposure in the field. The increment of total N was significant with urea (100–800 g/m^2) application, whereas no significant difference in total N content was observed among 100, 200, and 400 g/m^2 of urea application. This accumulation of nitrogen in the litter is most likely caused by inorganic nitrogen

Table 6. Content of each nutrient element in leaf litter of *Eucalyptus marginata* after different amounts of urea application. Leaf litter of *Eucalyptus marginata* packed in litterbags were placed between coarse and fine litters of a mixed forest of *E. marginata* and *E. calophylla*, and then harvested after 10 and 23 months of embedding.

Embedded period (month)	Element		Content (%) of each element in the litter at each urea application (g/m^2)				
		Control*	**0**	**100**	**200**	**400**	**800**
Control*	Total N	0.3463 ± 0.011					
10			0.5340 ± 0.0062^a	0.7448 ± 0.0226^b	0.7640 ± 0.0364^b	0.7364 ± 0.0210^b	
23			0.7013 ± 0.0410^a	0.8778 ± 00433^b	0.9499 ± 0.0348^b	0.9550 ± 0.0288^b	1.2088 ± 0.0544^c
Control*	Total P	0.0050 ± 0.003					
10			0.0088 ± 0.0002^{a,b}	0.0077 ± 0.0004^b	0.0070 ± 0.0002^b	0.0070 ± 0.0002^b	0.0110 ± 0.0013^a
23			0.0155 ± 0.0009^{a,b}	0.0142 ± 0.0007^{a,b}	0.0130 ± 0.0004^b	0.0141 ± 0.0005^{a,b}	0.0169 ± 0.0010^a
Control*	K	0.1401 ± 0.062					
10			0.0384 ± 0.0017^a	0.0383 ± 0.0020^a	0.0351.8 ± 0.0036^a	0.0351 ± 0.0014^a	0.0561 ± 0.0067^b
23			0.0561 ± 0.0020^a	0.0619 ± 0.0056^a	0.0522 ± 0.0024^a	0.0550 ± 0.0015^a	0.0575 ± 0.0044^a
Control*	Na	0.1934 ± 0.009					
10			0.0195 ± 0.0007^a	0.0243 ± 0.0012^a	0.0273 ± 0.0024^a	0.0244 ± 0.0018^a	0.0418 ± 0.0057^b
23			0.0290 ± 0.0016^a	0.0324 ± 0.0015^{a,b}	0.0342 ± 0.0028^{a,b}	0.0335 ± 0.0008^{a,b}	0.0364 ± 0.0012^b
Control*	Ca	0.8920 ± 0.048					
10			0.9708 ± 0.0375^a	0.9192 ± 0.0455^a	0.9396 ± 0.0268^a	0.9524 ± 0.0194^a	1.0374 ± 0.0449^a
23			1.2250 ± 0.0448^a	1.2925 ± 0.0287^a	1.1838 ± 0.0302^a	1.3063 ± 0.0558^a	1.5025 ± 0.05311^b
Control*	Mg	0.3370 ± 0.008					
10			0.2056 ± 0.0057^a	0.2652 ± 0.0080^b	0.2684 ± 0.0092^b	0.2790 ± 0.0083^b	0.3200 ± 0.0100^c
23			0.1030 ± 0.0179^a	0.2069 ± 0.0193^b	0.1999 ± 0.0148^b	0.2241 ± 0.0105^b	0.2107 ± 0.0309^b

Values in the table indicate average with standard error (n = 5 for 10 months; n = 8 for 23 months embedded samples and control samples stored in a room).
Superscripts in the same row are significantly different at $p < 0.05$ based on Tukey-Kramer method.
*: Control samples were stored in a dark room at room temperature for 23 months.

immobilized mainly in the mycelia of saprotrophic ammonia fungi. Total P and Ca contents in the leaf litter slightly increased by exposure to the forest floor. This tendency became slightly stronger in Ca by treatment with 800 g/m^2 of urea for 23 months. Calcium content of the Jarrah Forest litter increased slightly by 800 g/m^2 of urea application, but not by 100–400 g/m^2 of urea application. The amounts of Na > Mg > K were released from the *Eucalyptus* leaf litter in that order by exposure in the field (Table 6). An application of 800 g/m^2 of urea induced suppression of the release of Na, K and Mg from the leaf litter in that order during decomposition. The suppression of Mg release was also observed in the litter treated with 100–400 g/m^2 of urea, but that of K and Na was not observed (Table 6). Increases in P in the litter by exposure in the field have been frequently reported and attributed partly to microbial immobilization (Baker and Attiwill, 1985).

Cromack, Jr. et al. (1975) reported that Ca, K, Na and Sr were significantly concentrated in the rhizomorphs and fruiting bodies of basidiomycetous fungi. Qiu et al. (2012) reported that the loss of Na, K, and Mg would be largely leach through structural damages during the decomposition of the litter. In the Jarrah Forest, N, P, and Ca are likely to be accumulated in the mycelia of fungi, especially in those of ammonia fungi. In contrast, K, Na, and Mg release into the soil are likely to be superior to the immobilization abilities of saprotrophic ammonia fungi grown in the litter. Among six nutrient elements, changes in N contents in the litter corresponded only to changes in decomposition rates of the leaf litter, suggesting that the increased amount of endogenous nitrogen contributed to stimulative decomposition of the litter through sustaining vegetative growth of saprotrophic fungi.

Conclusion

Excrements and dead bodies of terrestrial vertebrates and invertebrates, especially decomposition of those from vertebrates, generate a large amount of various organic and or inorganic nitrogen, such as amines, amino acids and ammonia. Those nitrogenous products (derived mostly from bacterial activities) induce colonization of post-putrefaction fungi including ammonia fungi in various habits of different climate zones. The disturbance caused by enriched nitrogen input from decomposing animals occurs more sporadically and in smaller sized areas compared with the impacts of larger-scale events such as fires, but never the less contributes to significant ecological processes.

Changing patterns of NH_4-N concentration and pH after urea application are similar, irrespective of the type of habitat, but those of NO_3-N concentration, moisture contents, and decomposition rates of plant organic matter depend on the type of vegetation. Changes in soil properties over time after NH_4-N enrichment observed in forests of both the Cfa region and the Csa Mediterranean region suggest that successive colonization of saprotrophic and ectomycorrhizal ammonia fungi, both endemic and ubiquitous species, successively contributes to the recovery of the forest after nitrogen disturbance, irrespective of vegetation type. The recovery period of changed soil properties after nitrogen impact spans over a period of one to two years in forests in the Cfa region in Japan (Yamanaka, 1995a, b; Sagara et al., 2008), and over two to three years in the Jarrah Forest in a Csa region in Australia. In comparison, the recovery period of soil properties after fire impact in a range of forest types is three years depending on the indicator assessed (e.g., gross N mineralization) (Wang et al., 2014). Relatively slower recovery of the soil properties after fire events is hampered by the erosion of topsoil, which is likely to be at least partly attributable to the elimination of binding hyphae and other microbes. Such a high degree of soil erosion typical after fire events does not occur on the forest floor after nitrogen disturbance, even if the latter is of high intensity.

Little is known about how ecological groups of fungi such as ammonia fungi and pyrophilous fungi colonize and or survive in the field prior to and after the disturbance (Sagara, 1976b, 1992). Recently, a 'body snatchers' hypothesis for pyrophilous fungi has been supported by the isolation of pyrophilous fungi from plants and lichens as endophytes (Raudabaugh et al., 2019). Hypotheses about how ammonia fungi inhabit the forest during non-nitrogen disturbed periods are yet to be proven. Ammonia fungi may survive as spores and or fragments of mycelia until they are exposed to nitrogen disturbance.

Research on the effect of urea at different intensities of nitrogen input has been undertaken mostly by incubation of mixtures of litter and soil collected from the field (Sagara and Hamada, 1965; Sagara, 1975; 1976a, b; Sagara et al., 2008). Currently, there is little data available about changes in fungal communities associated with changes in soil properties following low levels of nitrogen input, which would be the typical situation that occurs with the decomposition of excrements and dead bodies of the vast numbers of terrestrial invertebrates. It seems possible that species of fungi observed after the low intensity of nitrogen disturbance have been recorded as non-post-putrefaction fungi or not as ammonia fungi - perhaps recorded as wood rot fungi and or litter decomposing fungi. Therefore, further research based on different doses of nitrogen input in various vegetation sites is required to assess the role of ammonia fungi in the recovery of plant communities after disturbance. Surveys utilizing metatranscriptomic approaches in analyzing the population of various ecological groups of fungi in each vegetation site may prove to be indispensable.

Acknowledgements

Previously unpublished work presented in this paper was partly supported financially by the Japan Society for the Promotion of Science (JSPS), Bilateral Exchange Program (Australia) and the Australian Academy of Science (AAS) (Fiscal, 1996). We are thankful to the former CSIRO Division of Forestry and Forest Products group in Western Australia, and to Alcoa World Alumina Australia for their support and for making the experimental sites and analyses possible.

References

Attiwill, P.M., Polglase, P.J., Weston, C.J. and Adams, M.A. (1996). Nutrient cycling in forests of south-eastern Australia. pp. 191–227. *In*: Attiwill, P.M. and Adams, M.A. (eds.). Nutrition of Eucalypts. CSIRO Publishing, Collingwood.

Baker, T.G. and Attiwill, P.M. (1985). Loss of organic matter and elements from decomposing litter of *Eucalyptus obliqua* L'Herit. and *Pinus radiata* D. Don. Aust. For. Res., 15(3): 309–319.

Barua, B.S., Suzuki, A. and Pham H.N.-D. (2012a). Effects of different nitrogen sources on interactions between ammonia fungi and non-ammonia fungi. Mycology, 3(1): 36–53.

Barua, B.S., Suzuki, A., Pham, H.N.-D. and Inatomi, S. (2012b). Adaptation on ammonia fungi to urea enrichment environment. Int. J. Agric., 8(1): 173–89.

Certini, G. (2005). Effects of fire on properties of forest soils: A review. Oecologia, 143(1): 1–10.

Cochrane, V.W. (1958). Nitrogen nutrition and metabolism. *In*: Physiology of Fungi. John Wiley & Sons, New York, pp. 241–299.

Cooke, R.C. and Rayner, A.D.M. (1984). Uncomminuted terrestrial litter. *In*: Ecology of Saprotrophic Fungi. Longman, London, pp. 238–284.

Cromack, Jr., K., Todd, R.L. and Monk, C.D. (1975). Patterns of basidiomycete nutrient accumulation in conifer and deciduous forest litter. Soil Biol. Biochem., 7(4-5): 265–268.

Egli, S. (2011). Mycorrhizal mushroom diversity and productivity—An indicator of forest health? Ann. For. Sci., 68(1): 81–88.

Enokibara, S., Suzuki, A., Fujita, C., Kashiwagi, M., Mori, N. and Kitamoto, Y. (1993). Diversity of pH spectra of cellulolytic enzymes in Basidiomycetes. Trans. Mycol. Soc. Japan, 34(2): 221–228 (In Japanese with English summary).

Faustino, L.I., Moretti, A.P. and Graciano, C. (2015). Fertilization with urea, ammonium and nitrate produce different effects on growth, hydraulic traits and drought tolerance in *Pinus taeda* seedlings. Tree Physiol., 35(10): 1062–1074.

Flint, C.M., Harrison, R.B., Strahm, B.D. and Adams, A.B. (2008). Nitrogen leaching from Douglas-fir forests after urea fertilization. J. Environ. Qual., 37(5): 1781–1788.

Fukiharu, H. and Hongo, T. (1995). Ammonia fungi of Iriomote Island in the southern Ryukyus, Japan and a new ammonia fungus, *Hebeloma luchuense*. Mycoscience, 36(4): 425–430.

Fukiharu, T., Sato, Y. and Suzuki, A. (1997). The occurrence of ammonia fungi and changes in soil conditions and decay rate of bamboo in response to application of a large amount of urea in a bamboo grove in Chiba Prefecture, central Japan. The Bulletin of the Faculty of Education, Chiba University (III : Natural Sciences), 45(3): 61–67.

Fukiharu, T., Bougher, N.I., Buchanan, P.K., Suzuki, A., Tanaka, C. and Sagara, N. (2011). *Coprinopsis austrophlyctidospora* sp. nov., an agaric ammonia fungus from Southern Hemisphere plantations and natural forests. Mycoscience, 52(2): 137–142.

Fukiharu, T., Shimizu, K., Li, R., Raut, J.K., Yamakoshi, S., Horie, Y. and Kinjo, N. (2013). *Coprinopsis novorugosobispora* sp. nov., an agaric ammonia fungus from Beijing, China. Mycoscience, 54(3): 226–230.

Garraway, M.O. and Evans, R.C. (1984). Nitrogen nutrition. *In*: Fungal Nutrition and Physiology, John Wiley & Sons, New York, pp. 96–123.

He, X.M. and Suzuki, A. (2004). Effects of urea treatment on litter decomposition in *Pasania edulis* forest soil. J. Wood Sci., 50(3): 266–270.

Hilton, R.N. (1978). The ghoul fungus, *Hebeloma* sp. ined. Trans. Mycol. Soc. Japan, 19(4): 418.

Ho, B.T.Q., Pham, N.-D. H., Shimizu, K., Fukiharu, T., Truong, N. and Suzuki, A. (2013). The first record of *Hebeloma vinosophyllum* (Strophariaceae) in southeast Asia. Mycotaxon, 128: 25–36.

Hora, F.B. (1959). Quantitative experiments on toadstool production in woods. Trans. Br. Mycol. Soc., 42(1): 1–14.

Hora, F.B. (1972). Productivity of toadstools in coniferous plantations – natural and experimental. Mycopathol. Mycol. Appl., 48(1): 35–42.

Imamura, A. and Yumoto, T. (2004). The time of urea treatment and its effects on the succession of ammonia fungi in two warm temperate forests in Japan. Mycoscience, 45(1): 123–130.

Imamura, A., Yumoto, T. and Yanai, J. (2006). Urease activity in soil as a factor affecting the succession of ammonia fungi. J. For. Res., 11(2): 131–135.

Käärik, A.A. (1974). Decomposition of wood. pp. 129–174. *In*: Dickinson, C.H. and Pugh, G.J.F. (eds.). Biology of Plant Litter Decomposition, Volume I, Academic Press, London.
Laiho, O. (1970). *Paxillus involutus* as a mycorrhizal symbiont of forest trees. Acta Forestalia Fennica, 106: 1–72.
Lehmann, P.F. (1976). Unusual fungi on pine leaf litter induced by urea and urine. Trans. Mycol. Soc., 67(2): 251–253.
Lehmann, P.F. and Hudson, H.J. (1977). The fungal succession on normal and urea-treated pine needles. Trans. Mycol. Soc., 68(2): 221–228.
Licyayo, D.C. and Suzuki, A. (2006). Growth responses of ammonia fungi to different concentrations of ammonium-nitrogen. Mush. Sci. Biotechnol., 14(3): 145–156.
Licyayo, D.C.M., Suzuki, A. and Matsumoto, M. (2007). Interactions among ammonia fungi on MY agar medium with varying pH. Mycoscience, 48(1): 20–28.
Lloid, A.B. and Sheaffe, M.J. (1973). Urease activity in soils. Pl. Soil, 39(1): 71–80.
Manusweeraporn, S., Raut, J.K., Aramsirirujiwet, Y., Kitpreechavanich, V. and Suzuki, A. (2013). The variation of litter decomposing abilities of *Coprinopsis cinerea* from nitrogen-enriched environments in Thailand. Thai J. Bot., 5 (special issue): 89–98.
Marsh, K.L., Sims, G.K. and Mulvaney, R.L. (2005). Availability of urea to autotrophic ammonia-oxidizing bacteria as related to the fate of ^{14}C- and ^{15}N-labeled urea added to soil. Biol. Fert. Soils, 42(2): 137–145.
Miller Jr., O.K. and Hilton, R.N. (1986/1987). New and interesting Agarics from Western Australia. Sydowia, 39: 126–137.
Nagao, H., Udagawa, S., Bougher, N.L., Suzuki, A. and Tommerup, I.C. (2003). The genus *Thecotheous* (Pezizales) in Australia: *T. urinamans* sp. from urea-treated Jarrah (*Eucalyptus marginata*) forest. Mycologia, 95(4): 688–693.
Nguyen, P.T., Pham, N.D.H., Suzuki, A., Shimizu, K. and Fukiharu, T. (2019). *Coprinopsis neocinerea* sp. Nov., an ammonia fungus from Southern Vietnam. Mycoscience, 60(5): 307–312.
O'Connell, A.M. (1994). Decomposition and nutrient content of litter in a fertilized eucalypt forest. Biology and Fertility of Soils, 17(2): 159–166.
O'Connell, A.M. and Grove, T.S. (1996). Biomass production, nutrient uptake and nutrient cycling in the Jarrah (*Eucalyptus marginata*) and karri (*Eucalyptus diversicolor*) forests of south-western Australia. pp. 155–189. *In*: Attiwill, A.T. and Adams, M.A. (eds.). Nutrition of Eucalypts, CSIRO Publishing, Collingwood.
Ohenoja, E. (1978). Mushrooms and mushroom yields in fertilized forests. Ann. Bot. Fenn., 15(1): 38–46.
Ohenoja, E. (1988). Behaviour of mycorrihizal fungi in fertilized forests. Karstena, 28(1): 27–30.
Pham, N.D.H., Ho, B.T.Q., Pham, N.P.T., Nguyen, M.P.L. and Suzuki, A. (2019). Investigation of microfungi from urea plots at pine forest in Bidoup – Nui Ba National Park, Lam Dong Province, Vietnam. Int. J. Agric. Technol., 15(4): 625–636.
Qiu, S., McComb, A.J. and Bell, R.W. (2012). Leaf litter decomposition and nutrient dynamics in woodland and wetland conditions along a forest to wetland hillslope. International Scholarly Research Network ISRN Soil Science, Volume 2012, 346850. 10.5402/2012/346850.
Raudabaugh, D.B., Matheny, P.B., Hughes, K.W., Iturriaga, T., Sargent, M. and Miller, A.N. (2019). Where are they hiding? Testing the body snatchers hypothesis in pyrophilous fungi. Fungal Ecol., 43: 100870.
Raut, J.K., Fukiharu, T., Shimizu, K., Kawamoto, S., Takeshige, S. et al. (2015). *Coprinopsis novorugosobispora* (Basidiomycota, Agaricales), an ammonia fungus new to Canada. Mycosphere, 6(5): 612–619.
Raut, J. K., Suzuki, A., Fukiharu, T., Shimizu, K., Kawamoto, S. and Tanaka, C. (2011). *Coprinopsis neophlyctidospora* sp. A new ammonia fungus from boreal forests in Canada. Mycotaxon, 115: 227–238.
Sagara, N. (1973). Proteophilous fungi and fireplace fungi. Trans. Mycol. Soc. Japan, 14(1): 41–46.
Sagara, N. (1975). Ammonia fungi – A chemoecological grouping of terrestrial fungi. Contributions from the Biological Laboratory Kyoto University, 24(4): 205–276.
Sagara, N. (1976a). Growth and reproduction of the ammonia fungi. pp. 153–178. *In*: Biseibutsu-seitai, K. (ed.). Ecology of Microorganisms, Volume 3, University of Tokyo Press. Tokyo. (In Japanese).

Sagara, N. (1976b). Supplements to the studies of ammonia fungi (1). Trans. Mycol. Soc. Japan, 17(3/4): 418–428 (In Japanese with English summary).
Sagara, N. (1992). Experimental Disturbances and epigeous fungi. pp. 427–454. *In*: Carroll, G.C. and Wicklow, D.T. (eds.). The Fungal Community its Organization and Role in the Ecosystem, Mycology Volume 9, 2nd Edition, Marcel Dekker, Inc., New York.
Sagara, N. (1995). Association of ectomycorrhizal fungi with decomposed animal wastes in forest habitats: a cleaning symbiosis? Can. J. Bot., 73 (Suppl. 1): S1423–S1433.
Sagara, N. and Hamada, M. (1965). Responses of higher fungi to some chemical treatment of forest ground. Trans. Mycol. Soc. Japan, 6(3): 72–74.
Sagara, N., Yamanaka, T. and Tibbett, M. (2008). Soil fungi associated with graves and latrines: Toward a forensic mycology. pp. 67–107. *In*: Tibbett, M. and Carter, D.O. (eds.). Soil Analysis in Forensic Taphonomy Chemical and Biological Effects of Buried Human Remains. CRC Press, Boca Raton.
Salo, K. (1979). Mushrooms and mushroom yield on transitional peatlands in central Finland. Ann. Bot. Fenn., 16(3) 181–192.
Sato, Y. and Suzuki, A. (1997). The occurrence of ammonia fungi, and changes in soil conditions and wood decay rate in response to application of a large amount of urea in a *Quercus serrata* dominated mixed forest in Meguro, Tokyo. The Bulletin of the Faculty of Education, Chiba University (III: Natural Sciences), 45: 53–59.
Soponsathien, S. (1998a). Some characteristics of ammonia fungi 1. In relation to their ligninolytic enzyme activities. J. Gen. Microbiol., 44(5): 337–345.
Soponsathien, S. (1998b). Study on the production of acetyl esterase and side-group cleaving glycosidases of ammonia fungi. J. Gen. Microbiol., 44(6): 389–397.
Sullivan, T.P. and Sullivan, D.S. (2018). Influence of nitrogen fertilization on abundance and diversity of plants and animals in temperate and boreal forests. Environ. Rev., 26(1): 26–42.
Suzuki, A. (2000). A fixed point time survey of species assemblage of ammonia fungi. Natural History of Plants (Planta), 68: 27–35 (In Japanese).
Suzuki, A. (2006). Experimental and physiological ecology of ammonia fungi: studies using natural substrates and artificial media. Mycoscience, 47(1): 3–17.
Suzuki, A. (2009). Fungi from high nitrogen environments - Ammonia fungi: Eco-physiological aspects. pp. 189–218. *In*: Misra, J.K. and Deshmukh, S.K. (eds.). Fungi from Different Environments. Science Publishers, Enfield, USA.
Suzuki, A. (2017). Various aspects of ammonia fungi. pp. 39–58. *In*: Deshmukh, S.K., Johri, B. N. and Satyanarayana, T. (eds.). Developments of Fungal Biology and Applied Mycology. Springer-Verlag, New York.
Suzuki, A., Fukiharu, T., Tanaka, C., Ohono, T. and Buchanan, P.K. (2003). Saprobic and ectomycorrhizal ammonia fungi in the Southern Hemisphere. New Zealand J. Bot., 41(3): 391–40.
Suzuki, A., Tommerup, I.C. and Bougher N.L. (1998). Ammonia fungi in the Jarrah forest of Western Australia and parallelism with other geographic regions of the world. Second International Conference on Mycorrhiza, Uppsala, Sweden, 5–10 July, 1998.
Suzuki, A., Uchida, M. and Kita, M. (2002). Experimental analyses of successive occurrence of ammonia fungi in the field. Fungal Diversity, 10: 141–165.
Veverka, K., Štolcová, J. and Růžek, P. (2007). Sensitivity of fungi to urea, ammonium nitrate and their equimolar solution UAN. Pl. Prot. Sci., 43(4): 157–164.
Wang, Y., Xu, Z. and Zhou, Q. (2014). Impact of fire on soil gross nitrogen transformations in forest ecosystems. J. Soils Sed., 14(6): 1030–1040.
Wallenstein, M.D., McNulty, S., Fermandez, I.J., Boggs, J. and Schlesinger, W.H. (2006). Nitrogen fertilization decreases forest soil fungal and bacterial biomass in three long-term experiments. For. Ecol. Managem., 222(1-3): 459–468.
Williams, S.T. and Gray, T.R.G. (1974). Decomposition of litter on the soil surface. pp. 611–632. *In*: Dickinson, C.H. and Pugh, G.J.F. (eds.). Biology of Plant Litter Decomposition, Volume 2. Academic Press, London.
Wright, S.H.A., Berch, S.M. and Berbee, M.L. (2009). The effect of fertilization on the below-ground diversity and community composition of ectomycorrhizal fungi associated with western hemlock (*Tsuga heterophylla*). Mycorrhiza, 19(4): 267–276.

Yamanaka, T. (1993). Microbial and chemical properties of soil under the successive occurrence of the ammonia fungi. Proceedings of the First International Symposium of the Mycological Society of Japan, pp. 42–45.

Yamanaka, T. (1995a). Nitrification in a Japanese red pine forest soil treated with a large amount of urea. J. Jap. For. Soc., 77(3): 232–238.

Yamanaka, T. (1995b). Changes in organic matter composition of forest soil treated with a large amount of urea to promote ammonia fungi and the abilities of these fungi to decompose organic matter. Mycoscience, 36(1): 17–23.

Yamanaka, T. (1999). Utilization of inorganic and organic nitrogen in pure cultures by saprotrophic and ectomycorrhizal fungi producing sporophores on urea-treated forest floor. Mycol. Res., 103(7): 811–816.

Yamanaka, T. (2003). The effect of pH on the growth of saprotrophic and ectomycorrhizal ammonia fungi *in vitro*. Mycologia, 95(4): 584–589.

Young, A.M. (2002). Brief notes on *Hebeloma aminophilum* R.N. Hilton and O.K. Miller ('Ghoul fungus') from Northern Queensland and Tasmania. Austr. Mycol., 21(2): 79–80.

4

Ecology of Wild Mushrooms

Hoda M El-Gharabawy

1. Introduction

Mushrooms represent the fleshy delicate sporulating fruiting bodies belonging to various taxonomic classes of Ascomycetes and Basidiomycetes (Rashid et al., 2016). These fungi produce vegetative microscopic mycelium, which grows on different substrates and produce fruit bodies. They have diverse morphology forming toadstools, puffballs, hoofed fungi, coral fungi and other types of configurations (Govorushko et al., 2019). These species can grow as saprophytes on organic debris degrading lignocellulosic and pectic compounds, while in mycorrhizal association with plant roots (Jaradat, 2010). Under favorable seasonal conditions like temperature, moisture, nutrient source and light, the vegetative mycelia grow into a small pinhead and finally transformed into a mature fruit body. The fully grown mature basidiomycete mushroom consists of three main parts: stipe, pileus and gills. The gills of a mushroom bears spores that spread by air and different biotic agents into the atmosphere. These spores subsequently germinate into a new mycelium to reiterate the new life cycle under appropriate substrates (Dias and de Brito, 2017). Mushrooms may have unusual growth patterns as "fairy rings" in open spaces of fields and meadows. Most mushrooms are presented in the basidiomycete division and known as basidiocarps which formed specialized mycelium with clamp connections as a short-lived stage of their life cycle (Hibbett and Binder, 2002). A few species are presented in the division Ascomycetes including morels and truffles (Pegler, 2003).

Botany and Microbiology Department, Faculty of Science, Damietta University, New Damietta, 34517, Egypt.
Email: hoda_mohamed@du.edu.eg/hoda_elgharabawy@yahoo.com

Wild mushrooms are essential biota in the ecosystem and they play a vital role as saprotrophs or symbiotics with plant roots and animals (Dwivedi et al., 2012). Their diversity and distribution are dependent on the vegetation, landscape and environmental aspects (e.g., rainfall, light and temperature). They grow only in a specific part of the year mainly during the rainy season (Hu et al., 2022). These mushrooms occupy a specific niche determining their morphology including colour, shape, and other features. They grow on various substrates like fallen leaves, dead and decaying woods, organic matter and mycorrhizal association with the roots of higher plants (Karwa and Rai, 2010). Mushrooms are identified and taxonomically characterized through conventional approaches such as reactions of fungal mycelium to different chemicals, morphology, biochemistry and physiology (Hyde et al., 2020a). Recent trends in mushroom identification use morphological characterization combined with molecular phylogenetic analysis, ecology and chemotaxonomy (Samarakoon et al., 2020; Wibberg et al., 2021). Applying these advanced techniques provided insight into studying the diversity, ecology, taxonomy and phylogenetic relationships between fungal species (Hyde et al., 2020b).

Mushrooms have a worldwide distribution, some of them are edible, tasty and healthy food, while others are deadly poisonous (Ukwuru et al., 2018). They are an excellent source of nutrients like proteins and vitamins (Kumar et al., 2021). The wild edible mushroom provides an alternative nutritional and source of income for the rural population. They are consumed and sold in more than 80 countries across the world with an economic turnover of up to US $2.0 billion (Boa, 2004). The most frequently collected wild mushrooms that have commercial perspectives are *Armillaria ponderosa*, *Boletus edulis*, *Cantharellus cibarius*, *Dentinum repandum*, *Morchella esculenta* and *Tricholoma matsutake*. Other species of wild mushrooms as *Agaricus campestris*, *Caprinus commatus*, *Laetiporus sulphurcus*, *Lyophyllum multiceps* and *Morchella augusticeps* are also foraged (Chelela et al., 2014). Local folks owing to their knowledge of edibility, wild mushrooms are domesticated for commercial purposes.

Human activities have been threatening the biodiversity of wild mushrooms and their natural niches for the last several decades at rapid rate. Climatic changes increased damage to forests, drought, and damage by pests and pathogens are the impediments to wild mushroom conservation (Peñuelas et al., 2017; Huuskonen et al., 2021).

The aim of the current chapter is to highlight the ecology of wild mushrooms, the influence of environmental factors, soil habitats and microhabitats. Their mutualism, parasitism, dispersal, bioremediation and ecosystem services have been discussed.

2. Environmental Factors

Mushrooms are cosmopolitan in distribution tropical, sub-tropical and temperate regions. Geographic region, climatic situation, ecological niches, type of substrates, fauna in detritus and vegetation are the important factors responsible for the diversity and distribution of mushrooms (Srivastava, 2021). Different variables that influence the mushrooms include abiotic factors, soil variables, biotic factors, pests and pathogens (Kewessa et al., 2022).

2.1 Temperature

Temperature is a critical factor affecting the vegetative growth and fruitification of mushrooms. Generally, optimum fruitification temperatures are lower compared to the optimum necessary for the vegetative growth temperature, which ranges between 5 and 33°C. However, this optimum temperature varies with mushroom species and strain. Congenial temperature and humidity are necessary for the mycelium to build large biomass (Balai and Ahir, 2013). The formation of primordia and its further differentiation is controlled by environmental aspects especially rainfall and temperature. Mushroom production was positively correlated with rainfall, while the low humidity retards the growth rate during primordial development (Krebs et al., 2008; Kim et al., 2013).

The diverse climatic conditions and environmental features facilitate different mushrooms to grow and flourish (Feng and Yang, 2018). High temperature kills mycelia and spores, which reduces the yield. Oyster mushroom is a warm climate preferring mushroom, its vegetative growth and fruit body formation develops under a wide range of temperature (10–31°C) (Shen et al., 2014). Oyster mushrooms are divided into two groups based on temperature profile asL (i) Winter species (10–20°C); (ii) Summer species (16–30°C). Temperature also influences the pileus colour, for instance, *Pleurotus florida* has light-brown when fruit is at low temperature (10–15°C), however, transforms into white pale to yellowish at the increase in temperature (20–25°C). Likewise, the colour of *P. sajor-caju* fruit at 15–19°C is white to dull-white with increased dry matter and it transforms into whitish-brown to dark-brown with the low dry matter at 25–30°C (Barh et al., 2019).

2.2 Precipitation

Precipitation is one of the most important factors controlling mushroom occurrence. It will raise the soil water content, and facilitate the resting spores to get adequate water for germination and growth. Sufficient water softens the cell wall, triggers the hydrolysis of enzymes and enables the spore to germinate more efficiently (Hu et al., 2022). Water imbibition will dissolve the inhibitory substances and permits spore germination by breaking the dormancy (Feofilova et al., 2012). Moisture also promotes spore respiration, sugar conversion, provides energy for growth and stimulates spores for secretion of enzymes towards cessation of spore dormancy.

2.3 Relative Humidity

Relative humidity (RH) is a critical factor affecting the dominance of mushrooms in the ecosystem (Rai et al., 2015). It has a major influence on the dispersal of spores and the high humidity reduces the range of dispersal of spores (Davarzani et al.,

2014). During fruit body production, the RH varies from species to species and should be between 80 and 95%. The best RH for stimulating the primordial formation of *Flammulina velutipes* was reported as 60–70% (Rezaeian and Pourianfar, 2017), while 90–95% was optimal for fructification (Sangkaew and Koh, 2017).

2.4 *Shade and Light*

The dense shady trees provide moist conditions, which are favorable for spore germination and the growth of many wild mushrooms (Srivastava, 2021). Light is an environmental variable necessary to induce primordial differentiation in mushrooms and regulate the growth and formation of fruit bodies. Mushroom mycelia usually grow in dark conditions, but the presence of light is necessary for morphogenesis, fruit body induction and viable spore production (Fuller et al., 2015; Rezaeian and Pourianfar, 2017). Light also influences the morphology, colour, stipe length and pileus dimension. Treatment with light accelerated the growth and fruit body formation in *Alnicola lactariolens* and *Hebeloma vinosophyllum* (Yue et al., 2022). Button mushrooms require darkness for growth and fructification, while oyster mushrooms failed to develop cap, but stipes form a coral-like structure in absence of light (Bellettini et al., 2019).

2.5 *Wind*

The wind is also another variable that controls mushroom growth and activities. It promotes the dispersal of spores, increases the range of dispersal and aids in the formation of the dominant population within the same plant community (Rieux et al., 2014). Wind also controls the oxygen supply to the plant-macrofungal communities, which helps in the morphogenesis of fruit bodies (Neira et al., 2015). Wind speed affects soil moisture and atmospheric humidity, which leads the formation of fruit bodies (Davarzani et al., 2014). However, owing to more evaporation of soil moisture by wind, hinder spore germination as well as vegetative hyphal growth that promotes dormancy.

3. Soil Factors

Mycelial colonization, spore accumulation and diversity of mushrooms especially symbiotic mushrooms were strongly related to soil properties. A mycorrhizal mushroom develops a mutualistic association with roots, which help plants to tolerate harsh conditions, especially in coastal plant communities (Wang et al., 2021a). Soil pH, moisture, organic matter and sand contents play a significant role in mushroom sporocarp community composition. Significant relationships were also found with other variables such as soil Na, S in litter and nitrogen mineralization rate. These soil variables have a strong influence on the aboveground mushroom community and drive sporocarp occurrence, diversity and richness (Ponce et al., 2022).

3.1 Organic Matter

Soil organic matter affects the occurrence and activities of mushrooms. Based on the organic matter availability, mushrooms can be categorized into four groups: (i) Wood-associated mushrooms on decayed wood; (ii) Organic matter decomposing mushrooms in soil; (iii) Biotrophs as fungi associated with algae and symbiotic mycorrhizae; (iv) Parasitic mushrooms on live host plants (Wang et al., 2021b; Mukhtar et al., 2021). Soil organic carbon can double the atmospheric carbon content and it is about two- to three-fold greater than living organisms in terrestrial ecosystems (Andrew et al., 2018).

Tree fires may produce incomplete oxidation or pyrolysis of compounds leading to the production of pyromorphic humus, and accumulation of new particulate carbon forms with enhanced resistance against chemical oxidation and biological degradation (González-Pérez et al., 2004). For example, a succession of fire-associated pyrophilous mushrooms was rapid compared to the succession of wood-associated aphyllophoroid mushrooms. According to Salo et al. (2019), post-fire young successional forests are important ecosystems with specialized and rare fungi. *Octospora* and *Pyronema* spp. were clear examples of mushrooms growing on post-fire habitats, that have been observed during the author's mushroom forays in orchards of Damietta in Egypt.

3.2 Soil pH

The pH of the substrates also affects the pH of mushrooms. Thus, the growth, absorptive capacity, membrane permeability and enzymatic activity of the mushrooms are also affected. The optimal soil pH for mushroom fruiting may differ from those for mycelial growth. The optimum pH required for mycelial growth should be between 4 and 7 (Fazenda et al., 2008). Soils with low acidity were characterized by decomposers, while the frequency of mycorrhiza increased steadily towards the most acid soils. The use of mushrooms as indicators in environmental studies of soil acidification needs future research (Tinya et al., 2021).

3.3 Soil Salinity

Salinity affects mushroom development, reduces spore germination, hyphal growth and host root colonization (Porcel et al., 2012). Saprotrophic as well as ectomycorrhizal mushrooms decrease on increased soil salinity (Ben-David et al., 2011). Low or less frequent ectomycorrhizal fungi were recorded in saline sites owing to unfavourable soil conditions (Hrynkiewicz et al., 2015). Johromi et al. (2008) opined that salinity induces morphological variations in hyphae and affects hyphal ramification, which influences the symbiotic capability.

Saprotrophic mushrooms (e.g., *Ganoderma lucidum*, *Laetiporus sulphureus*, *Pleurotus ostreatus* and *Trametes versicolor*) decreased their growth and hyphal extensions in saline soils (Bencherif et al., 2015). Hyperosmotic stress inhibits cell wall extension as well as cellular expansion, which leads to growth reduction. Excess of Cl− and or Na+ ions in fungal cells also change the enzymatic activities leading to the decreased potential of ectomycorrhizal fungal successful root colonization (Hrynkiewicz et al., 2015).

3.4 *Oxygen and Carbon Dioxide*

Mushrooms require oxygen for the growth of mycelia and to decompose the organic matter during vegetative growth, which results in the production of carbon dioxide. Ventilation is a very crucial factor for fruiting, which needs good air exchange and a low CO_2 level. Less than 0.3–0.5% CO_2 is necessary for vegetative growth as well as to stimulate the development of primordia in *Flammulina* spp. Its concentration of more than 0.1–0.2% leads to an increase in the elongation of the stipe, a reduction in pileus diameter and suppresses development of basidiospores (Bellettini et al., 2016). Increased ambient CO_2 levels an increase in vegetative growth, while it may also hinder fruit body initiation as well as its normal development (Nikšić et al., 2022). Under the high CO_2 level, mushrooms produce lengthy stipes with the small pileus, while short stipes with expanded caps are produced under low CO_2 levels or due to frequent ventilation regimes (Klein, 2022).

4. Wild Mushrooms in Egypt

Egypt spreads its geographical area at the junction of two large continents (Africa and Asia). It includes part of the Mediterranean basin, which has influenced the biota of the country. The Nile basin consists of the valley in the south, while the Nile delta in the north with riparian oases and dense farmlands. The Nile Delta of the Mediterranean climate receives little rainfall (100–200 mm/annum mainly in winter) (Fig. 1). This delta has its hottest temperature during summer (July and

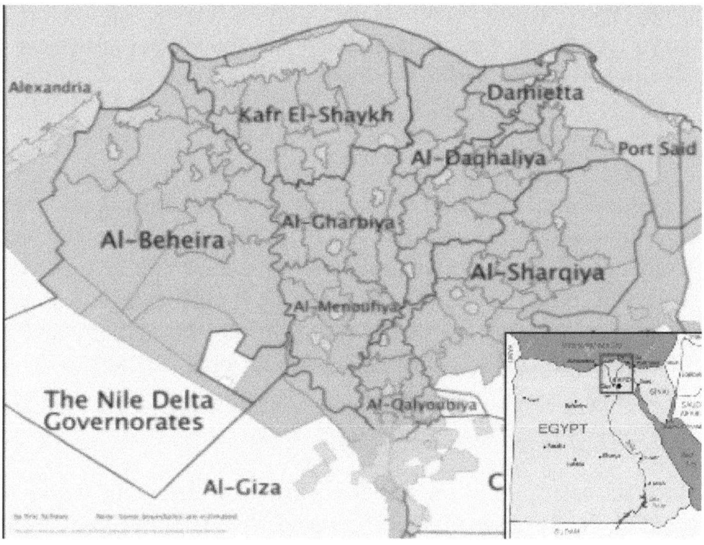

Fig. 1. Map showing the study area of the Nile Delta of Egypt, shown in the square (inset).

August: average, 30°C; maximum, ~ 42°C), while the winter temperatures range from 10–19°C. Thus, with the cooler temperatures with sporadic rain during winter, the Nile Delta region experiences humid conditions suitable for the wild mushrooms to bloom (Fishar, 2018).

Egypt in general, hosts a rich diversity of wild mushrooms. Mushrooms were reported growing in different habitats such as gardens, canal banks, wetlands, protected areas, and arid and semi-arid lands in Egypt. El-Fallal (2003) recorded several Agaricales from the countryside and grasslands during field surveys across the East Delta region. Up to 14 species belonging to 10 genera and six families were reported (*Agaricus*, *Agrocybe*, *Conocybe*, *Deconica*, *Lyophyllum*, *Paneolus*, *Pholiotina*, *Psathyrella*, *Strobilurus* and *Volvariella*). Mushroom forays during the last ten years (2010–2020) across the Nile Delta, several wild mushrooms were recorded during different fruiting seasons and climatic conditions (Table 1). During late summer and autumn, most mushrooms pop up in orchards, gardens and grassy areas except for some basidiomycetes as *Coprinus* and *Volvariella* spp. and ascomycetes including morel and puffballs, which start fruiting by Winter under cooler temperature and higher humidity degrees. Mushroom fruiting during the early season of spring was rare, and for the summer, some species were fruiting in a fascinating manner (*Amylospora* sp. and *Ganoderma mbrekobenum*). On the other hand, *Crystallicutis damiettenses* occurrence was recorded during the whole year on surviving different climate conditions.

5. Microhabitats of Mushrooms

Most of the mushroom biomass consists of fine thread-like hyphae, which extend profusely throughout the organic matter in their habitats. They periodically develop spore-producing fruit bodies under congenial environmental conditions using their nutrient reserves. It may require years to expose to favourable circumstances in forests, prairies, fields and other habitats to occur in abundance (Janaki, 2022). These fungi exist in various types of microhabitats such as soil surfaces, surfaces of organic matter, logs, woody litter and herbivore dung.

5.1 *Surface of Soil and Organic Matter*

The hyphae of many mushrooms grow extensively in soil and organic matter on the forest floor of prairies and savannas (Hu et al., 2022). These hyphal networks on the soil urface constitute an important component of the decomposer food web. Several mushrooms were recorded in the Nile Delta of Egypt growing in soil microhabitat and organic matter (*Lepista sordida*, *Leucocoprinus birnbaumii* and *Morchella galilaea*) (El-fallal et al., 2017, 2019; El-Gharabawy et al., 2019).

5.2 *Wood Microhabitat*

Many mushrooms as saprophytes prefer to grow on decaying wood (e.g., logs, branches, standing dead trees including rotting heartwood of live trees. Some of these are responsible for significant economic loss owing to dry-rot in the forests as well

Table 1. Wild mushrooms recorded in the Nile Delta of Egypt and their fruiting seasons.

No.	Order	Mushroom species	Collection season											
			Autumn			Winter			Spring			Summer		
			Sep	Oct	Nov	Dec	Jan	Feb	Mar	Apr	May	Jun	July	Aug
1.	*Polyporales*	*Ganoderma mbrekobenum*												
2.		*Ganoderma resinaceum*												
3.		*Bjerkandera adusta*												
4.		*Amylospora* sp.												
5.		*Inonotus rickii*												
6.		*Oxysporus latemarginatus*												
7.		*Megasporoporia minor*												
8.		*Crystallicutis damiettensis*												
9.	*Agaricales*	*Leucocoprinus cretaceus*												
10.		*Leucocoprinus birnbaumii*												
11.		*Coprinopsis cinerea*												
12.		*Coprinopsis urticicola*												
13.		*Coprinus comatus*												
14.		*Conocybe* sp.												
15.		*Psathyrella candolleana*												
16.		*Panaeolus antillarum*												
17.		*Panaeolus papilionaceus*												
18.		*Podaxis pistillaris*												
19.		*Volvariella speciosa*												

Table 1 contd. ...

...Table 1 contd.

No.	Order	Mushroom species	Collection season											
			Autumn			Winter			Spring			Summer		
			Sep	Oct	Nov	Dec	Jan	Feb	Mar	Apr	May	Jun	July	Aug
20.		*Ossicaulis lachnopus*			■		■							
21.		*Marasmius winnei*					■							
22.		*Agaricus punjabensis*			■									
23.		*Agaricus bisporus*			■	■	■	■						
24.		*Agaricus bitorquis*			■	■	■	■						
25.		*Agrocybe aegerita*			■	■	■	■	■					
26.		*Pleurotus fuscosquamulosus*			■	■	■						■	■
27.		*Chlorophyllum molybdites*		■	■	■								
28.		*Hygrocybe* sp.				■								
29.		*Lepista sordida*				■								
30.	Ascomycota	*Pyronema* sp.		■										
31.		*Octospora* sp.		■										
32.		*Peziza visculosa*				■	■							
33.		*Vascellum pratense*				■					■			
34.		*Mycenastrum corium*		■	■									
35.		*Morchella galilaea*				■								

as wooden components of buildings. The annual loss of wood has been estimated up to one-third of the timber cut per annum in the United States by fungal decay, fire, insects and other natural catastrophic events (Schmidt, 2006).

The fungal hyphae invade the wood through cut edges and wounds to spread through the lumens and adjacent cells. Mushrooms secrete several enzymes to digest cell wall components and polysaccharides as a source of energy. Occurrence of mushrooms at the base of the tree trunks or branches (e.g., bracket and shelf-like structures) shows indications of decay (Goodell and Nielsen, 2023) (Fig. 2). However, fruit bodies of some mushrooms during wood decay will not appear until attaining the advanced stage of decay and masks the visible indications.

In Egypt, highly economic trees suffer from wood-decay fungi, which has an adverse effect on tree quality and strength with a serious economic loss. The problem of wood decay is spreading through numerous trees such as palms, citrus, grapes, lemon, pear, guava, mulberry, *Ficus*, *Casuarina*, willow and poinciana. Consequently, it has a serious impact and significant economic losses (El-Fouly et al., 2011; Shetafa and El-Wahab, 2013; El-Gharabawy, 2016). Three main types of wood-degrading fungi such as white-rot, brown-rot and soft-rot fungi have been recognized.

Fig. 2. Wood decay mushrooms growing on trees in the North Nile Delta of Egypt, observed and recorded by the author during the years of 2013–2014: (A) *Inonotus rickii* on casuarina tree, (B) *Irpex latemarginatus* on Mulberries tree, (C) *Ganoderma mbrekobenum* on Lemon tree, and (D) *Ganoderma resinaceum* on a date palm tree.

5.3 *White-Rot Fungi*

White-rot fungi destroy almost all the components of plant cell walls by breaking down lignin leaving the light-coloured cellulose behind, while some of them are efficient in breakdown lignin as well as cellulose (Rayner and Boddy, 1988). White-rot fungi usually degrade hardwoods of deciduous trees, which are usually resistant to brown-rot fungi. Rotted wood is usually moist, soft and spongy with white pockets or streaks (Goodell and Nelsen, 2023). *Armillaria* spp. are white-rot fungi, which are well-known for attacking live trees. Commonly cultivated white-rot fungi include *Pleurotus ostreatus P. sajor-caju* and others (Wesenberg et al., 2003). Many white-rot fungi cause selective rot and they initially degrade starch and pectin followed by hemicellulose and lignin (e.g., *Tramates versicolor*) (Qi et al., 2022). *Crystallicutis damiettenses* is a rare and novel species of white-rot fungus reported for the first time growing on the decayed trunks of date palms (*Phoenix dactylifera*) in the orchards of the Nile Delta (Egypt) (El-Gharabawy et al., 2021) (Fig. 3).

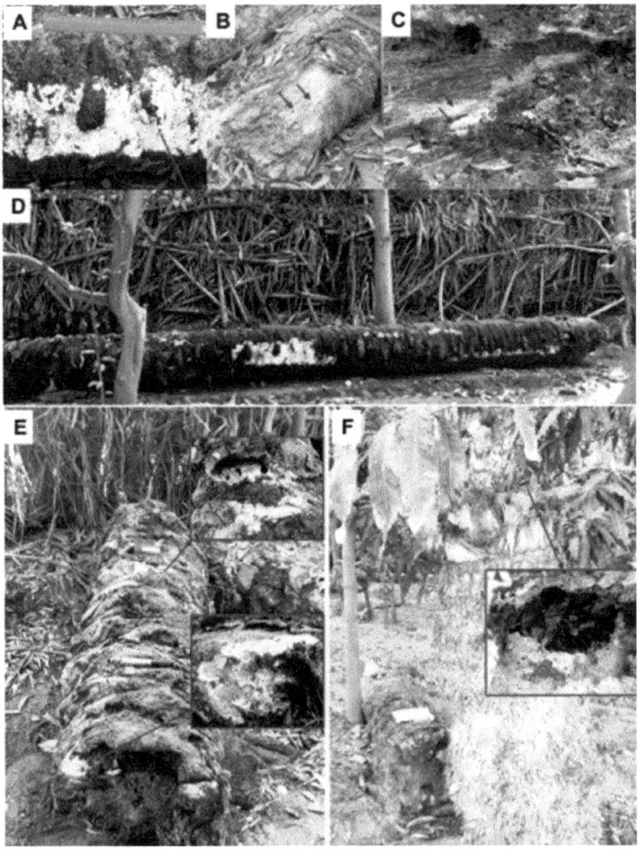

Fig. 3. *Crystallicutis damiettensis* grow on date palm trunks, recorded during the author's field surveys in different governorates of the Nile Delta of Egypt during 8 years of study (2013–2021). Honey-yellow basidiomes on the bark (A), occupy the boreholes of red weevil (B, arrow) causing white-rot decay (C). Sometimes the basidiome formed in the long decay columns (D), at the trunk ends (E) and within the hollowed dead stumps (F).

5.4 *Brown-Rot Fungi*

Brown-rot fungi represent less than 10% of wood-rot basidiomycetes, the preferred attack of softwoods breaking down hemicellulose and cellulose removing most of thre wood carbohydrates, with little or without metabolizing lignin (Pournou, 2020). These fungi can mediate lignin demethylation non-enzymatically through the Fenton reaction to access complex polysaccharides (Venkatesagowda and Dekker, 2021). Extensive depolymerization of the cellulose within the wood cell wall occurs at early stages of decay with very low weight loss. These fungi include *Coniophora puteana, Fibroporia vaillantii, Lentinus* spp., *Lenzites* spp. and *Serpula lacrymans*, which may attack indoor timber in buildings (Durmaz et al., 2016).

5.5 *Soft-Rot Fungi*

The soft-rot fungi have more similarity with brown-rots due for their preference to cellulose degradation and relatively low lignin degradation (Goodell et al., 2020). Soft-rot occurs not only on wet wood, but also in dry conditions. Although exceeding wet or dry conditions inhibit the growth of wood decaying basidiomycetes, these adverse conditions will not limit the colonization soft-rot fungi (Goodell et al., 2020). *Ustulina deusta* is an extensively degrades wood even in the central parts of the live tree. It causes a characteristic pattern of soft-rot during the early stage of decay on different hosts (Schwarze et al., 1995).

6. Mutualism and Parasitism

Macrofungi are well known for mutualistic association, pathogenicity and parasitism on a wide variety of plants and animals. The following sections emphasize the mutualism and parasitism with plants as well as animals in different ecosystems.

6.1 *Mutualism with Plants*

Many mushrooms have a mutualistic association with roots of higher plants are called ectomycorrhizae. Such mutualism is very important to fetch many nutrients to the host plant especially phosphate (Grose, 2018). They support the growth of forest trees through mycorrhizal associations with roots and plant pathogens. Their absence can decline the population of trees to the brink of extinction (Wojciechowska, 2017; Jones et al., 2018).

6.2 *Parasitism with Plants*

Some mushrooms are parasites on higher plants causing severe damage to the health of trees, which leads to creating new habitats for other organisms. There are many types of parasitism, which depend on the plant species and its health. Some kind of parasitism attack plant host without killing it, while some attack only unhealthy hosts and hastens death (Doehlemann et al., 2017). Parasitic mushrooms are also edible especially honey mushrooms (e.g., *Armillaria mellea*). Some saprophytic mushrooms are weakly parasitic and are known as facultative saprophytes. *Pleurotus ostreatus* is an example of a classic saprophytic mushroom that often grows on dying trees and appears to be operating parasitically. In Asia, *Ganoderma* spp. are the most widely distributed pathogens associated with root rot diseases of various tropical plant species especially acacia, coconut and tea (Wahab and Aswad, 2015).

6.3 Mutualism with Animals

Symbiosis involving mushrooms and animals, specific insects is an important relationship due to the inability of animals to digest cellulose. Some animals consume mycelium in well-decomposed plant material as their source of food (Tang et al., 2015). The relationship between *Termitomyces* and fungus-growing termites (Macrotermitinae) is a well-known mutualism. The termites offer suitable microclimatic conditions as well as a substrate to the fungus, selectively inhibit fungal competitors and control microbial pests by secretion of specific inhibitory substances. Termitomycetes provide nitrogen-rich nutrients and digestive enzymes as aid to termites (Rouland-Lefèvre et al., 2006). Termites disperse the associated fungi to different locations (e.g., hyphal fragments, spores and spherules). Such vertical transmission results in the co-phylogenetic pattern of evolution between the termites and mutualistic fungi (Vreeburg, 2020).

6.4 Parasitism with Animals

Mushrooms parasitizing animals are rare, *Cordyceps* and allied species parasitize insects and other arthropods (Zhang et al., 2013). About 400 species of *Cordyceps* and related fungi have been reported and several of them serve traditional Chinese medicines for centuries. They are assigned to five genera (*Cordyceps*, *Elaphocordyceps*, *Metacordyceps*, *Ophiocordyceps* and *Tyrannicordyceps*) (Li et al., 2019). The well-known species are *Cordyceps militaris* and *Ophiocordyceps sinensis*, they are also called the Chinese caterpillar fungi. The *Cordyceps* spp. infects insects, grows and produces fruit bodies by sprouting from the flesh of insects. They possess potential medicinal value owing to many bioactive compounds, especially cordycepin. Cordycepin is therapeutically valued as anti-proliferation and anti-metastasis to treat different cancer cell lines (e.g., renal, ovarian, lung and others) owing to its apoptotic action (Lee et al., 2022). It is also believed that it is effective against pathogenic viruses including COVID-19 (Johri et al., 2022).

Ophiocordyceps parasitizes the larvae of ghost moths (*Hepialus humuli*), which usually live in underground tunnels (Zhang et al., 2013). The anamorph of *Ophiocordyceps* lives around plant roots, which helps to infect the larvae of the ghost moths parasitic to plant roots (Zhong et al., 2014). *Ophiocordyceps* infections occur in three phases: (i) Penetration of the fungal hyphae into the spiracles of larvae; (ii) Adherence of ascospores to the cuticles of larvae; (iii) Ingestion of ascospores by the larvae. Eventually, the *Ophiocordyceps* develops inside the parasitized larva. During the summer (~ 2–3 years), the fungal stroma emerges from the head region of the dead larva (Zhang et al., 2013). *Ophiocordyceps nutans* is also parasitic on the stick bugs (*Halyomorpha halys*), which is a sap-sucking pentatomid bugs of the bark of evergreen trees (*Cassine glauca*) in the Western Ghats of India (Karun and Sridhar, 2013). A tripartite association of a basidiomycete *Auruculoscypha* with trees belonging to Anacardiaceae with a scale insect (*Neogreenia zeylinaca*) has been reported Manimohan (2019).

7. Dispersal of Mushrooms

Mushrooms and animals although they live independently, at certain conditions interact with each other leading to beneficial effects. Mushrooms serve as the nutritional sources for the animal, while mushrooms spread spores through animal feeding and their appendages (Tang et al., 2015). Associated animals expand the opportunities for mushrooms to widen its horizon. Such macrofungal interactions with animal could be generally classified into three groups: (i) Hyogeous mushrooms; (ii) Stinkhorns; (iii) Coprophilous fungi.

7.1 *Hypogeous Mushrooms*

Most of the hypogeous fungi are ectomycorrhizal with plants to obtain carbohydrates, while these mushrooms facilitate plants to get water, essential mineral nutrients and protection from root pathogens. Owing to the limited ability of dispersal in hypogeous mushrooms, their spores will be transmitted to different habitats by soil invertebrates as well as small mammals (Magyar et al., 2016). The mycophagous mammals benefit from the nutritional value of hypogeous fungi and disperse their propagules/spores on their surface as well as through defecation. Mycophagous mammals have a significant influence on the transmission, reproduction and genetic exchange of hypogeous mushrooms (Elliott et al., 2022).

7.2 *Stinkhorns*

Members of the order Phallales (stinkhorns) are mainly saprophytic grown in soils of woodlands and well-rotted woody material in many regions of Africa, America, Asia and Australia (Begum, 2021). Fruit body formation of stinkhorns initiated underground and basidiospores developed within the epigeic gleba. On the breakdown of gleba, the gelatinous mass of basidiospores attracts insects owing to the production of foul odour as well as volatile substances. Insects feed on gelatinous matrix and transfer live basidiospores to different habitats through their appendages and excrements (Magyar et al., 2016).

7.3 *Coprophilous Mushrooms*

Coprophilous mushrooms a special group adapted to grow on herbivore dung and transmitted to different regions. They withstand the conditions of the digestive tract, which leads to the cessation of spore dormancy. In addition, dung attracts several insects and soil invertebrates, thus, coprophilous mushrooms will be further transmitted to different habitats. Typical examples of copropholous mushrooms are cup and flask ascomycetes, they spread spores by themselves, herbivores and insects

Tang et al., 2015). *Coprinus* sp. (basidiomycete) also known as inky caps are famous examples of coprophilous fungi that commonly growing on decaying substrates in grasslands as well as herbivorous dung (Nagy et al., 2013). *Coprinus* get transmitted by the coprophilous insects, which carry spores on their bodies as well as a digestive system from place to place (Gomes and Wartchow, 2018). *Panaeolus antillarum* and *Psathyrella* spp. are other examples of coprophilous mushrooms, which grow on dung (Amandeep et al., 2015). They were also recorded growing on dung in different microhabitats during our mushroom forays in the Nile Delta of Egypt (Fig. 4).

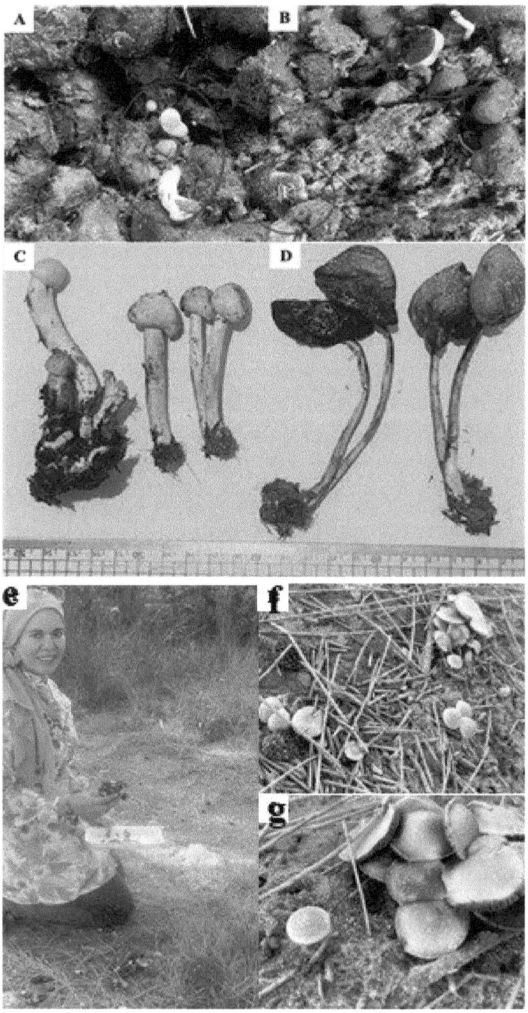

Fig. 4. Coprophilous mushroom (*Panaeolus antillarum*) recorded by the author in Damietta farm during a mushroom foray in August 2022: A) and B) showing young and mature fruit bodies in rings on the dung of a domestic donkey; C) and D) showing collected young and mature mushrooms at the laboratory, respectively; Coprophilous mushroom (*Psathyrella* sp.) recorded by the author in Alexandria Botanical Garden during a mushroom foray in October 2022: E) fruit bodies growing in groups on dung; F) and G) magnified pictures of the mushrooms.

8. Ecological Role of Wild Mushrooms

Wild mushrooms are the key component of natural ecosystems (e.g., as symbionts, parasites and as pathogens) and perform a significant role in ecosystem functioning. They maintain the fertility of the soil by recycling nutrients through decomposing organic matter and uptake nutrients to the roots through mycorrhizal associations. Parasitic as well as pathogenic mushrooms fetch nutrients for their growth and perpetuation by exploiting the resources of plants and animals. Wild mushrooms possess the ability to remediate pollutants through their extracellular enzymes. Wild mushrooms are also an important source of food for wildlife. In addition, other species such as *Cordyceps* and *Pleurotus* are used as biological control agents against phytopathogenic pests (i.e., insects and nematodes) (Moosavi and Zare, 2020).

8.1 Mycorrhizae

These mushrooms facilitate the uptake of water and nutrients, especially in nitrogen- and phosphorus-limited habitats to enhance carbon sequestration as well as storage. They are capable to regulate the forest structure and functions in varied climate changes. The ecological relevance of mycorrhizal mushrooms, the pattern of association with roots, and mechanism of nutrient transport has extensively drawn the attention of the scientific community to understand the structure and role in forest ecosystem dynamics (Pennisi and Cornwall, 2020).

8.2 Decomposers

Decomposers are the saprotrophic mushrooms involved in the degradation of organic matter of plant origin and play an essential role in nutrient cycling as well as carbon sequestration in forest ecosystems. They are exploited for the decomposition of litter, cellulose, hemicelluloses, lignin compounds and the biodegradation of environmentally hazardous materials. These mushrooms can breakdown the lignin and other related constituents like hemicelluloses and cellulose into a simpler form by secretion of laccases, phenoloxidases and peroxidases enzymes (Malik et al., 2021). The pace of degradation generally depends upon the nutrients present in the soil. Several mushrooms (e.g., *Agaricus* sp., *Ganoderma* sp., *Irpex lacteus*, *Lentinula* sp., *Lentinus squarrosulus*, *L. tigirinus*, *Nematolana prowardii*, *Pleurotus* sp. and *Stropharia coronilla*) interact with obstinate waste litter such as chitin, keratin, lignin, fats and disintegrate them into simple molecules (i.e., cellulose and sugars) (Barh et al., 2019).

8.3 Mycoremediation

Mycoremediation is one of the means to transform or degrade environmental pollutants (Sahrawat et al., 2018). Mushrooms are known to produce a variety of extracellular enzymes (e.g., oxidases, cellulases, laccases, lignin peroxidase, peroxidases, manganese-dependent peroxidase, pectinases and xylanases) (Nyanhongo et al., 2007). These enzymes involve in the complete mineralization of several pollutants into simpler and more safe products. They can degrade non-polymeric and many recalcitrant pollutants (e.g., nitrotoluenes, organic dyes, pentachlorophenol, polyaromatic hydrocarbons and synthetic dyes) (Kumar et al., 2018). Moreover, it is known that mushrooms are also capable to degrade polymers like plastics (da Luz et al., 2013). Several wood-decaying basidiomycetes such as *Abortiporus biennis*, *Bjerkandera adusta*, *Ganoderma* sp., *Inonotus hispidus*, *Irpex lacteus*, *Lentinus tigrinus*, *Panellus stipticus* and *Pleurotus* spp. are potent agents to degrade or detoxify a broad range of pollutants (chlorinated compounds, textile or industrial effluents, aromatic hydrocarbons, preservatives, fungicides, herbicides and insecticides) into non-toxic simple organic compounds (Ntougias et al., 2012). Other species like *Ceriporiopsis subvermispora*, *Phanerochaete chrysosporium* and *Trametes versicolor* have been utilized in bio-pulping and these species have also been used to degrade some of the toxic wastes present in municipal dumps (Kumar et al., 2020).

Wild mushrooms are also capable to tolerate heavy metals and participate in the biosorption of metallic ions, pollutants and xenobiotics (de Jesus Menk et al., 2019). Such functions of mushrooms are simple, more economical, environment-friendly means of removal of heavy metals from industrial effluents. *Pleurotus ostreatus* and its spent substrates are also efficient in the biosorption of lead (II) from aqueous solutions (Eliescu et al., 2020). Mushroom composts are used successfully to remove copper (II) and iron (II) from the wastewater (Kamarudzaman et al., 2022).

Recently, mushrooms have been used to produce ecofriendly mycelium-bound composite materials in the construction industry for building and insulations (Xing et al., 2018). The bio-based materials can be used as alternatives for conventional bricks, polystyrene with the same physical and mechanical characteristics but enhanced biodegradability. Three Egyptian mushrooms were tested by Xing et al. (2018), their study includes *Ganoderma resinaceum*, *Megasporoporia minor* and *Oxyporus latermarginatus*. Substrates such as coconut husk, rice bran and sawdust are useful in the production of sturdy mycelium composites. The quality of such bricks and composites can be enhanced by increasing the water absorption and thermal insulation capacities. Amending mycelium with other products such as silver grass (*Miscanthus*) improves its quality as non-conventional building material (Sharma and Sumbria, 2022).

Conclusion

Wild mushrooms play a vital role in maintaining ecosystem integrity. They are nutritionally, medicinally and industrially valuable. They possess a high potential to degrade organic matter, are efficient in nutrient cycling and involve in ecofriendly bioremediation. They are under the influence of various abiotic and biotic factors in different ecological niches. Wild mushrooms mutualistically associated with a variety of tree species in forest ecosystems. Many wild mushrooms are pathogens and parasites of plants and animals. They are dependent on abiotic and biotic factors to disperse their spores and occupy varied ecological niches. They are responsible for several ecosystem services as mutualists, decomposers of organic matter, nutrient cycling, nutrient distribution and involve in ecofriendly bioremediation of environmental pollutants. Wild mushrooms have become valuable tools in modern research involves in environment-friendly innovations.

Acknowledgments

Many thanks to Prof. Mamdouh S. Serag, Professor of Plant Ecology, Faculty of Science, Damietta University, Egypt for his efforts to draft this chapter, genuine guidelines and constructive suggestions.

References

Amandeep, K., Atri, N.S. and Munruchi, K. (2013). Two new coprophilous varieties of *Panaeolus* (Psathyrellaceae, Agaricales) from Punjab, India. Mycosphere, 4: 616–625. 10.5943/mycosphere/4/3/13.

Andrew, C., Halvorsen, R., Heegaard, E., Kuyper, T.W., Heilmann-Clausen, J. et al. (2018). Continental-scale macrofungal assemblage patterns correlate with climate, soil carbon and nitrogen deposition. J. Biogeog., 45: 1942–1953. 10.1111/jbi.13374.

Balai, L.P. and Ahir, R. (2013). Role of temperature and relative humidity on mycelial growth of *Alternaria alternata* infecting brinjal. Tr. Biosci., 6: 307–308.

Barh, A., Sharma, V.P., Kamal, S., Shirur, M., Annepu, S.K. et al. (2019). Speciation of cultivated temperate and tropical *Pleurotus* species—An *in silico* prediction using conserved sequences. Mushr. Res., 28: 31–37. 10.36036/MR.28.1.2019.91990.

Begum, N. (2021). Diversity of Macrofungi. Bhumi Publishing, Nigave Khalasa, Kolhapur, Maharashtra, India, p. 46.

Bellettini, M.B., Fiorda, F.A., Maieves, H.A., Teixeira, G.L., Ávila, S. et al. (2019). Factors affecting mushroom *Pleurotus* spp. Saudi J. Biol. Sci., 26: 633–646. 10.1016/j.sjbs.2016.12.005.

Bencherif, K., Boutekrabt, A., Fontaine, J., Laruelle, F., Dalpe, Y. and Sahraoui, A.L.H. (2015). Impact of soil salinity on arbuscular mycorrhizal fungi biodiversity and microflora biomass associated with *Tamarix articulata* Vahll rhizosphere in arid and semi-arid Algerian areas. Sci. Tot. Environ., 533: 488–494. 10.1016/j.scitotenv.2015.07.007.

Ben-David, E.A., Zaady, E., Sher, Y. and Nejidat, A. (2011). Assessment of the spatial distribution of soil microbial communities in patchy arid and semi-arid landscapes of the Negev Desert using combined PLFA and DGGE analyses. FEMS Microbiol. Ecol., 76: 492–503. 10.1111/j.1574-6941.2011.01075.

Boa, E. (2004). Wild edible fungi: A global overview of their use and importance to people. Non-Wood Forest Products # 17. Food and Agriculture Organization, Rome, p. 147.

Chelela, B.L., Chacha, M. and Matemu, A. (2014). Wild edible mushroom value chain for improved livelihoods in Southern Highlands of Tanzania. American J. Research Communication, 2: 1–14.

da Luz, J.M.R., Paes, S.A., Nunes, M.D., da Silva, M.D.C.S. and Kasuya, M.C.M. (2013). Degradation of oxo-biodegradable plastic by *Pleurotus ostreatus*. Plos One, 8: e69386.

Davarzani, H., Smits, K., Tolene, R.M. and Illangasekare, T. (2014). Study of the effect of wind speed on evaporation from soil through integrated modeling of the atmospheric boundary layer and shallow subsurface. Water Resour. Res., 50: 661–680.

de Jesus Menk, J., do Nascimento, A.I.S., Leite, F.G., de Oliveira, R.A., Jozala, A.F. et al. (2019). Biosorption of pharmaceutical products by mushroom stem waste. Chemosphere, 237: 124515.

Dias, E.S. and de Brito, M.R. (2017). Mushrooms: Biology and life cycle. pp. 15–33. *In*: Deigo, C.Z. and Pardo-Gaménez, A. (eds.). Edible and Medicinal Mushrooms: Technology and Applications, John Wiley & Sons, Inc. 10.1002/9781119149446.ch3.

Doehlemann, G., Ökmen, B., Zhu, W. and Sharon, A. (2017). Plant pathogenic fungi. 703–726. *In*: Heitman, J., Howlett, B., Crous, P., Stukenbrock, E., James, T. and Gow, N. (eds.). The Fungal Kingdom. ASM Press, Washington DC.

Durmaz, S., Ozgenc, O., Boyaci, I.H., Yildiz, U.C. and Erisir, E. (2016). Examination of the chemical changes in spruce wood degraded by brown-rot fungi using FT-IR and FT-Raman spectroscopy. Vib. Spectrosc., 85: 202–207.

Dwivedi, S., Tiwari, M.K., Chauhan, U.K. and Pandey, A.K. (2012). Biodiversity of mushrooms of Amarkantak Biosphere Reserve Forest of Central India. Int. J. Pharm. Life Sci., 3: 1363–1367.

El-Fallal, A.A., Elsayed, A.K.A. and El-Gharabawy, H.M. (2017). First Record of *Lepista sordida* (Schumach) Singer in Eastern North Africa. Eg. J. Bot., 57: 111–118. 10.21608/ejbo.2017.980.1087.

El-Fallal, A.A., El-Sayed, A.K. and El-Gharabawy, H.M. (2019). *Podaxis pistillaris* (L.) Fr. and *Leucocoprinus birnbaumii* (Corda) Singer; new addition to macrofungi of Egypt. Eg. J. Bot., 59: 413–423. 10.21608/ejbo.2019.5990.1255.

El-Fouly, M.Z., Shahin, A.A.F.M. and El-Bialy, H.A.A. (2011). Biological control of sapstain fungi in Egyptian wood stores and infected trees. Ann. Microbiol., 61: 789–799.

El-Gharabawy, H.M. (2016). Wood Decay of Trees by Basidiomycete Fungi in the North East Nile Delta Region, PhD Dissertation, Faculty of Science, Damietta University, Egypt, p. 287.

El-Gharabawy, H.M., El-Fallal, A.A. and El-Sayed, K.A.A. (2019). Description of a yellow morel from Egypt using morphological and molecular tools. Nova Hedw., 109: 95–110. 10.1127/novahedwigia/2019/0535.

El-Gharabawy, H.M., Leal-Dutra, C.A. and Griffith, G.W. (2021). *Crystallicutis* gen. nov. (Irpicaceae, Basidiomycota), including *C. damiettensis* sp. nov., found on *Phoenix dactylifera* (date palm) trunks in the Nile Delta of Egypt. Fungal Biol., 125: 447–458. 10.1016/j.funbio.2021.01.004.

Eliescu, A., Georgescu, A.A., Nicolescu, C.M., Bumbac, M., Cioateră, N. et al. (2020). Biosorption of Pb (II) from aqueous solution using mushroom (*Pleurotus ostreatus*) biomass and spent mushroom substrate. Anal. Lett., 53: 2292–2319.

Elliott, T.F., Truong, C., Jackson, S.M., Zúñiga, C.L., Trappe, J.M. and Vernes, K. (2022). Mammalian mycophagy: A global review of ecosystem interactions between mammals and fungi. Fun. Syst. Evol., 9: 99–159.

Fazenda, M.L., Seviour, R., McNeil, B. and Harvey, L.M. (2008). Submerged culture fermentation of "higher fungi": The macrofungi. Adv. App. Microbiol., 63: 33–103. 10.1016/S0065-2164(07)00002-0.

Feng, B. and Yang, Z. (2018). Studies on diversity of higher fungi in Yunnan, southwestern China: A review. Plant Divers., 40: 165–171. 10.1016/j.pld.2018.07.001.

Feofilova, E.P., Ivashechkin, A.A., Alekhin, A.I. and Sergeeva. Y.E. (2012). Fungal spores: Dormancy, germination, chemical composition, and role in biotechnology. Appl. Biochem. Microbiol., 48: 1–11. 10.1134/S0003683812010048.

Fishar, M. (2018). Nile Delta (Egypt). *In*: Finlayson, C., Milton, G., Prentice, R. and Davidson, N. (eds.). The Wetland Book, Springer, Dordrecht. 10.1007/978-94-007-4001-3_216.

Fuller, K.K., Loros, J.J. and Dunlap, J.C. (2015). Fungal photobiology: Visible light as a signal for stress, space and time. Curr. Genet., 61: 275–288. 10.1007/s00294-014-0451-0.

Gomes, A.R.P. and Wartchow, F. (2018). Notes on two coprinoid fungi (Basidiomycota, Agaricales) from the Brazilian semiarid region. Edin. J. Bot., 75: 285–295. 10.1017/S0960428618000094.

González-Pérez, J.A., González-Vila, F.J., Almendros, G. and Knicker, H. (2004). The effect of fire on soil organic matter—A review. Environ. Int., 30: 855–870. 10.1016/j.envint.2004.02.003.

Goodell, B. and Nielsen, G. (2022). Wood biodeterioration. pp. 1–39. *In*: Niemz, P., Teischinger, A. and Sandberg, D. (eds.). Springer Handbook of Wood Science and Technology, p. 2060. 10.1007/978-3-030-81315-4_4.

Goodell, B., Winandy, J.E. and Morrell, J.J. (2020). Fungal degradation of wood: Emerging data, new insights and changing perceptions. Coatings, 10: 1210–1229. 10.3390/coatings10121210.

Govorushko, S., Rezaee, R., Dumanov, J. and Tsatsakis, A. (2019). Poisoning associated with the use of mushrooms: A review of the global pattern and main characteristics. Food Chem. Toxicol., 128: 267–279. 10.1016/j.fct.2019.04.016.

Grose, T.K. (2018). Engineered symbiosis. ASEE Prism, 28: 12–12.

Hibbett, D.S. and Binder, M. (2002). Evolution of complex fruiting-body morphologies in homobasidiomycetes. Proc. R. Soc. Lond. B, 269: 1963–1969.

Hrynkiewicz, K., Złoch, M., Kowalkowski, T., Baum, C., Niedojadło, K. and Buszewski, B. (2015). Strain-specific bioaccumulation and intracellular distribution of Cd^{2+} in bacteria isolated from the rhizosphere, ectomycorrhizae, and fruitbodies of ectomycorrhizal fungi. Environ. Sci. Pollut. Res., 22: 3055–3067.

Hu, J.-J., Zhao, G.-P., Tuo, Y.-L., Qi, Z.-X., Yue, L., Zhang, B. and Li, Y. (2022). Ecological factors influencing the occurrence of macrofungi from eastern mountainous areas to the central plains of jilin Province, China. J. Fungi, 8: 871–921. 10.3390/jof8080871.

Huuskonen, S., Domisch, T., Finér, L., Hantula, J., Hynynen, J. et al. (2021). What is the potential for replacing monocultures with mixed-species stands to enhance ecosystem services in boreal forests in Fennoscandia?. For. Ecol. Manage., 479: 118558–118579. 10.1016/j.foreco.2020.118558.

Hyde, K.D., Chethana, K.W.T., Jayawardena, R.S., Luangharn, T., Calabon, M.S. et al. (2020a). The rise of mycology in Asia. Sci. Asia, 46: 1–11. 10.2306/scienceasia1513-1874.2020.S001.

Hyde, K.D., Dong, Y., Phookamsak, R., Jeewon, R., Bhat, D.J. et al. (2020b). Fungal diversity notes 1151–1276: Taxonomic and phylogenetic contributions on genera and species of fungal taxa. Fungal Divers., 100: 5–277. 10.1007/s13225-020-00439-5.

Jahromi, F., Aroca, R., Porcel, R. and Ruiz-Lozano, J.M. (2008). Influence of salinity on the *in vitro* development of *Glomus intraradices* and on the *in vivo* physiological and molecular responses of mycorrhizal lettuce plants. Microb. Ecol., 55: 45–53. 10.1007/s00248-007-9249-7.

Janaki, T. (2022). Fungi. pp. 83–106. *In*: Singh, H.K. (ed.). Current Research and Innovations in Plant Pathology, AkiNik Publications, New Delhi.

Jaradat, A.A. (2010). Genetic resources of energy crops: Biologica systems to combat climate change. Aust. J. Crop. Sci., 4: 309–323.

Johri, S., Nair, Y.V. and Selvapandiyan, A. (2022). Cultivation and medicinal uses of *Cordyceps militaris* (L.) link: A revolutionary entomopathogenic fungus. pp. 579–595. *In*: Arya, A. and Rusevska, K. (eds.). Biology, Cultivation and Applications of Mushrooms. Springer, Singapore. 10.1007/978-981-16-6257-7_22.

Jones, M., Bhat, T., Huynh, T., Kandare, E., Yuen, R., Wang, C.H. and John, S. (2018). Waste-derived low-cost mycelium composite construction materials with improved fire safety. Fire Mat., 42: 816–825.

Kamarudzaman, A.N., Adan, S.N.A.C., Hassan, Z., Wahab, M.A., Makhtar, S.M.Z. et al. (2022). Biosorption of copper (II) and iron (II) using spent mushroom compost as biosorbent. Biointerface Res. Appl. Chem., 12: 7775–7786.

Karun, N.C. and Sridhar, K.R. (2013). The stink bug fungus *Ophiocordyceps nutans* – a proposal for conservation and flagship status in the Western Ghats of India. Fungal Conserv., 3: 43–49.

Karwa, A. and Rai, M.K. (2010). Tapping into the edible fungi biodiversity of Central India. Biodiversitas, 11: 97–101. 10.13057/biodiv/d110209.

Kewessa, G., Dejene, T., Alem, D., Tolera, M. and Martín-Pinto, P. (2022). Forest type and site conditions influence the diversity and biomass of edible macrofungal species in Ethiopia. J. Fungi, 8: 1023–1039. 10.3390/jof8101023.

Kim, S.Y., Kim, M.-K., Im, C.H., Kim, K.-H., Kim, D.S. et al. (2013). Optimal relative humidity for *Pleurotus eryngii* cultivation. J. Mushroom, 11: 131–136. 10.14480/JM.2013.11.3.131.

Klein, J. (2022). Role of Pde_2 in the CO_2 sensing during fruiting in *Schizophyllum commune*. Ph.D. Dissertation, Utrecht University, Netherlands. p. 31. https://studenttheses.uu.nl/handle/20.500.12932/43274.

Krebs, C., Carrier, P., Boutin, S., Boonstra, R. and Hofer, E. (2008). Mushroom crops in relation to weather in the southwestern Yukon. Botany, 86: 1497–1502.

Kumar, A., Gautam, A. and Dutt, D. (2020). Bio-pulping: An energy saving and environment-friendly approach. Phys. Sci. Rev., 5(10): 1–9. 10.1515/psr-2019-0043.

Kumar, K., Mehra, R., Guiné, R.P., Lima, M.J., Kumar, N., Kaushik, R., Ahmed, N., Yadav, A.N. and Kumar, H. (2021). Edible Mushrooms: A comprehensive review on bioactive compounds with health benefits and processing aspects. Foods, 10(12): 2996.

Kumar, M., Ansari, R.A. and Ashraf, S. (2018). Does mycoremediation reduce the soil toxicant? pp. 423–431. *In*: Kumar, V., Kumar, M. and Prasad, R. (eds.). Phytobiont and Ecosystem Restitution, Springer, Singapore. 10.1007/978-981-13-1187-1_21.

Lee, Y.H., Tsai, K.W., Lu, K.C., Shih, L.J. and Hu, W.C. (2022). Cancer as a dysfunctional immune disorder: pro-tumor th1-like immune response and anti-tumor thαβ immune response based on the complete updated framework of host immunological pathways. Biomed., 10: 2497. 10.3390/biomedicines10102497.

Li, X., Liu, Q., Li, W., Li, Q., Qian, Z. et al. (2019). A breakthrough in the artificial cultivation of Chinese cordyceps on a large-scale and its impact on science, the economy, and industry. Crit. Rev. Biotechnol., 39: 181–191.

Magyar, D., Vass, M. and Li, D.W. (2016). Dispersal strategies of microfungi. pp. 315–371. *In*: Li, D.-W. (ed.). Biology of Microfungi, Springer Cham. 10.1007/978-3-319-29137-6.

Malik, N.A., Kumar, J., Wani, M.S., Tantray, Y.R. and Ahmad, T. (2021). Role of mushrooms in the bioremediation of soil. pp. 77–102. *In*: Dar, G.H., Bhat, R.A., Mehmood, M.A. and Hakeem, K.R. (eds.). Microbiota and Biofertilizers, Volume 2: Springer Cham.

Manimohan, P. (2019). *Auroculoscypha*: A fascinating instance of fungus-insect-plant interactions. pp. 34–39. *In*: Sridhar, K.R. and Deshmukh, S.K. (eds.). Advances in Macrofungi - Diversity, Ecology and Biotechnology. CRC Press, Boca Raton.

Moosavi, M.R. and Zare, R. (2020). Fungi as biological control agents of plant-parasitic nematodes. pp. 333–384. *In*: Mérillon, J.M. and Ramawat, K.G. (eds.). Plant Defence: Biological Control, Volume 22, Springer Cham.

Mukhtar, H., Lin, C.M., Wunderlich, R.F., Cheng, L.C., Ko, M.C. and Lin, Y.P. (2021). Climate and land cover shape the fungal community structure in topsoil. Sci. Tot. Environ., 751: 141721. 10.1016/j.scitotenv.2020.141721.

Nagy, L.G., Vágvölgyi, C. and Papp, T. (2013). Morphological characterization of clades of the Psathyrellaceae (Agaricales) inferred from a multigene phylogeny. Mycol. Progr., 12: 505–517. 10.1007/s11557-012-0857-3.

Neira, J., Ortiz, M., Morales, L. and Acevedo, E. (2015). Oxygen diffusion in soils: Understanding the factors and processes needed for modeling. Chil. J. Agric. Res., 75: 35–44.

Nikšić, M., Podgornik, B.B. and Berovic, M. (2022). Farming of Medicinal Mushrooms. Adv. Biochem. Eng. Biotechnol. 10.1007/10_2021_201.

Ntougias, S., Baldrian, P., Ehaliotis, C., Nerud, F., Antoniou, T. et al. (2012). Biodegradation and detoxification of olive mill wastewater by selected strains of the mushroom genera *Ganoderma* and *Pleurotus*. Chemosphere, 88: 620–626.

Nyanhongo, G.S., Gűbitz, G., Sukyai, P., Leitner, C., Haltrich, D. and Ludwig, R. (2007). Oxidoreductases from *Trametes* spp. in biotechnology: A wealth of catalytic activity. Food Technol. Biotechnol., 45(3): 250–268.

Pegler, D. (2003). Useful fungi of the world: Morels and truffles. Mycologist, 17: 174–175. 10.1017/S0269915X04004021.

Pennisi, E. and Cornwall, W. (2020). Hidden web of fungi could shape the future of forests. Science, 369: 1042–1043. 10.1126/science.369.6507.1042.

Peñuelas, J., Sardans, J., Filella, I., Estiarte, M., Llusià, J. et al. (2017). Impacts of global change on Mediterranean forests and their services. Forests, 8: 463–500. 10.3390/f8120463.

Porcel, R., Aroca, R. and Ruiz-Lozano, J.M. (2012). Salinity stress alleviation using arbuscular mycorrhizal fungi. A review. Agron. Sustain. Dev., 32: 181–200.

Pournou, A. (2020). Wood Deterioration by Terrestrial Microorganisms. Biodeterioration of Wooden Cultural Heritage, pp. 345–424. Springer Cham. 10.1007/978-3-030-46504-9_6.

Qi, J., Jia, L., Liang, Y., Luo, B., Zhao, R. et al. (2022). Fungi's selectivity in the biodegradation of *Dendrocalamus sinicus* decayed by white and brown rot fungi. Ind. Crops Prod., 188: 115726.

Rai, A., Rai, P.K., Singh, S. and Sharma, N.K. (2015). Environmental factors affecting edible and medicinal mushroom production. pp. 67–81. *In*: Thakur, T. and Thakur, M.P. (eds.). Production Techniques of Tropical Mushrooms in India. Nirmal Publications (India), New Delhi, India.

Rashid, S.N., Aminuzzaman, F.M., Islam, M.R., Rahaman, M. and Rumainul, M.I. (2016). Biodiversity and distribution of wild mushrooms in the southern region of Bangladesh. J. Adv. Biol. Biotechnol., 9: 1–25. 10.9734/JABB/2016/27711.

Rayner, A.D.M. and Boddy, L. (1988). Fungal Decomposition of Wood: Its Biology and Ecology. John Wiley & Sons, Great Britain, p. 587.

Rezaeian, S. and Pourianfar, H.R. (2017). A comparative study on bioconversion of different agrowastes by wild and cultivated strains of *Flammulina velutipes*. Waste Biomass Valorization, 8: 2631–2642. 10.1007/s12649-016-9698-7.

Rieux, A., Soubeyrand, S., Bonnot, F., Klein, E.K., Ngando, J.E. et al. (2014). Long-distance wind-dispersal of spores in a fungal plant pathogen: Estimation of anisotropic dispersal kernels from an extensive field experiment. PLoS ONE, 9: e103225.

Rouland-Lefèvre, C., Inoue, T. and Johjima, T. (2006). *Termitomyces*/Termite interactions. pp. 335–350. *In*: König, H. and Verma, A. (eds.). Intestinal Microorganisms of Termites and other Invertebrates. Springer, Berlin and Heidelberg.

Sahrawat, A., Sharma, J., Tiwari, S. and Rahul, S.N. (2018). A comparative study of nutritional and non-nutritional composition of mushroom capable of growing on the different waste. Int. J. Chem. Stud., 6: 223–229.

Salo, K., Domisch, T. and Kouki, J. (2019). Forest wildfire and 12 years of post-disturbance succession of saprotrophic macrofungi (Basidiomycota, Ascomycota). Forest Ecol. Manag., 451: 117454. 10.1016/j.foreco.2019.117454.

Samarakoon, M.C., Thongbai, B., Wang, K.D., Brönstrup, M., Beutling, U. et al. (2020). Elucidation of the life cycle of the endophytic genus *Muscodor* and its transfer to *Induratia* in Induratiaceae fam. nov., based on a polyphasic taxonomic approach. Fungal Div., 101: 177–210. 10.1007/s13225-020-00443-9.

Sangkaew, M. and Koh, K. (2017). The cultivation of *Flammulina velutipes* by using sunflower residues as mushroom substrate. J. Adv. Agric. Technol., 4: 140–144. 10.18178/joaat.4.2.140-144.

Schmidt, O. (2006). Wood and Tree Fungi. Springer-Verlag, Berlin Heidelberg, p. 334.

Schwarze, F.W.M.R., Lonsdale, D. and Mattheck, C. (1995). Detectability of wood decay caused by *Ustulina deusta* in comparison with other tree-decay fungi. Eur. J. For. Pathol., 25: 327–341.

Sharma, R. and Sumbria, R. (2022). Mycelium bricks and composites for sustainable construction industry: a state-of-the-art review. Innov. Infrastruct. Solut., 7: 298. 10.1007/s41062-022-00903-y.

Shen, Y., Gu, M., Jin, Q., Fan, L., Feng, W. et al. (2014). Effects of cold stimulation on primordial initiation and yield of *Pleurotus pulmonarius*. Sci. Horticul., 167: 100–106. 10.1016/j.scienta.2013.12.021.

Shetafa, A.G.S. and El-Wahab, A. (2013). First record of *Ptychogaster cubensis* causing canker and decay on citrus and grape trees in Egypt. Eg. J. Phytopathol., 41: 215–216.

Srivastava, M. (2021). Biodiversity of wild mushrooms and their future perspectives. Int. J. Pl. Environ., 7: 164–168. 10.18811/ijpen.v7i02.07.

Tang, X., Mi, F., Zhang, Y., He, X., Cao, Y. et al. (2015). Diversity, population genetics, and evolution of macrofungi associated with animals. Mycol., 16: 94–109. 10.1080/21501203.2015.

Tinya, F., Kovács, B., Bidló, A., Dima, B., Király, I. et al. (2021). Environmental drivers of forest biodiversity in temperate mixed forests–A multi-taxon approach. Sci. Tot. Environ., 795: 148720. 10.1016/j.scitotenv.2021.148720.

Ukwuru, M.U., Muritala, A. and Eze, L.U. (2018). Edible and non-edible wild mushrooms: Nutrition, toxicity and strategies for recognition. J. Clin. Nutr. Metab., 2: 1–9.

Venkatesagowda, B. and Dekker, R.F. (2021). Microbial demethylation of lignin: Evidence of enzymes participating in the removal of methyl/methoxyl groups. Enz. Microb. Tech., 147: 109780.

Vreeburg, S.M.E. (2020). On conflict and resolution in the termite-fungus symbiosis. Ph.D. Dissertation, Wageningen University, Netherlands.

Wahab, M.A.A. and Aswad, M. (2015). *Ganoderma* stem rot of oil palm: epidemiology, diversity and pathogenicity, Ph.D. Dissertation, University of Bath, England.

Wang, J., Ma, S., Wang, G.G., Xu, L., Fu, Z. et al. (2021a). Arbuscular mycorrhizal fungi communities associated with wild plants in a coastal ecosystem. J. For. Res., 32: 683–695. 10.1007/s11676-020-01127-5.

Wang, J.Q., Shi, X.Z., Zheng, C.Y., Suter, H. and Huang, Z.Q. (2021b). Different responses of soil bacterial and fungal communities to nitrogen deposition in a subtropical forest. Sci. Tot. Environ., 755: 142449. 10.1016/j.scitotenv.2020.142449.

Wesenberg, D., Kyriakides, I. and Agathos, S.N. (2003). White rot fungi and their enzymes for the treatment of industrial dye effluents. Biotechnol. Adv., 2003: 161–187. 10.1016/j.biotechadv.2003.08.011.

Wibberg, D., Stadler, M., Lambert, C., Bunk, B., Spröer, C. et al. (2021). High quality genome sequences of thirteen Hypoxylaceae (Ascomycota) strengthen the phylogenetic family backbone and enable the discovery of new taxa. Fungal Diversity, 106: 7–28. 10.1007/s13225-020-00447-5.

Wojciechowska, I. (2017). The leather underground: biofabrication offers new sources for fabrics. AATCC Review, 17: 18–23.

Xing, Y., Brewer, M., El-Gharabawy, H., Griffith, G. and Jones, P. (2018). Growing and testing mycelium bricks as building insulation materials. In IOP Conference Series: Earth and Environmental Science 121(2): 022032. IOP Publishing. 10.1088/1755-1315/121/2/022032.

Yue, Z., Zhang, W., Liu, W., Xu, J., Liu, W. and Zhang, X. (2022). Effect of different light qualities and intensities on the yield and quality of facility-grown *Pleurotus eryngii*. J. Fungi, 8: 1244. 10.3390/jof8121244.

Zhang, S., Zhang, Y., Shrestha, B., Xu, J., Wang, C. and Liu, X. (2013). *Ophiocordyceps sinensis* and *Cordyceps militaris*: research advances, issues and perspectives. Mycosystema, 32: 577–597.

Zhong, X., Peng, Q.-Y., Li, S.-S., Chen, H., Sun, H.-X. et al. (2014). Detection of *Ophiocordyceps sinensis* in the roots of plants in alpine meadows by nested-touchdown polymerase chain reaction. Fungal Biol., 118: 359–363. 10.1016/j.funbio.2013.12.005.

5

Ecology of Macrofungi in Southwest India

Kandikere R Sridhar

1. Introduction

Fungi are the largest biological entity after insects evolved ~ 1,800 mya with complex morphology and lifestyles (Kirk et al., 2008; Hawksworth, 1991). The global conservative estimate of fungi ranges from 2.2–3.8 million, while the macrofungi range between 0.14 and 1.25 million (Hawksworth and Lucking, 2017; Azeem et al., 2020). Another prediction of macrofungal estimate predicts the range from 0.053–0.11 million based on the plant-macrofungal ratio (Mueller et al., 2007). However, our knowledge is limited on the significance of fungi owing to the exploration of less than 10% of the total estimate. Macrofungi are non-conventional resources to cater to the needs of the human diet, health, agriculture and industry. Macrofungi are known to colonize different substrates (soil, leaf litter, woody litter, tree roots and animal remains) and in turn responsible for a variety of ecosystem services (decomposition, erosion prevention, plant growth promotion, nutraceutical source and increased carrying capacity) (Fig. 1). Evaluation of macrofungi in different ecosystems and their favourable substrates expands our knowledge on their applications and future prospects.

Southwestern India consists mainly of the Western Ghats, coastal plains and maritime habitats. Each of these ecosystems could be further classified into several ecological niches that sustain macrofungi. The Western Ghats is one of the major hotspots of biodiversity, characterized by a wide range of mountains stretch of about 1,600 km (about 160,000 km^2) in the states mainly Maharashtra, Goa, Karnataka,

Department of Biosciences, Mangalore University, Mangalagangotri, Mangalore, Karnataka, India.
Email: kandikere@gmail.com

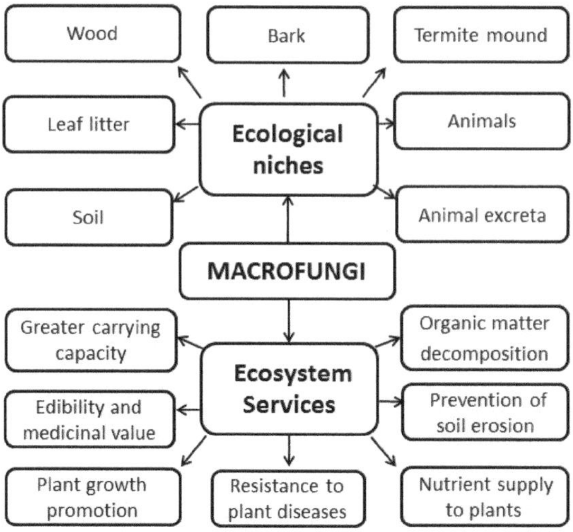

Fig. 1. Schematic presentation of ecological niches and ecosystem services of macrofungi.

Kerala and Tamil Nadu. The vegetation pattern, as well as the ecosystems, differ from the plains of the Deccan plateau with several peaks in the Western Ghats followed by the mid-altitude, foothill (~ 500 to 1,200 m asl) and coastal plains.

In spite of reports of macrofungi from the Western Ghats on distribution, taxonomy and phylogeny, quantitative investigations are sporadic (e.g., Natarajan et al., 2005; Brown et al., 2006; Swapna et al., 2008; Karun and Sridhar, 2014, 2016b). Checklists have documented 1175 species (190 genera) of wood rot fungi (Aphyllophorales) from the Indian subcontinent, 616 species of agarics (112 genera) in Maharashtra, 550 species (166 genera) in Kerala, 315 species (101 genera) in Karnataka, 178 species of agarics (68 genera) in Kerala, 135 species (56 genera) in Karnataka; 99 species (48 genera) of polypores in Kerala, 81 species (39 genera) of non-gilled fleshy macrofungi in Kerala, 50 species of *Mycena* (fifteen sections) in Kerala, 48 species (34 genera) of ectomycorrhizas in southwest India and 30 species of *Inocybe* (four clades) in Kerala (Swapna et al., 2008; Mohanan et al., 2011; Ranadive, 2013; Farook et al., 2013; Kumar et al., 2019; Senthilarasu, 2014; Usha and Janardhan, 2014; Aravindakshan and Manimohan, 2015; Latha and Manimohan, 2017; Sridhar and Karun, 2019; Vijusha and Kumar, 2022). Selected contributions of macrofungal diversity, distribution, phylogeny and ecology have been compiled in Table 1. This study provides an overview of the diversity, distribution and ecological aspects of macrofungi in southwest India mainly based on the quantitative studies carried out in the recent past.

2. Ecosystems

Southwest India is endowed with a variety of ecological niches those support growth, perpetuation and dissemination of macrofungi. The Western Ghats consists of various vegetation types such as evergreen forests, semi-evergreen forests, deciduous forests, moist-dry deciduous forests, shola forests, agroforests and swamps. The agroforests

Table 1. Selected contributions on the diversity and ecology of macrofungi in southwest India.

	References
Diversity, distribution and phylogeny	Sathe and Daniel (1980); Sathe and Deshpande (1980); Bhavanidevi (1995); Natarajan (1995); Leelavathy and Ganesh (2000); Brown et al. (2006); Leelavathy et al. (2006); Manimohan et al. (2007); Riviere et al. (2007); Swapna et al. (2008); Bhosle et al. (2010); Pradeep and Vrinda (2010); Mohanan (2011, 2014); Ranadive et al. (2013); Farook et al. (2013); Karun and Sridhar (2013, 2014, 2015a, b, 2016b, 2017); Senthilarasu (2014, 2015); Usha and Janardhan (2014); Aravindakshan and Manimohan (2015); Borkar et al. (2015); Ghate and Sridhar, 2016a, b); Greeshma et al. (2016); Pavithra et al. (2016); Senthilarasu and Kumaresan (2016); Akash et al. (2017); Latha and Manimohan (2017); Karun et al. (2018, 2022); Thulasinathan et al. (2018); Jagadeesh et al. (2019); Kumar et al. (2019); Datttaraj et al. (2000); Bijeesh et al. (2022); Vinjusha and Kumar (2022)
Ecology	Brown et al. (2006); Karun and Sridhar (2014, 2016b, 2017); Ghate and Sridhar, 2016a, b); Pavithra et al. (2015, 2016); Greeshma et al. (2016); Karun et al. (2018, 2022); Thulasinathan et al. (2018); Jagadeesh et al. (2019); Datttaraj et al. (2000)

and plantations developed in different climatic conditions of the Western Ghats as well as the west coast have a major influence on the dynamics of macrofungi. The pattern of plantation management especially specific tree/plant species and litter strata (leaf litter and woody litter) will influence the macrofungal activity. The coastal plains offer ecosystems such as grasslands, scrub jungles, plantations, agroforests and swamps. Maritime zones consist of estuaries, islets (freshwater and saltwater), mangroves, bays and lagoons, while the coastal sand dunes (mid- and hind-dunes) possess natural as well as cultivated commercial plantations (e.g., *Acacia* and *Casuarina*).

The climatic conditions drastically vary in the Western Ghats, foothills, plains of the west coast and maritime habitats. Besides vegetation, climatic conditions will have a major impact on the occurrence of macrofungi. For instance, the Western Ghats receive pre-monsoon showers (before June), which support a specific group of macrofungi. Similarly, on the west coast pre-monsoon showers result in an eruption of fruit bodies mainly gasteromycetes. Maritime habitats supply a variety of organic matter along with minerals for the growth of macrofungi. The vegetation in maritime habitats has a major influence on the species richness, distribution and diversity of macrofungi.

3. Diversity

Species richness and diversity of macrofungi in the Western Ghats have been studied by various workers and some major checklists include Aphyllophorales, agarics, non-gilled fleshy macrofungi and ectomycorrhizas (Swapna et al., 2008; Farook et al., 2013; Ranadive, 2013; Senthilarasu, 2014; Kumar et al., 2019; Sridhar and Karun, 2019). Some of the studies carried out the quantitative studies on macrofungal species/sporocarp richness, seasonal fluctuations and core-group species in southwest India (Karun and Sridhar, 2014; Ghate et al., 2016a, b; Greeshma et al., 2016; Pavithra et al., 2016). The gist of some of these studies has been highlighted here.

3.1 Species and Sporocarp Richness

Macrofungi occur in wide ecological niches in southwest India. In comparison in eight ecological niches, the coastal sand dunes possess the highest species richness followed by mangroves, reserve forests and shola forests (Fig. 2). Fully identified macrofungi were the highest in Shola forest (85.4%) followed by sacred grove (81.1%), arboretum (76.7%) and reserve forest (70.7%), while in the rest six ecosystems the known macrofungi were below 70%. Unidentified macrofungi up to species level was the highest in coastal sand dunes (40.6%) followed by botanical garden (38.9%), scrub jungle (35.3%) and mangroves (32.6%), while in the rest six habitats it was

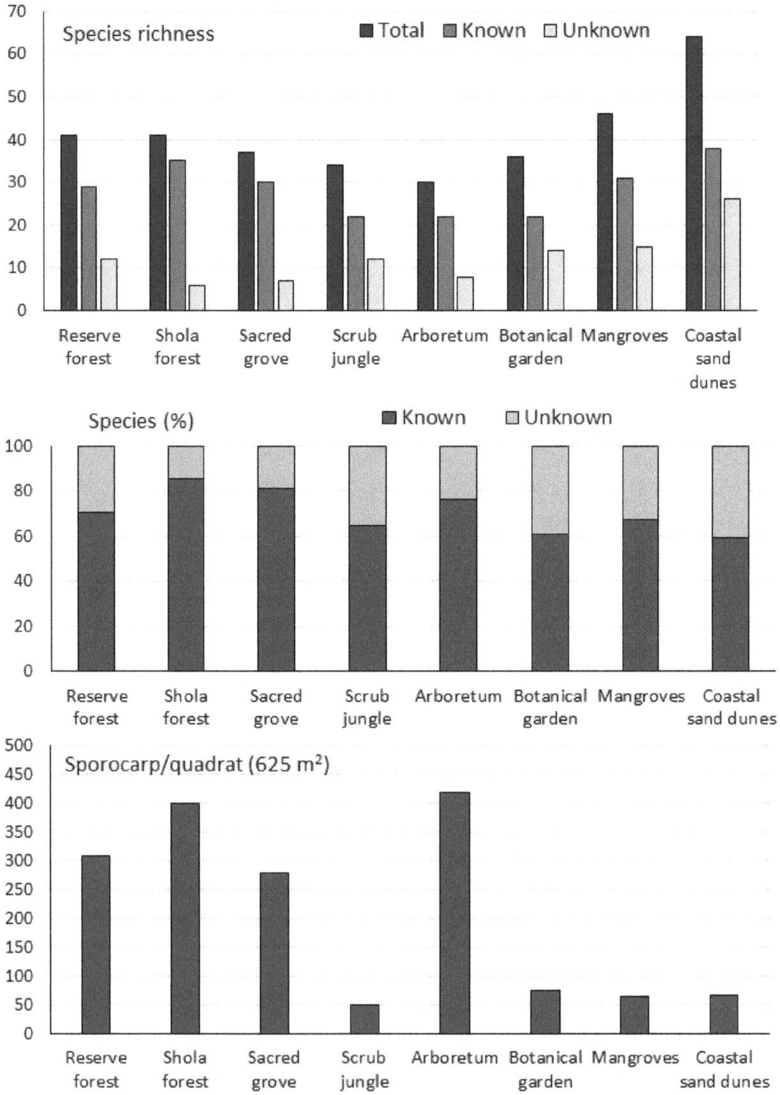

Fig. 2. Species and sporocarp richness of macrofungi in eight ecosystems of southwest India.

below 30%. The sporocarp richness did not match with the species richness in eight ecosystems. The arboretum showed the highest number of sporocarps per quadrat (625 m^2) (418) followed by the shola forest (400), reserve forest (308) and sacred grove (279), while the rest four habitats possess below 100 sporocarps per quadrat.

The species richness is not matching with the sporocarp richness. Although scrub jungle, botanical garden, mangroves and coastal sand dunes possess higher species richness than other habitats, they are represented by low sporocarp density (48–74/625 m^2) (Fig. 2). While the arboretum possesses the least species richness but the sporocarp density was the highest (418/625 m^2). It is likely the sporocarp density depends on the vegetation pattern especially in the reserve forest, shola forest sacred groves (Western Ghats) and arboretum (west coast). The arboretum was established in southwest India and endowed with a variety of endemic and endangered species growing in the Western Ghats. Thus, there seems to be a strong link between sporocarp richness and with vegetation type of a specific habitat. However, the unidentified species was relatively higher in scrub jungle, botanical garden, mangroves and coastal sand dunes compared to the other four habitats indicating their uniqueness in supporting different macrofungi.

Species richness and overlap of macrofungi in three regions have been represented by a Venn diagram (Fig. 3). In three habitats of the Western Ghats, the

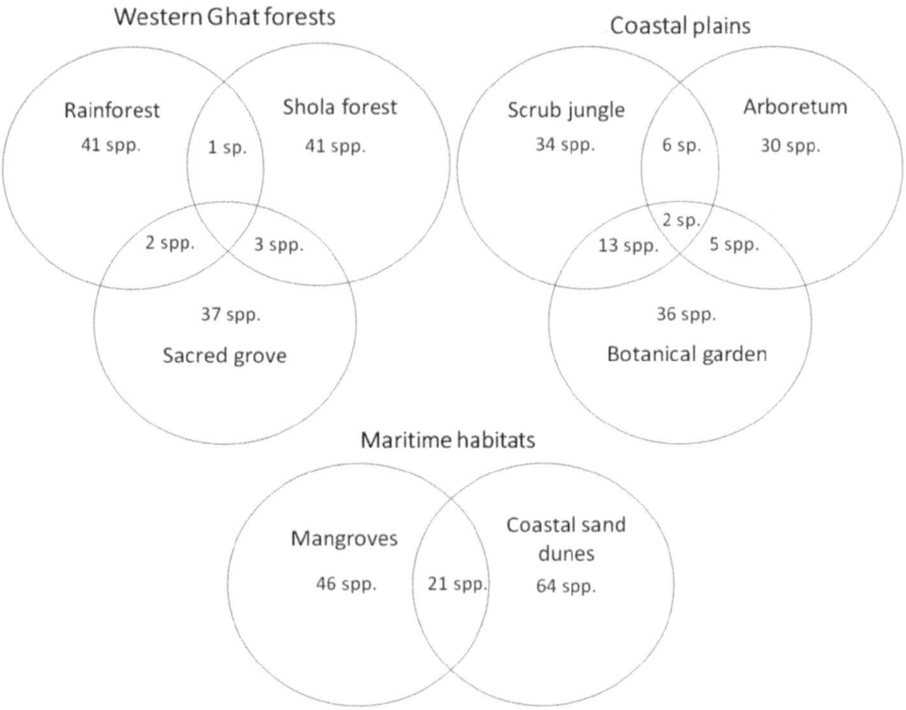

Fig. 3. Venn diagram showing distribution and overlap of macrofungi in three ecosystems of southwest India.

overlap of species was 1–3, while none of the species was common to all habitats indicating the uniqueness of macrofungi. Although the species richness was lower in the coastal plains than in the Western Ghats, the overlap of species was 5–13 and two species of *Marasmius* were common to three habitats. Two maritime habitats possess higher species richness than the Western Ghats as well as coastal plains, showing 21 species of macrofungi in common.

3.2 Seasonal Variation in Species Richness

Generally, the macrofungal species richness showed two peaks in all habitats with one highest peak during the wet season (June to November) (Fig. 4). In the Western Ghats (reserve forest, shola forest and sacred grove), the highest peak reached up to 40 species (in September), the second peak with about 30 species (in June) and

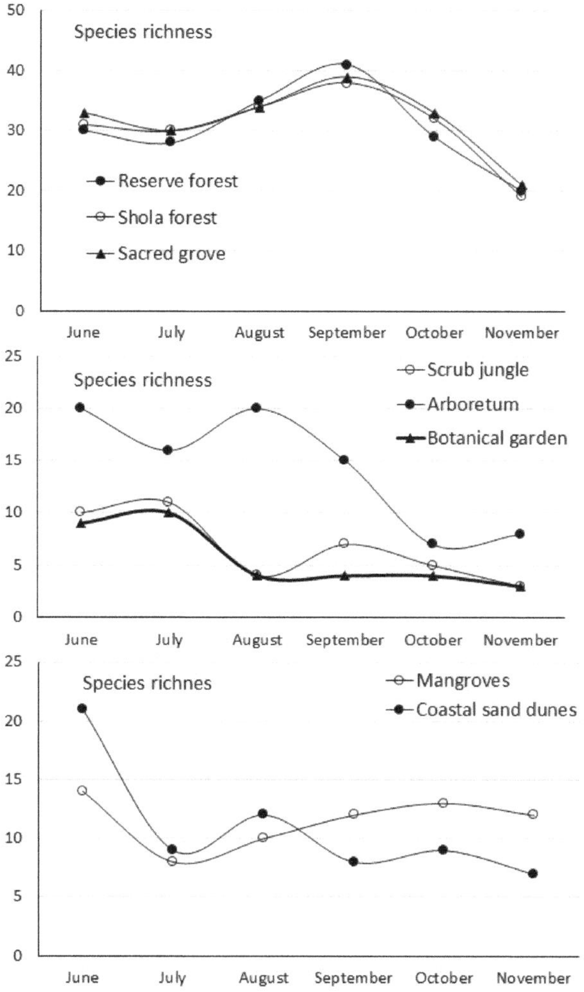

Fig. 4. Seasonal variation in species richness in three ecosystems of southwest India.

attained minimum species of about 20 species (in November). The species richness in the wet season on the west coast and maritime habitats ranged between 3 and 21. On the west coast, the arboretum possesses a higher number of species throughout the wet season as an outlier compared to the scrub jungle and botanical garden. In maritime habitats, coastal sand dunes showed the highest peak in June compared to mangroves.

There is a shift in the highest peak in species richness from September in the Western Ghats to July on the west coast (except for arboretum: August) and June in the maritime habitats (Fig. 4). The second peak in species richness was during June in the Western Ghats as well as west coast (except for arboretum: equal in June and August). Mangroves showed second species richness during August like an arboretum, while it was during October in coastal sand dunes. The arboretum stands as an outlier in species richness almost throughout the west season compared to other habitats on the west coast.

3.3 *Core-Group Species*

A total of 44 species have been identified as core-group species in southwest India (Table 2). The highest core-group species was found in shola forest (11 spp.) followed by reserve forest (10 spp.), arboretum (9 spp.), sacred grove, botanical garden (5 spp. each) and mangroves (3 spp.). The least number of core-group species was recorded in scrub jungle and coastal sand dunes (1 spp. each).

Table 2. Core-group macrofungi (> 10 sporocarp/quadrat: 625 m^2) in eight ecosystems of southwest India.

	Core-group species	**References**
Reserve forest	*Amylosporus campbellii, Cyclomyces setiporus, Ganoderma oregonense, Gymnopilus bryophilus, Hypholoma* sp., *Microporus vernicipes, Phellinus gilvus, Phellinus* sp., *Stereum hirsutum* and *Xylaria multiplex*	Karun and Sridhar (2016b)
Shola forest	*Clavulinopsis laeticolor, Cookeina tricholoma, Coprinus disseminates, Entoloma theekshnagandhum, Microporus vernicipes, Pleurotus djamor, Pleurotus flabellatus, Pluteus* sp., *Royoporus spathulatus, Termitomyces microcarpus* and *Xylaria filiformis*	Karun and Sridhar (2016b)
Sacred grove	*Inocybe viridiumbonata, Irpex lacteus, Pleurotus cornucopiae, Psathyrella lucipeta* and *Xylaria obovata*	Karun and Sridhar (2016b)
Scrub jungle	*Xylaria hypoxylon*	Greeshma et al. (2016)
Arboretum	*Crepidotus uber, Hygrocybe* sp., *Ileodictyon gracile, Marasmius guyanensis, Marasmius spegazzinii, Microcarpus xanothopus, Mycena vitilis, Trametes verscolar* and *Unknown* sp. (basidiomycete)	Karun and Sridhar (2014)
Botanical garden	*Clathrus delicatus, Marasmius* sp., *Nectria cinnabarina, Termitomyces microcarpus* and *Xylaria* sp.	Pavithra et al. (2016)
Mangroves	*Dacryopinax spathularia, Marasmiellus stenophyllus* and *Thelephora palmata*	Ghate and Sridhar (2016b)
Coastal sand dunes	*Dacryopinax spathularia*	Ghate and Sridhar (2016a)

The core-group macrofungi play a major role in the breakdown of organic matter in different habitats. The total species richness in the Western Ghats (reserve forest, shola forest and sacred grove) and arboretum coincide with the richness of core-group species. It is interesting to note that there is only one overlap of core-group species between reserve forest and shola forest (*Microporus vernicipes*). Such dominance and non-overlap of core-group macrofungi indicate the uniqueness of different habitats to support the dominant macrofungi. Similar to the genus *Termitomyces* (22 spp.), another dominant genus *Xylaria* in southwest India with a richness of up to 24 species (Karun and Sridhar, 2013, 2015b, 2017).

4. Substrates

The macrofungal occurrence will be usually investigated in soil, leaf litter and woody litter (see Fig. 1). These three major habitats could be further classified into various niches. The type of soil varies and the growth of macrofungi depends on the content of organic matter in the soil. In addition, moisture, organic matter, humus content, minerals and animal activity (invertebrates and vertebrates) also influence the soil quality. Leaf litter quality, extent and decomposition depend on the colonization of microbes under specific climatic conditions. The growth of macrofungi depends on the type of woody litter (twigs, stumps and logs). Besides, the bark serves as a unique habitat that supports many macrofungi. Leaf litter, humus and partially decomposed woody material in the tree canopy also serve as habitats for macrofungal perpetuation (Karun and Sridhar, 2016).

Macrofungal growth and dissemination also depend on the phenology of vegetation in a specific habitat, which may determine the uniqueness of macrofungi. Vegetation supporting ectomycorrhizal fungi also determines the dynamics of macrofungi in a specific habitat. The dung of polygastric and monogastric animals supports many macrofungi of medicinal significance. Live and dead insects also support many unique macrofungi of medicinal and industrial importance.

4.1 Entomopathogens

The endoparasitic genus *Cordyceps* (Ascomycota, Cordycipitaceae) with morphologically diverse perithecia has worldwide distribution consisting of about 500 species (Sung et al., 2007; Kepler et al., 2012). The high-altitude Himalayan region is well known for *Cordyceps* and allied species (Shrestha and Bawa, 2013, 2014; Quan et al., 2014; Baral et al., 2015; Negi et al., 2015). Besides the Himalayas, the Western Ghats and southwest India (Maharashtra, Karnataka and Kerala) are known to support up to 15 species of *Cordyceps* and allied species (Dattaraj et al., 2018). These species are known to be pathogenic with a variety of insect cocoons and larvae (e.g., Coleoptera, Formicide, Pentatomidae and Scarabaeidae). Interestingly, the root infecting coconut grub (*Leucopholis coneophora*) in Kerala and Karnataka were infected by a *Cordyceps* sp. seems to be a natural biological control agent (Kumar and Aparna, 2014; Prathibha, 2015). Similarly, the parasitic pentatomid bug *Halyomorpha halys* in the trees of *Casinae glauca* was attacked by the *Ophiocordyceps nutans* in Western Ghats (Sridhar and Karun, 2017). A one-time survey in a narrow stretch of scrub jungles on the west coast revealed the occurrence of six *Cordyceps* and allied species (Dattaraj et al., 2018). Such studies on entomopathogenic macrofungi will be valuable to following the tripartite relationship (plant, insect and macrofungus) and may be helpful to practice natural biological control measures.

4.2 *Coprophilous Macrofungi*

The herbivore dung of wild and domestic animals worldwide serves as an important substrate for the growth of macrofungi (e.g., Manimohan et al., 2007, Doveri, 2010, Kaur et al., 2014; Karun and Sridhar, 2015a). The elephant dung provides a variety of processed plant substrate, which is suitable for macrofungal growth including hallucinogenic mushrooms. So far a total of 35 species (16 genera) of macrofungi have been reported from the elephant dung of southwest India mainly from the states of Kerala and Karnataka (Karun and Sridhar, 2015a) (Table 3). *Psilocybe* is represented by a maximum of nine species, followed by *Panaeolus* (6 spp.), *Conocybe* (4 spp.) and *Stropharia* (3 spp.), while the rest 12 genera are represented by one or two species. Although *Conocybe pubescens* usually grow on elephant dung, it was also reported in the soil, grasslands, and evergreen and deciduous forests of Kerala and Maharashtra (Sathe and Deshpande, 1980; Mohanan, 2011). Similarly, *Agrocybe pediades, Canthrocybe virosa* and *Chlorophyllum molybdites* were grown on cattle dung in Kerala (Mohanan, 2011; Bijeesh et al., 2022). Elephant dung was investigated in many forests and national parks (Silent Valley) and wildlife sanctuaries (Mathunga) in Kerala (Thomas et al., 2001, Manimohan et al., 2007, Latha and Manimohan, 2012). Similarly, several elephant corridors provide ample scope to evaluate dung inhabiting macrofungi in the Western Ghats region of Karnataka, Kerala and Tamil Nadu (Karun and Sridhar, 2015a).

5. Mutualistic Association

Macrofungi have a variety of mutualistic associations with plant species and animals. Among such mutualistic associations, ectomycorrhizal association with tree species has drawn more attention. The bipartite (macrofungi vs. animals) and tripartite association (macrofungi vs. plants vs. animals) are still in the emerging stage. Some of these associations are discussed here.

5.1 *Ectomycorrhizae*

Among the macrofungi, ectomycorrhizal (EM) fungi have global concerns owing to their association with a variety of trees and involvement in the nutrition of tree species. The global estimate exposes the existence of 20,000–25,000 EM fungi in a mutualistic association with 6,000 plant species (Rinaldi et al., 2008; Tedersoo et al., 2010). The diversity of EM fungi occurring in the Western Ghats has been related to tree diversity (Riviere et al., 2007). Up to 148 species (34 genera) of EM fungi colonized about 60 host tree species (40 genera) have been reported from southwest India (Sridhar and Karun, 2019). Among them, the four dominant genera of EM

Table 3. Revised list of macrofungi reported on the elephant dung in southwest India.

	State	**References**
Agrocybe guruvayoorensis	Kerala	Manimohan et al. (2007)
Agrocybe pediades	Kerala	Mohanan (2011)
Bolbitius coprophilus	Kerala	Thomas et al. (2001); Manimohan et al. (2007)
Conocybe brunneoaurantiaca	Kerala	Manimohan et al. (2007)
Conocybe pseudopubescens	Kerala	Thomas et al. (2001); Manimohan et al. (2007)
Conocybe pubescens	Karnataka	Karun and Sridhar (2015a)
Conocybe volvata	Kerala	Thomas et al. (2001); Manimohan et al. (2007)
Copelandia cyanescens	Kerala	Manimohan et al. (2007)
Coprinus patouillardii	Karnataka	Karun and Sridhar (2015a)
Entoloma anamikum	Kerala	Manimohan et al. (2007); Noordeloos et al. (2007)
Panaeolina rhombisperma	Kerala	Manimohan et al. (2007)
Panaeolus acuminatus	Kerala	Vrinda et al. (1999); Bijeesh et al. (2022)
Panaeolus africanus	Kerala	Bijeesh et al. (2022)
Panaeolus antillarum	Kerala	Manimohan et al. (2007); Mohanan (2011); Bijeesh et al. (2022)
Panaeolus cyanescens	Kerala	Bijeesh et al. (2022)
Panaeolus fimicola	Karnataka	Karun and Sridhar (2015a)
Panaeolus rickenii	Kerala	Manimohan et al. (2007); Bijeesh et al. (2022)
Pholiotina indica	Kerala	Thomas et al. (2001); Manimohan et al. (2007); Mohanan (2011)
Podosordaria elephanti	Kerala	Latha and Manimohan (2012); Karun and Sridhar (2015a)
Poronia pileiformis	Kerala	Latha and Manimohan (2012)
Psathyrella conopilus	Kerala	Vrinda et al. (1999)
Psilocybe argentina	Kerala	Mohanan (2011)
Psilocybe caeruleoannulata	Kerala	Bijeesh et al. (2022)
Psilocybe coprophila	Kerala and Karnataka	Manimohan et al. (2007); Karun and Sridhar (2015a)
Psilocybe cubensis	Kerala	Bijeesh et al. (2022)
Psilocybe fimetaria	Karnataka	Karun and Sridhar (2015a)
Psilocybe inquilina	Kerala	Mohanan (2011)
Psilocybe pegleriana	Kerala	Manimohan et al. (2007); Mohanan (2011)
Psilocybe subaeruginascens	Kerala	Manimohan et al. (2007)
Psilocybe subcubensis	Kerala	Manimohan et al. (2007); Mohanan (2011); Bijeesh et al. (2022)
Stropharia bicolor	Kerala	Manimohan et al. (2007)
Stropharia rugosoannulata	Kerala	Manimohan et al. (2007)
Stropharia semiglobata	Kerala	Mohanan (2011)
Macrocybe gigantea	Kerala	Manimohan et al. (2007)
Volvariella volvacea	Kerala	Manimohan et al. (2007)

fungi include *Inocybe* (36 spp.), *Russula* (31 spp.), *Amanita* (13 spp.) and *Boletus* (8 spp.). It is interesting to note that among 148 species of EM fungi, 114 species (77%) were in association with eight tree species belonging to Dipterocarpaceae. In eight species of Dipterocarpaceae, *Inocybe*, *Russula*, *Amanita* and *Boletus* were represented by 33, 31, 10 and 7 species, respectively. The EM fungi were highest in *Vateria indica*, *Hopea ponga*, *H. parviflora*, *Diospyros malabarica*, *Myristica malabarica* and *Dipterocarpus indicus* represented by 69, 50, 48, 37, 10 and 10 species, respectively. Natarajan et al. (2005) also reported the highest number of EM fungi associated with *Vateria indica*. Similarly, the members of the Dipterocarpaceae possess the highest species of ectomycorrhizal *Inocybe* in Kerala (Latha and Manimohan, 2017). From 160 collections from evergreen, deciduous and exotic forests Pradeep and Vrinda (2010) reported the highest EM fungi in the evergreen forest. Up to 30 new species of EM fungi have been reported from Kerala state (about 20% of EM fungi known in the Western Ghats).

Many EM fungi reported in southwest India are valued as edible and medicinal. They have been recognized as edible and or medicinal based on ethnic knowledge (host tree species and substrates) (Pavithra et al., 2015; Karun and Sridhar, 2017). The edible EM fungi identified by the tribals on their association with the host tree species include *Boletinellus merulioides*, *Boletus edulis*, *B. reticulatus*, *Lycoperdon utriforme*, *Phlebopus marginatus*, *P. portentosus*, *Rubroboletus caespitosus*, *Scleroderma citrinum*, *Suillus brevipes*, *S. placidus*, and *S. tomentosus* (Karun and Sridhar, 2017). *Scleroderma* is one of the important genera of ectomycorrhizal fungi owing to their wide distribution, capability to withstand wide climatic conditions and broad host range (Mark et al., 2017; Zhang et al., 2020; Vaario and Matsushita, 2021). Their ability to endure extreme habitats like mining, coal waste, xeric and sand dunes qualify them to be considered for rehabilitation purposes (Mark et al., 2017). Up to eight species of *Scleroderma* are known from southwestern India as native as well as exotic tree species (Karun et al., 2022). *Amanita konajensis* has mutualistic association with *Acacia auriculiformis*, *A. mangium*, *Anacardium occidentale* and *Eucalyptus globulus* (Crous et al., 2022). Based on the literature, up to 80 tree species are associated with about 240 species of ectomycorrhizal fungi revealing about a 1:3 ratio of tree species with ectomycorrhizal fungi in southwestern India.

5.2 *Macrofungi in Termite Mounds*

One of the fascinating ecological niches that support the mutualistic association of termitophilic fungi is the termite mounds. The lignocellulosic enzymes produced by microorganisms (fungi and bacteria) are utilized by termites as seen in rumen symbiosis (Norbe and Aanen, 2012). Nearly 30 species of *Termitomyces* are known globally and about 20 species of *Termitomyces* have been reported from southwestern India (Karun and Sridhar, 2013). The highest number of *Termitomyces* have been reported from Karnataka (18 spp.) followed by Kerala (15 spp.) and Goa (10 spp.). Wide distribution was seen in *Termitomyces micorcarpus* followed by *T. eurrhizus* / *T. heimii* and *T. clypeatus* in southwest India (Table 4). In addition to *Termitomyces*

Table 4. Revised list of *Termitomycetes* reported from the termite mounds in southwest India.

	State	References
Termitomyces clypeatus	Goa	Nandkumar (2013)
	Karnataka	Pahlevanlo and Janardhana (2012); Karun and Sridhar (2013, 2014, 2016b, 2017); Greeshma et al. (2016)
	Kerala	Leelavathy et al. (1983); Pradeep and Vrinda (2007); Varghese et al. (2010); Mohanan (2011)
Termitomyces cylindricus	Karnataka	Pahlevanlo and Janardhana (2012)
Termitomyces entolomoides	Goa	Nandkumar (2013)
	Kerala	Leelavathy et al. (1983); Pradeep and Vrinda (2007)
Termitomyces eurrhizus	Maharashtra	Sathe and Deniel (1980); Sathe and Deshpande (1980)
	Goa	Nandkumar (2013)
	Karnataka	Pahlevanlo and Janardhana (2012); Karun and Sridhar (2013, 2014, 2016b, 2017)
	Kerala	Sathe and Deniel (1980); Leelavathy et al. (1983); Sankaran and Florence (1995); Florence and Yesodharan (1997); Florence and Yesodharan (2000); Florence (2004); Pradeep and Vrinda (2007); Varghese et al. (2010); Mohanan (2011)
Termitomyces fulginosus	Karnataka	Karun and Sridhar (2017); Karun et al. (2018); Dattaraj et al. (2020)
	Goa	Nandkumar (2013)
Termitomyces globulus	Goa	Nandkumar (2013)
	Karnataka	Pahlevanlo & Janardhana (2012); Sudheep & Sridhar (2014); Karun & Sridhar (2017)
	Kerala	Leelavathy et al. (1983); Mohanan (2011)
Termitomyces heimii	Maharashtra	Sathe & Deniel (1980)
	Goa	Nandkumar (2013)
	Karnataka	Pahlevanlo and Janardhana (2012); Karun and Sridhar (2013, 2014, 2016b, 2017)
	Kerala	Sathe and Deniel (1980); Leelavathy et al. (1983); Sankaran and Florence (1995); Florence and Yesodharan (2000); Florence (2004); Pradeep and Vrinda (2007); Varghese et al. (2010)
	Tamil Nadu	Natarajan (1979); Johnsy et al. (2011); Sargunam et al. (2012)
Termitomyces indicus	Karnataka	Pahlevanlo and Janardhana (2012); Karun and Sridhar (2017)
Termitomyces lanatus	Karnataka	Karun and Sridhar (2017)
Termitomyces le-testui	Kerala	Mohanan (2011); Karun and Sridhar (2017)
Termitomyces mammiformis	Goa	Nandkumar (2013)
	Karnataka	Pahlevanlo and Janardhana (2012); Karun and Sridhar (2017)
	Kerala	Bhavanidevi and Nair (1983); Leelavathy et al. (1983); Florence (2004); Mohanan (2011)

Table 4 contd. ...

...Table 4 contd.

	State	References
Termitomyces medius	Goa	Nandkumar (2013); Karun and Sridhar (2017)
Termitomyces microcarpus	Gujarat	Sidhu (2009); Lahiri et al. (2010)
	Maharashtra	Sathe and Deshpande (1980)
	Goa	Nandkumar (2013)
	Karnataka	Pahlevanlo and Janardhana (2012); Karun and Sridhar (2013, 2016b, 2017); Pavithra et al. (2016)
	Karnataka and Kerala	Sathe and Deniel (1980); Leelavathy et al. (1983); Sankaran and Florence (1995); Florence and Yesodharan (1997); Florence and Yesodharan (2000); Mohanan (2003); Pradeep and Vrinda (2007); Yesodharan and Sujana (2007); Kavishree et al. (2008); Varghese et al. (2010); Mohanan (2011); Karun and Sridhar (2013)
	Tamil Nadu	Johnsy et al. (2011); Sargunam et al. (2012)
Termitomyces perforans	Kerala	Leelavathy et al. (1983)
Termitomyces robustus	Karnataka	Karun and Sridhar (2017)
	Kerala	Bhavanidevi and Nair (1983); Bilgrami et al. (1991); Mohanan (2011)
Termitomyces sagittiformis	Kerala	Vrinda and Pradeep (2009)
Termitomyces schimperi	Karnataka	Karun and Sridhar (2014, 2017); Ghate and Sridhar (2016a)
	Kerala	Mohanan (2011)
Termitomyces spiniformis	Karnataka	Karun and Sridhar (2017)
Termitomyces striatus	Goa	Nandkumar (2013)
	Karnataka	Ghate and Sridhar (2016a, 2017); Greeshma et al. (2016)
	Kerala	Leelavathy et al. (1983); Mohanan (2011)
Termitomyces titanicus	Karnataka	Karun and Sridhar (2017)
Termitomyces tylerianus	Gujarat	Lahiri et al. (2010)
	Karnataka	Karun and Sridhar (2016b, 2017)
	Kerala	Mohanan (2011)
Termitomyces umkowaan	Karnataka	Karun and Sridhar (2013, 2017)
	Kerala	Vrinda et al. (2002); Pradeep and Vrinda (2007); Mohanan (2011); Karun and Sridhar (2014)

spp., some of the *Scleroderma* as well as *Xylaria* spp. were also associated with the termite mounds in southwest India (e.g., *Scleroderma dictyosporum*, *Xylaria acuminatilongissima*, *X. escharoidea*, *X. longipes* and *X. nigripes*) (Latha and Manimohan, 2012; Karun and Sridhar, 2015b; Sridhar et al., 2022).

5.3 *Auriculoscypha*

A fascinating tripartite relationship of macrofungus-insect-plant interactions has been reported by Manimohan from Kerala (2019). The basidiomycete *Auriculoscypha* is symbiotic with sap-feeding scale-insect *Neogreenia zeylanica* the minor pest of cashew (*Anacardium occidentale*) and mango (*Mangifera indica*) infest the cracks or underneath the bark. *Auriculocypha* controls the scale-insect population in the juvenile stage as a unique biological control. *Auriculocypha* seems to be endemic to southwest India and widely distributed in Kerala and parts of Karnataka and Goa.

6. Ecosystem Services

Colonization, growth and dissemination of macrofungi on different substrates lead to several ecosystem services (see Fig. 1). One of the major functions of macrofungi is the breakdown of organic matter, which supports soil-dependent organisms as a nutritional source and supplies minerals to plants. Owing growth of macrofungal binding of soil and substrates associated with soil and colonization of tree roots by ectomycorrhizae are responsible for a slowdown or prevention of soil erosion. The ectomycorrhizal association helps to prevent diseases in plants.

The uniqueness of macrofungi (bioprospect potential: metabolites and therapeutics) depends on the substrate on which it establishes (e.g., insects and dung). Edible mushrooms serve as an alternate food source other than conventional plant and animal foods. The entomopathogenic nature of macrofungi serves as natural biological control of insects in the ecosystem. The complex interaction of macrofungi with substrates, plants and animals in an ecosystem ultimately improves the carrying capacity. Based on the ethnic knowledge of tribals, 51 edible mushrooms (23 genera) have been documented by Karun and Sridhar (2017) in the Western Ghats. It is interesting to note the skill of tribals in the identification of poisonous and edible mushrooms based on the colonized substrates.

7. Climatic and Human Impacts

Studies on macrofungal diversity, distribution and interaction reveal the positive and negative influence of climatic conditions as well as human interference. Soil erosion, floods, drought and invasion of exotic plant or tree species have several influences on macrofungi, which leads to removal or elimination of substrates, increased temperature or xeric conditions (e.g., wildfire) and influence of monoculture and exotic tree species (e.g., *Acacia* and *Eucalyptus*). Such climatic conditions are partially due to human interference such as tree felling, plantation activities (e.g., use of pesticides and agricultural chemicals), removal of substrates (leaf and woody litter), the impact of fire and expansion of exotic tree plantations. Organic farming, prevention of soil erosion, ecofriendly plantations/agroforests, minimizing agricultural chemicals, retention of termite mounds and prevention of excessive removal of tree canopies are the possible measures to support the conditions required for the growth and perpetuation of macrofungi in an ecosystem.

Mild fire-impacted scrub jungle in southwest India possesses 25 species of macrofungi against 34 species in the control region (Greeshma et al., 2016). There was a shift in the peak of species and sporocarp richness in the fire-impacted region to the fourth week (second week in the control region). As fire has eliminated leaf litter, leaf litter inhabiting macrofungi was not found, while only the soil and wood-inhabiting fungi were recorded. The common species found in control and fire-impacted regions were: *Chlorophyllum molybdites, Lepiota* sp., *Leucocoprinus birnbaumii, Marasmius* sp., *Polyporus* sp., *Schizophyllum commune* and *Tetrapyrgos nigripes*. Only one species showed dominance (*Xylaria hypoxylon*) in control region, while *Cyathus striatus* as well as *Lentinus squarrosulus* were dominant in the fire-impacted region. Severn species of EM fungi in control region (*Amanita angustilamellata, Geastrum triplex, Hygrocybe astatogala, H. aurantioalba, Pisolithus albus, Scleroderma citrinum* and *Telephora palmata*) have been reduced to two species in the fire-impacted region (*Astraeus odoratus* and *Lycoperdon utriforme*).

Conclusion

There are several knowledge gaps in understanding the macrofungal dynamics in southwest India. Firstly, the diversity, distribution and ecology of macrofungi have not been assessed in different ecological niches (e.g., national parks, sanctuaries, swamps, grasslands, mangroves, coastal sand dunes, exotic plantations, urban ecosystems and aquatic/semi-aquatic habitats). The species and sporocarp richness were highest in the forests of the Western Ghats compared to the coastal ecosystems. Similarly, the seasonal pattern of species richness also differs among these ecosystems. Although soil, leaf litter and woody substrates were explored for macrofungi, other substrates like termite mounds, animals (e.g., insects, larvae, millipedes, molluscs and others), animal excrements and exotic substrates are not studied extensively. Such studies facilitate to development strategies of conservation of macrofungi in southwest India.

During many forays in southwest India, species belongs to genera *Amanita, Amauroderma, Dictyophora, Geastrum, Phallus, Simblum* and *Xylaria* were in association with many tree species. In spite of several studies on ectomycorrhizal fungi, their hosts are not identified more precisely. Unpublished studies revealed that *Geastrum fimbriatum* has an association with tree species like *Artocarpus heterophyllus, Coffea robusta* and *Mangifera indica* grown in coffee plantations in the Western Ghats. Macrofungal socioeconomic relevance was least considered especially the food and health security of tribals who are dependent on mushrooms during the wet season. Traditional knowledge of tribals needs an appraisal to follow macrofungal significance as the nutritional and medicinal source. Sustainable approaches need to be followed to support wild macrofungi in natural and human-influenced ecosystems of southwest India.

Acknowledgements

The author is grateful to the Department of Biosciences, Mangalore University for extending the facilities. Thanks are also due to Dr. S. Mahadevakumar, Kerala Forest Research Institute, Kerala for helpful discussion and technical assistance.

References

Akash, D., Eranna, N. and Subramanya, S. (2017). Mushroom diversity in the Biligiri Rangana hills of Karnataka (India). J. Appl. Nat. Sci., 9: 1381–1387.

Aravindakshan, D. and Manimohan, P. (2015). Mycenas of Kerala. SporePrint Books, Calicut, Kerala, India, p. 213.

Azeem, U., Hakkem, K.R. and Ali, M. (2020). Fungi for Human Health – Current Knowledge and Further Perspectives. Springer Nature, Switzerland, p. 113.

Baral, B., Shrestha, B. and Da Silva, J.A.T. (2015). A review of Chinese *Cordyceps* with special reference to Nepal, focusing on conservation. Environ. Exp. Biol., 13: 61–73.

Bhavanidevi, S. (1995). Mushroom flora of Kerala. pp. 277–316. *In*: Chadha, K.L. and Sharma, S.R. (eds.). Advances in Horticulture – Mushrooms, Volume 13. Malhotra Publishing House, New Delhi.

Bhavanidevi, S. and Nair, M.C. (1983). Addition to Indian Agaricales. pp. 271–274. *In*: Kaul, T.N. and Kapoor, B.M. (eds.). Indian Mushroom Science II, Council of Scientific and Industrial Research, New Delhi.

Bhosle, S., Ranadive, K., Bapat, G. et al (2010). Taxonomy and diversity of *Ganoderma* from the Western parts of Maharashtra (India). Mycosphere, 1: 249–262.

Bijeesh, C., Pradeep, C.K. and Vrinda, K.B. (2022). Poisonous Mushrooms of Kerala, India: An Illustrated Handbook. Jawaharlal Nehru Tropical Botanic Garden and Research Institute, Trivandrum, Kerala, p. 187.

Bilgrami, K.S., Jamaluddin, S. and Rizwi, M.A. (1991). Fungi of India, Volume 2: List and References. Today and Tomorrows Printers and Publishers, New Delhi, p. 798.

Borkar, P., Doshi, A. and Navathe, S. (2015). Mushroom diversity of Konkan region of Maharashtra, India. J. Threat. Taxa, 7: 7625–7640.

Brown, N., Bhagwat, S. and Watkinson, S. (2006). Macrofungal diversity in fragmented and disturbed forests of the Western Ghats of India. J. Appl. Ecol., 43: 11−17.

Crous, P.W., Boers, J., Holdom, D., Osieck, E.R., Steinrucken, T.V. et al. (2022). Fungal planet description sheets 1383–1435. Persoonia, 48: 261–371.

Dattaraj, H.R., Jagadish, B.R., Sridhar, K.R. and Ghate, S.D. (2018). Are the scrub jungles of southwest India potential habitats of *Cordyceps*? KAVAKA – Trans. Mycol. Soc. India, 51: 20–22.

Dattaraj, H.R., Sridhar, K.R., Jagadish, B.R. and Pavithra, M. (2020). Bioactive potential of the wild edible mushroom *Ramaria versatilis*. Stud. Fungi, 5: 73–83.

Doveri, F. (2010). Occurrence of coprophilous *Agaricales* in Italy, new records, and comparisons with their European and extraeuropean distribution. Mycosphere, 1: 103–140.

Farook, V.A., Khan, S.S. and Manimohan, P. (2013). A checklist of agarics (gilled mushrooms) of Kerala State, India. Mycosphere, 4: 97–131.

Florence, E.J.M. (2004). Biodiversity Documentation for Kerala, Part 2: Microorganisms (Fungi). Kerala Forest Research Institute, Handbook #17, Peechi, Kerala, India, p. 293.

Florence, E.J.M. and Yesodharan K. (2000). Macrofungal Flora of Peechi-Vazhani Wildlife Sanctuary. Kerala Forest Research Institute Research Institute, Report # 191, Peechi, Kerala, India, p 41.

Florence, E.J.M. and Yesodharan, K. (1997). Status survey of macrofungal flora of Peechi-Vazhani wildlife sanctuary. pp. 477–479. *In*: Iyengar P.K. (ed.). Proceedings of Ninth Kerala Science Congress, Thiruvananthauram, Kerala, India.

Ghate, S.D. and Sridhar. K.R. (2016a). Spatiotemporal diversity of macrofungi in the coastal sand dunes of southwestern India. Mycosphere, 7: 458–472.

Ghate, S.D. and Sridhar, K.R. (2016b). Contribution to the knowledge on macrofungi in mangroves of the Southwest India. Pl. Biosyst., 150: 977–986.

Greeshma, A.A., Sridhar, K.R., Pavithra, M. and Ghate, S.D. (2016). Impact of fire on the macrofungal diversity of scrub jungles of Southwest India. Mycology, 7: 15–28.

Hawksworth, D.L. (1991). The fungal dimension of biodiversity: Magnitude, significance, and conservation. Mycol. Res., 95: 641–655.

Hawksworth, D.L. and Lücking, R. (2017). Fungal diversity revisited: 2.2 to 3.8 million species. Microbiol. Spectr., 5: FUNK-0052-2016.

Jagadish, B.R., Sridhar, K.R. and Dattaraj, H.R. (2019). Macrofungal assemblage with two tree species in scrub jungles of south-west India. Stud. Fungi, 4: 72–82.

Johnsy, G., Sargunam, S.D., Dinesh, M.G. and Kaviyarasan, V. (2011). Nutritive value of edible wild mushrooms collected from the Western Ghats of Kanyakumari District. Bot. Res. Int., 4: 69–74.

Karun, N.C. and Sridhar, K.R. (2013). Occurrence and distribution of *Termitomyces* (Basidiomycota, Agaricales) in the Western Ghats and on the west coast of India. Czech Mycol., 65: 233–254.

Karun, N.C. and Sridhar, K.R. (2014). A preliminary study on macrofungal diversity in an arboretum and three plantations of the southwest coast of India. Curr. Res. Environ. Appl. Mycol., 4: 173–187.

Karun, N.C. and Sridhar, K.R. (2015a). Elephant dung-inhabiting macrofungi in the Western Ghats. Curr. Res. Environ. Appl. Mycol., 5: 60–69.

Karun, N.C. and Sridhar, K.R. (2015b). *Xylaria* complex in the South Western India. Pl. Pathol. Quar., 5: 83–96.

Karun, N.C. and Sridhar, K.R. (2016a). Two new records of hydnoid fungi from the Western Ghats of India. Stud. Fungi, 1: 135–141.

Karun, N.C. and Sridhar, K.R. (2016b). Spatial and temporal diversity of macrofungi in the Western Ghat forests of India. Appl. Ecol. Environ. Res., 14: 1–21.

Karun, N.C. and Sridhar, K.R. (2017). Edible wild mushrooms in the Western Ghats: Data on the ethnic knowledge. Data in Brief, 14: 320–328.

Karun, N.C., Bhagya, B. and Sridhar, K.R. (2018). Biodiversity of macrofungi in Yenepoya Campus, Southwest India. Microb. Biosyst., 3: 1–11.

Karun, N.C., Mahadevakumar, S. and Sridhar, K.R. (2022). On the *Scleroderma* in southwest India. Species, 23: 297–312.

Kaur, A., Atri, N.S. and Kaur, M. (2014a). Diversity of coprophilous species of *Panaeolus* (Psathyrellaceae, Agaricales) from Punjab, India. Biodiversitas, 15: 115–130.

Kavishree, S., Hemavathy, J., Lokesh, B.R., Shashirekha, M.N. and Rajarathnam, S. (2008). Fat and fatty acids of Indian edible mushrooms. Food Chem., 106: 597–602.

Kepler, R.M., Sung, G.H., Ban, S., Nakagiri, A., Chen, M.-J. et al. (2012). New teleomorph combinations in the entomopathogenic genus *Metacordyceps*. Mycologia, 104: 82–197.

Kirk, P., Cannon, P., Minter, D. and Stalpers, J. (2008). Dictionary of the Fungi, 10th edn. CABI, Wallingford, UK.

Kumar, T.K.A., Thomas, A., Kuniyil, K., Nanu, S. and Nellipunth, V. (2019). A checklist of non-gilled fleshy fungi (Basidiomycota) of Kerala State, India. Mycotaxon, 134: 221–229.

Kumar, T.S. and Aparna, N.S. (2014). *Cordyceps* species as a bio-control agent against coconut root grub, *Leucopholis coneophora* Burm. J. Environ. Res. Develop., 8: 614–618.

Lahiri, S.S., Shukla, M.D., Shah, M.B. and Modi, H.A. (2010). Documentation and analysis of certain macrofungal traditional practices from Western India (Gujarat). Ethnobot. Leaflets, 14: 626–641.

Latha, D.K.P. and Manimohan, P. (2012). Two remarkable xylariaceous ascomycetes associated with elephant dung. Mycosphere, 3: 261–265.

Latha, K.P.D. and Manimohan, P. (2017). Inocybes of Kerala. SporePrint Books, Calicut, Krala, India, p. 181.

Leelavathy, K.M., Flower, L. and Suja, C.P. (1983). The genus *Termitomyces* in India. pp. 402–407. *In*: Kaul, T.N. and Kapoor, B.M. (eds.). Indian Mushroom Science II, Council of Scienentific and Industrial Research, New Delhi.

Leelavathy, K.M. and Ganesh, P.N. (2000). Polypores of Kerala. Daya Publishing House, New Delhi, pp. 402–407.

Leelavathy, K.M., Manimohan, P. and Arnolds, E.J.M. (2006). *Hygrocybe* in Kerala state, India. Persoonia, 19: 101–151.

Manimohan, P. (2019). *Auriculoscypha*: A fascinating instance of fungus-plant interactions. pp. 34–39. *In*: Sridhar, K.R. and Deshmukh, S.K. (eds.). Advances in Macrofungi: Diversity, Ecology and Biotechnology, CRC Press, Boca Raton, USA.

Manimohan, P., Thomas, K.A. and Nisha, V.S. (2007). Agarics on elephant dung in Kerala State, India. Mycotaxon, 99: 147–157.

Mohanan, C. (2014). Macrofungal diversity in the Western Ghats, Kerala, India: Members of Russulaceae. J. Threat. Taxa, 6: 5636–5648.

Mohanan, C. (2003). Mycorrhizae in forest plantations: Association, diversity and exploitation in planting improvement. Kerala Forest Research Institute, Research Report # 252, Peechi, Kerala, India, pp. 133.

Mohanan, C. (2011). Macrofungi of Kerala. Kerala Forest Research Institute, Peechi, Kerala, India, p. 597.

Mrak, T., Kühdorf, K., Grebenc, T., Štraus, I., Münzenberger, B. and Kraigher, H. (2017). *Scleroderma areolatum* ectomycorrhizae on *Fagus sylvatica* L. Mycorrhiza, 27: 283–293.

Mueller, G.M., Schmit, J.P., Leacock, P.R., Buyck, B., Cifuentes, J. et al. (2007). Global diversity and distribution of macrofungi. Biodivers. Conserv., 16: 37–48.

Nandkumar, K. (2013). Ecoconservation of Goa's *Termitomyces* biodiversity. https://www.researchgate.net/publication/280623912.

Natarajan, K. (1995). Mushroom flora of South India (except Kerala). pp 381–397. *In*: Chadha, K.L. and Sharma, S.R. (eds.). Advances in Horticulture. Malhotra Publishing House, New Delhi.

Natarajan, K. (1979). South Indian agaricales V: *Termitomyces heimii*. Mycologia, 71: 853–855.

Natarajan, K., Narayanan, K., Ravindran, C. and Kumaresan, V. (2005). Biodiversity of agarics from Nilgiri Biosphere Reserve, Western Ghats, India. Curr. Sci., 88: 1890–1893.

Negi, C.S., Joshi, P. and Bohra, S. (2015). Rapid vulnerability assessment of Yartsa Gunbu *Ophiocordyceps sinensis* [Berk.] G.H. Sung et al.) in Pithoragarh District, Uttarakhand, India. Mountain Res. Develop., 35: 382–391.

Nobre, T., Eggleton, P. and Aanen, D.K. (2010). Vertical transmission as the key to the colonization of Madagascar by fungus-growing termites? Proc. Royal Soc. B, Biol. Sci., 277: 359–365.

Noordeloos, M.E., Vrinda, K.B. and Manimohan, P. (2007). On two remarkable brown-spored agarics from Kerala State, India. Fungal Diversity, 27: 145–155.

Pahlevanlo, A. and Janardhana, G.R. (2012). Diversity of *Termitomyces* in Kodagu and need for conservation. Journal of Advanced Laboratory Research in Biology, 3: 54–57.

Pavithra, M., Greeshma, A.A., Karun, N.C. and Sridhar, K.R. (2015). Observations on the *Astraeus* spp. of Southwestern India. Mycosphere, 6: 421–432.

Pavithra, M., Sridhar, K.R., Greeshma, A.A. and Karun, N.C. (2016). Spatial and temporal heterogeneity of macrofungi in the protected forests of Southwestern India. Int. J. Agric. Technol., 12: 105–124.

Pradeep, C.K. and Vrinda K.B. (2007). Some noteworthy agarics from Western Ghats of Kerala. J. Mycopathol. Res., 45: 1–14.

Pradeep, C.K. and Vrinda, K.B. (2010). Ectomycorrhizal fungal diversity in three different forest types and their association with endemic, indigenous and exotic species in the Western Ghat forests of Thiruvananthapuram District, Kerala. J. Mycopathol. Res., 48: 279–289.

Prathibha, P.S. (2015). Behavioral Studies of Palm White Grubs *Leucopholis* spp. (Coleoptera: Scarabaeidae) and Evaluation of New Insecticides for their Management. Ph.D. Thesis in Agricultural Entomology, University of Agricultural Sciences, Bangalore.

Quan, Q.-M., Chen, L.-L., Wang, X., Li, S., Yang, X.-L. et al. (2014). Genetic diversity and distribution patterns of host insects of caterpillar fungus *Ophiocordyceps sinensis* in the Qinghai-Tibet Plateau. PLoS ONE, 9: e92293.

Ranadive, K.R. (2013). An overview of Aphyllophorales (wood rotting fungi) from India. Int. J. Curr. Microbiol. Appl. Sci., 2: 91–114.

Rinaldi, A.C., Comandini, O. and Kuyper, T.W. (2008). Ectomycorrhizal fungal diversity: Separating the wheat from the chaff. Fungal Ecol., 33: 1–45.

Riviere, T.R., Diedhiou, A.G., Diabate, M., Senthilarasu, G., Natarajan, K. et al. (2007). Genetic diversity of ectomycorrhizal basidiomycetes from African and Indian tropical rain forests. Mycorrhiza, 17: 145–248.

Sankaran, K.V. and Florence, E.J.M. (1995). Macrofungal flora and checklist of plant diseases of Malayattoor forests (Kerala). pp. 147–168. *In*: Mukherjee, S.K. (ed.). Advances in Forestry Research in India, Volume 12, Dehra Dun, India.

Sargunam, S.D., Johnsy, G., Samuel, A.S. and Kaviyarasan, V. (2012). Mushrooms in the food culture of the Kaani tribe of Kanyakumari District. Ind. J. Trad. Knowl., 11: 150–153.

Sathe, A.V. and Daniel, J. (1980). Agaricales (mushrooms) of Kerala State. pp. 75–108. *In*: Sathe, A.V. (ed.). Agaricales (Mushrooms) of South West India, Monograph # 1, Part # 3. Maharashtra Association of Cultivation of Science, Pune, Maharashtra, India.

Sathe, A.V. and Deshpande, S. (1980). Agaricales (Mushrooms) of Maharashtra State. pp. 9–42. *In*: Agaricales (Mushrooms) of South West India. Maharashtra Association for the Cultivation of Science, Research Institute, Pune, Maharashtra, India.

Senthilarasu, G. (2014). Diversity of agarics (gilled mushrooms) of Maharashtra, India. Curr. Res. Environ. Appl. Mycol., 4: 58–78.

Senthilarasu, G. and Kumaresan, V. (2016). Diversity of agaric mycota of Western Ghats of Karnataka, India. Curr. Res. Environ. Appl. Mycol., 6: 75–101.

Sentihlarasu, G. (2015). The letinoid fungi (*Lentinus* and *Panus*) from Western Ghats, India. IMA Fungus, 6: 119–128.

Shrestha, U.B. and Bawa, K.S. (2014). Impact of climate change on potential distribution of Chinese caterpillar fungus (*Ophiocordyceps sinensis*) in Nepal Himalaya. PLoS ONE, 9: e106405.

Sidhu, A.S. (2009). Annual Report 2008–09, Indian Institute of Horticultural Research, Bangalore, India, p. 125.

Sridhar, K.R. and Karun, N.C. (2017). Observations on *Ophiocordyceps nutans* in the Western Ghats. J. New Biol. Rep., 6: 104–111.

Sridhar, K.R. and Karun, N.C. (2019). Diversity and ecology of ectomycorrhizal fungi in the Western Ghats. pp. 479–507. *In*: Singh, D.P., Gupta, V.K. and Prabha, R. (eds.). Microbial Interventions in Agriculture and Environment, Volume 1, Research Trends, Priorities and Prospects. Springer Nature, Singapore.

Sudheep, N.M. and Sridhar, K.R. (2014). Nutritional composition of two wild mushrooms consumed by tribals of the Western Ghats of India. Mycology, 5: 64–72.

Sung, G.H., Hywel-Jones, N.L., Sung, J.M., Luangsa-Ard, J.J., Shrestha, B. and Spatafora, J.W. (2007). Phylogenetic classification of *Cordyceps* and the clavicipitaceous fungi. Stud. Mycol., 57: 5–59.

Swapna, S., Abrar, S. and Krishnappa, M. (2008). Diversity of macrofungi in semi-evergreen and moist deciduous forest of Shimoga District, Karnataka, India. J. Mycol. Pl. Pathol., 38: 21–26.

Tedersoo, L., May, T.W. and Smith, M.E. (2010). Ectomycorrhizal lifestyle in fungi: Global diversity, distribution, and evolution of phylogenetic lineages. Mycorrhiza, 20: 217–263.

Thomas, K.A., Hausknecht, A. and Manimohan, P. (2001). Bolbitiaceae of Kerala State, India: New species and new and noteworthy records. Österreichische Zeitschrift für Pilzkunde, 10: 87–114.

Thulasinathan, B., Kulanthaisamy, M., Nagarajan, A., Soorangakattan, S., Muthuramalingam, J. et al. (2018). Studies on the diversity of macrofungus in Kodaikanal region of Western Ghats, Tamil Nadu, India. Biodiversitas, 19: 2283–2293.

Usha, N. and Janardhana, G.R. (2014). Diversity of macrofungi in the Western Ghats of Karnataka (India). The Indian Forester, 140: 531–536.

Vaario, L. and Matsushita, N. (2021). Conservation of edible ectomycorrhizal mushrooms: Understanding of the ECM fungi mediated carbon and nitrogen movement within forest ecosystems. pp. 1–24. *In*: Ohyama, T. and Inubushi, K. (eds.). Nitrogen in Agriculture - Physiological, Agricultural and Ecological Aspects. IntechOpen. https://doi.org/10.5772/intechopen.95 399.

Varghese, S.P., Pradeep, C.K. and Vrinda, K.B. (2010). Mushrooms of tribal importance in Wayanad area of Kerala. J. Mycopathol. Res., 48: 311–320.

Vinjusha, N. and Kumar, T.K.A. (2022). The Polyporales of Kerala. SporePrint Books, Calicut, Kerala, India, p. 229.

Vrinda, K.B. and Pradeep, C.K. (2009). *Termitomyces sagittiformis*—A lesser known edible mushroom from the Western Ghats. Mushroom Res., 18: 33–36.

Vrinda, K.B., Pradeep, C.K. and Abraham, T.K. (2002). *Termitomyces umkowaani* (Cooke & Mass.) Reid – an edible mushroom from the Western Ghats. Mushroom Res., 11: 7–8.

Vrinda, K.B., Pradeep, C.K., Mathew, S. and Abraham, T.K. (1999). Agaricales from Western Ghats–VI. Ind. Phytopathol., 52: 198–200.

Yesodharan, K. and Sujana, K.A. (2007). Wild edible plants traditionally used by the tribes in the Parambikulam wildlife sanctuary, Kerala, India. Nat. Prod. Radiance, 6: 74–80.

Zhang, Y.-Z., Sun, C.-Y., Sun, J., Zhang, K.-P., Zhang, H.-S. et al. (2020). *Scleroderma venenatum* sp. nov., *S. venenatum* var. *macrosporum* var. nov. and *S. suthepense* new to China. Phytotaxa, 438: 107–118.

Ectomycorrhizal Fungi

6

Mycogeography and Ecology of Ectomycorrhizal Fungi in Northern México

Fortunato Garza Ocañas,[1,*] *Miroslava Quiñonez Martínez,*[2] *Lourdes Garza Ocañas,*[3] *Artemio Carrillo Parra,*[4] *Jesús García Jiménez,*[5] *Gonzalo Guevara Guerrero,*[5] *Ricardo Valenzuela Garza,*[6] *Mario García Aranda,*[7] *Javier de la Fuente,*[8] *Gerardo Cuellar Rodríguez*[1] and *José Guadalupe Martínez Avalos*[9]

1. Introduction

Ectomycorrhizal fungi form associations with many hosts in different vegetation types of the world (e.g., oak and conifer forests) and both symbiotic partners obtain

[1] Universidad Autónoma de Nuevo León, Campus Linares, Facultad de Ciencias Forestales, Carretera Nacional km 145, Apdo. postal 41, 67700 Linares, Nuevo León, Mexico.
[2] Universidad Autónoma de Ciudad Juárez. Departamento de Ciencias Químico-Biológicas. Av. Benjamín Franklin No. 4650, Zona PRONAF, Cd. Juárez, Chihuahua, México. C. P. 32315.
[3] Facultad de Medicina, Departamento de Farmacología y Toxicología, Universidad Autónoma de Nuevo León, México.
[4] Instituto de Silvicultura e Industria de la Madera Universidad Juárez del Estado de Durango, México
[5] Tecnológico Nacional de México, Instituto Tecnológico de Ciudad Victoria, Boulevard Emilio Portes Gil número 1301, 87010 Cd. Victoria, Tamaulipas, Mexico.
[6] Laboratorio de Micología, Departamento de Botánica, Escuela Nacional de Ciencias Biológicas, Instituto Politécnico Nacional, Apartado Postal 63-351, 02800, México.
[7] Mar Caspio, 8212, Loma Linda, Monterrey, N.L. C.P. 64120, México.
[8] Colegio de Postgraduados, km 36.5, 56230 Montecillo, Texcoco, Estado de México, Mexico.
[9] Instituto de Ecología Aplicada, Universidad Autónoma de Tamaulipas, Cd. Victoria, Mexico
* Corresponding author: fortunatofgo@gmail.com

nutritional benefits (e.g., water, minerals and photosynthesis products). The nutrients are shared *via* the functional mycelial nets starting from the soil and moving inside and outside of the ectomycorrhizal roots of their hosts. These relationships have a long evolutionary history starting at the Devonian period and it is suggested that such partnerships helped both associates to conquest new territories reaching the current biogeographic distribution of forests in the world (Retallack et al., 2014; Heads et al., 2017). Hosts involved in this evolution process belong to families (e.g., Pinaceae, Fagaceae, Betulaceae, Salicaceae, Ericaceae, Myrtaceae, Dipterocarpaceae and Caesalpiniaceae) (Smith and Read, 1997). They are well distributed in the world in Boreal, Temperate, Tropical, and semiarid conditions and carry with them their fungal partners to survive obtaining better nutrition from the different soils where they grow. In every soil condition of the world where these mutualistic symbioses occur, the ectomycorrhizal fungi will produce fruiting bodies either on top (epigeous) or below the soil (hypogeous) during the rainy season (North, 2002).

A great number of ectomycorrhizal fungal species are involved in these processes, and yearly they contribute millions of new spores to the forest ecosystem (Villarreal and Luna, 2019). Spores are dispersed by millions of insects, mites, worms, and other fauna (e.g., rodents, deer, wild pigs and bear) as well as by rain and wind and each will germinate in the forest soil and will find plenty of new roots to colonize and form new ectomycorrhizas (Maser et al., 2008; Schiegel, 2012; Elliot et al., 2022). They will form agglomerates or complexes integrated by several species of ECM fungi with each tree species and they will be connected with a few other tree species via their mycelial nets, and they will also be exploring the soil for nutrients. These fungal nets contribute significantly to the nutrition of trees as well as to other organisms feeding on mycelium or fruiting bodies in the forest. Almost every terrestrial vegetation of the world functions associated with mycorrhizal fungi as they uptake and share nutrients with their hosts. Due to the nutritional contribution of the mycorrhizal associations to the forest of the world, their relevance is very high, and they also contribute to forest's health and biogeographical distribution. Understanding these nutritional processes from every vegetation type is a key subject that helps to understand their global contribution and generates local or regional knowledge from different geographical latitudes. The mutualistic associations are needed to carry out the management of specific associations (e.g., to improve the establishment of endangered host species *Picea*, *Abies* or *Pseudotsuga* spp. in Mexico, using inoculated seedlings or to carry out reforestation projects of burned sites) (Garza et al., 1985; Hall et al., 2019). Regarding taxonomical studies from different biogeographical regions, these studies are useful to know the diversity of species, their plant partners and some of the soil and abiotic conditions occurring in each region as well as some possible species of animals that might be contributing for their dispersion (Maser et al., 2008; Halling et al., 2008; Elliot et al., 2022). Some species of macromycetes might have had a wide distribution in the past and were separated in different continents. Nowadays they are forming ectomycorrhizal associations with oak forests in North America, Mexico and China (e.g., Boletales: *Harrya chromipes*, *Tylopilus cyaneotinctus*, *T. griseus*, *Tylopilus felleus* and *Sutorius eximius*) (Arora, 1986; Wu et al., 2016; Yan-Chun Li and Zhu L. Yang, 2021). Some populations of Boletes might have been isolated for long periods (Halling et al., 2008). According to the geological substrate, vegetation type, soils, and climatic conditions from every region of the world, the mycorrhizal associations form symbiosis with many hosts (Garza et al., 1985, 1986, 2002, 2019, 2022; García et al., 2014, 2016). Examples of the latter can be observed in the fungal diversity present in western and eastern North America and in the past, some of these associations moved with their partners (e.g., the genus *Pinus*, southwards into Mexico reaching the big Sierra Madre Occidental) (SMW) in the northwest and the Sierra Madre Oriental (SME). Some fungal species remained in north America in boreal and temperate climatic conditions associated with different tree species and they are not found in the north of Mexico. However, some fungal species grow in both, the south of North America and the north of Mexico associated with different host species (e.g., nut pines, other conifers and oaks species) (Bessette et al., 1997, 2016, 2019; Beug et al., 2014; García, 2016; Baroni, 2017). In Mexico, pine and oak forests are widely diverse and dispersed and the ectomycorrhizal fungi are associated to all the species, so far, no studies comparing the similarity of the diversity of ectomycorrhizal macromycetes species from both Sierras (SMW) and (SME) with those from the Pacific and Atlantic coast from USA and Canada have been attempted.

This study presents mycogeography of ectomycorrhizal fungi in the biogeographic regions in the western and eastern parts of Mexico and some of the animals that interact with them in the forests. An attempt to see similarities between the species found in the north of Mexico with some species reported from both the western and east southern regions of the USA, and it also includes some species reported from these zones in Canada.

2. Description of the Sierras at the North of Mexico

2.1 The Sierra Madre Occidental (SMW)

This biogeographic territory covers an area of 12.6% it has rocks belonging to the Paleozoic, Mesozoic, and Cenozoic eras mainly of extrusive volcanic origin and soils are represented mainly by litosol, regosol, fluvisol, rendzina and cambisol. The SMW adjoints the states of Durango, Coahuila, Sonora and Sinaloa in México, and in the north reaches the USA in Texas, New Mexico and Arizona (Fig. 1). It has a mean elevation of 2,500 m.a.s.l and the highest elevation reaches 3,711 m.a.s.l. at Cerro El Moinora at the municipality of Yecora. This Sierra has a 75% dry or arid (B) climate type, followed by temperate (C) in 13% semiarid, temperate, subhumid (A) with 12% A cold-temperate (C) climate is the predominant one at high altitudes. Mean rain ranges from 550–750 mm and temperate forests cover 25.3% of the state and it has pine, oak and mixed pine-oak forests, and the more open areas of these forests have *Arbutus* and *Juniperus*. January is the coldest month and lower temperatures are reached at the high Sierra with a mean of –5°C and June is the hottest and temperatures reach a mean of 35°C (Gonzalez et al., 2012; Reyes Gomez and Nuñez López, 2014).

Fig. 1. Location of both Sierras at the west and east of Mexico.

2.2 *The Sierra Madre Oriental (SME)*

This Sierra is a physiographic province and as in the SMW it has its own topographical, geological and climatological characteristics (Fig. 1). The SME covers an area of ca. 220,192.3 km^2 and is almost 1,250 km long and it is distributed in 11 states, it has a mean elevation of 1,313 m.a.s.l., and the highest elevation reaches 3,711 m.a.s.l. at Cerro El Potosí in the municipality of Galeana. The SME has rocks belonging to the Paleozoic, Mesozoic, and Cenozoic eras the Limestone the predominant, whereas the predominant soil type is Litosol and xerosol, rendzina, feozem, vertisol, luvisol, and yermosol are also present. Due to its size, the SME has 14 climate types and the predominant is dry (BS), in the northern and western regions, it is followed by the humid-warm type (A) (C) and sub-humid-warm type (Aw) that are in the eastern and southern regions. A cold-temperate (C) climate is the predominant one at high altitudes and main rain ranges from 600–700 mm (Cervantes-Zamora et al., 1990; Salinas et al., 2017).

3. Sampling Data and Vegetation

A sampling of ectomycorrhizal fungi was carried out at both Sierras: in the last 25 years. The main vegetation types are temperate forests with conifers (e.g., *Pinus*,

Pseudotsuga, *Abies*, *Picea* and *Cupressus* and Oaks *Quercus* spp.). Altitude at the different Sierras vary from 500 to 3750 m and vegetation is formed by plant communities that are integrated into many associations (e.g., *Pinus-Quercus*, *Pinus-Pseudotsuga*, *Pinus-Abies*, *Pinus-Picea*, *Quercus-Pinus* and *Pinus-Quercus*) and they also form mixed associations with other species like *Arbutus xalapensis*, *Arctostaphylos pungens* and *A. manzanitae*. Some oak species are very small and grow at the soil level forming small communities of decumbent plants (e.g., *Quercus depressipes*), while others, when growing in semiarid conditions, form shrub-like communities (e.g., *Quercus emory*). *Quercus grisea* may be found growing isolated in small hills surrounded by a semiarid environment at hundreds of kilometers away from the main forests, and they have exclusive ectomycorrhizal fungi associated. Visits at different locations in both Sierras were made in the rainy season to try to cover as much geography as possible, these Sierras are separated by more than a thousand kilometers. Collection of mushrooms was carried out from July to September at the SMW and in from September onwards at the SME. This is because the phenology of fruit bodies production and rainy seasons are different in each region (Garza, 1986; Garza et al., 2019; Quiñonez et al., 2008; Quiñonez et al., 2014).

3.1 Field Studies

In the field, the location was recorded with GPS, and the ecological information, such as the associated plant communities, the possible hosts of the mycorrhizal species, the habitat data, the date of the fruiting, and the characteristics of the fresh basidiocarps such as color, smell and size. were recorded for each species. To obtain the macro chemical reactions, reagents such as NH_3 25%, KOH 15% and $FeSO_4$ 10% were applied in the different parts of the basidiocarps. The specimens were taken to the laboratory for dehydration and final labeling. Specimens of the species and some strains obtained are deposited either at the mycological collection of Facultad de Ciencias Forestales, CFNL, Instituto de Ciencias Biomédicas, UACH or Instituto Tecnológico de Ciudad Victoria (ITCV). Animals shown here were observed in the field at both Sierras and photography were obtained *in situ*.

3.2 Laboratory Studies

In the laboratory, mature specimens of each species were selected and used to determine the microscopic characteristics, and fine handmade sections were made from the different parts of the basidiocarps, the fine cuts made were moistened with water on a slide and then mounted in a solution of KOH 5% and NH_3 3% or Congo red solution. All microscopic structures were measured to obtain a mean of length by width with a lens magnification of 1000X, a calibrated eye micrometer was used and the average sizes of each of the different structures were calculated for each species using a Carl Zeiss Axiostar microscope (Quiñonez et al., 2008).

4. Mycorrhizal Association with Vegetation

Apparently, the first evidence of the occurrence of Basidiomycetes forming ectomycorrhizal associations occurred in the Pennsylvanian during the Carboniferous 299 million years ago (ICCICS, 2014). It is speculated that ectomycorrhizal associations might have contributed to the survival of their associated partners and helped to the diversification and richness of species (Singer, 1986; Bruns et al., 1989; Binder and Bresinsky, 2002a; Binder and Hibbett, 2006; Wang and Qiu, 2006; Nuhn et al., 2013; Wu et al., 2014, 2016). Mirov (1967) mentioned that the genus *Pinus* might have migrated from boreal conditions existing in north America reaching its southwards distribution in Nicaragua with *Pinus caribaea* var. *caribaea*. Nowadays studies on Boletales from Mexico show that they are widely distributed and are intimately associated to *Pinus* and *Quercus* species (García and Garza, 2001; North, 2002). According to Haling et al. (2008), "the array of plant-associated distributions provides a potential handle for evaluating Bolete distribution on a global scale". The species from the genus *Suillus* might be the most widely distributed in the Mexican territory and they form mycorrhizal associations with many different conifers but especially with the genus *Pinus* and might also be associated with some oak species (Garcia and Garza, 2001). The observation of this kind of ecological relationships is important to produce seedlings of every host (e.g., *Pinus*) in the wide geography of Mexico. Thus, inoculating conifer seedlings with their native ectomycorrhizal species strains or spores under greenhouse and nursery or field conditions is particularly important to obtain good field establishment results (Hall et al., 2019). Some other interesting examples of very close relationship might occur with the association formed between *Suillus lakei* and *Pseudotsuga menziesi* or *Turbinellus floccosus* and *Abies vejarii* or *Leccinum insigne* and *Populus tremuloides* in both Sierras (Garcia and Garza, 2001; Garza et al., 2019). These associations are very much specific and close and so far, the fungal partner has not been found associated with other hosts. Considering these field observations, it might be necessary to use these ecological associations when the reproduction of these hosts for reforestation or restoration procedures is required. *Pisolithus tinctorius*, *Scleroderma verrucosum*, *S. areolatum*, *S. texense* and *S. cepa* are widely distributed forming ectomycorrhizas with many oak species in both SMW and SME. Species of *Russula*, *Lactiflus*, *Lactarius*, *Cortinarius*, *Boletus*, *Gyroporus*, *Amanita*, *Cantharellus*, *Inocybe* and *Astraeus* are widely distributed and form ectomycorrhizal associations with many oak species (Laferriere and Gilbertson, 1996; North, 2002; García et al., 2014; Olimpia et al., 2017; Flores et al., 2018). Interesting examples of mycorrhizal associations are found in either of the sierras (e.g., *Cortinarius magnivelatus* is associated to *Quercus sartorii* and *Q. depressipes* and *Boletus chippewaensis*, *B. rubripes*, *B. barrowsii* form associations with *Pinus chihuahuensis*, *P. herrerae* and *P. lumholtzii*) at the high Sierra of Chihuahua at the (SMW) (Perez et al., 1986; Quiñonez et al., 1999; 2008; 2015). Other interesting ectomycorrhizal associations formed by Boletales (e.g., *Suillelus amigdalinus*, *Rubroboletus eastwoodiae*, *R. pulcherrimus* and *Rhizopogon occidentalis* in forests with *Pinus quadrifolia*, *P. monophyla*, *P. coulterii* and *Quercus agrifolia*) in Baja California (Sánchez et al., 2015). At the SME several species of *Tuber*, *Sclerogaster*, *Rhizopogon*, *Hydnobolites*, *Hymenogaster*, *Gilkeya* and *Genea* form ectomycorrhizas with *Pinus culminicola* at a high altitude of 3700 m at cerro El Potosí at the (SME) and Tamaulipas (Garza et al., 1985; Garza et al., 2002; Guevara et al., 2014). It is interesting to mention that some Boletales (e.g., *Boletus luridellus, Boletus paulae, Hortiboletus rubellus, Boletus campestris, Boletinellus merulioides, Boletinellus rompelli, Phlebopus portentosus, P. brassiliensis, P. mexicanus* and *Xerocomellus* spp.) grow at low altitudes 500–600 m associated either with oaks in the submontane forests at the SME or at lower altitudes 260–360 m with thorn scrubs to others potential hosts (e.g., *Acacia rigidula, Vachelia farnesiana, Helietta parvifolia, Ebenopsis ebano*) and other species at the low coastal province. Table 1 complies the ectomycorrhizal tree species in Mexico.

5. Ectomycorrhizae of Vegetation and Relationship with Animals

A review of the species of animals from the different taxonomic groups shows that there is quite a number, and some are endemic to either of the sierras. Very many species of mycophagous animals from different families have been recorded around the world (Elliot et al., 2022). Ectomycorrhizal fungi are very abundant, and they form mycorrhizas with all the species of host mentioned before establishing nutritional nets through which nutrients move actively in the soil of temperate forest ecosystem (Halling et al., 2008; Garza et al., 2019; Rilling and Allen, 2019). These fungi produce millions of spores in their fruiting bodies and for many animals, the water and wind are their main dispersers. Multisymbiosis systems are established in the temperate forest and many species are interconnected in some part of the nutritional chain (Maser et al., 2008; Elliot et al., 2022). More attention on all trophic interactions occurring in the temperate forest should be considered. Minerals in the soils require microorganisms to be dissolved, ectomycorrhizal fungi secrete enzymes and uptake the dissolved minerals in their nets, and they are then translocated to the inner tissues of the roots of every host plant, and they even share these minerals through these nets, from one host to another thus participating actively in the forest growth (Villarreal and Luna, 2019). Every year during the rainy season the spores of the ectomycorrhizal fungi germinate and find their genetic counterparts to produce secondary mycelia which will initiate the production of fruiting bodies. These will grow and produce attractive colors and volatile compounds to attract animals that will eat and disperse their spores all over the forest (Elliot et al., 2022). Mycophagy takes place not only in mammals but in very many species of other animals like earthworms, slugs, snails, plain worms, mites, collembola and many other insects either eat or use the fruiting bodies as a place lay their eggs which in turn will grow and use their "nursery beds" as food (Maser et al., 2008). This study aims to show some examples of the diversity of ECM fungi growing at both Sierras and give some examples of animals interacting with them. Mycophagy of fruiting bodies of ECM fungi forms an important part of the diet of many animals during the rainy season. In the north of Mexico, some animals like the Peccary have incorporated ECM fruiting bodies in their diet and so they became active bioturbation species in the temperate forests in the north of Mexico, seeking actively for fruiting bodies of either epigeous or hypogeous ECM fungi. Fruit bodies are attractive by their colors and their volatile

Table 1. Ectomycorrhizal host tree species in Mexico.

	SMW-MEX	SME-MEX
Family Pinaceae		
Pinus pseudostrobus var. apulcensis		•
P. teocote		•
P. nelsonii		•
P. greggii		•
P. culminicola		•
P. cembroides	•	•
P. monophylla	•	
P. quadrifolia	•	
P. coulterii	•	
P. engelmannii	•	
P. lambertiana	•	
P. montezumae var. montezumae		•
P. johannis		•
P. hartwegii		•
P. leiophylla	•	
P. arizona var. cooperii	•	
P. duranguensis	•	
P. chihuahuana	•	
P. devoniana		•
P. discolor	•	
P. douglasiana	•	
P. patula		•
P. lumholtzii	•	
P. ayacahuite		•
P. flexilis	•	
P. herrerae	•	
P. strobiformis	•	
P. yecorensis	•	
P. arizonica var. arizonica		•
P. arizonica var. stormiae	•	
P. jeffreyi	•	
Abies duranguensis	•	
A. vejarii		•
A. concolor	•	
A. guatemalensis		•
Picea chihuahuana	•	
P. engelmanii ssp.	•	

Table 1 contd. ...

...Table 1 contd.

	SMW-MEX	SME-MEX
P. engelmanii	●	
Pseudotsuga mensiezii	●	●
Family Fagaceae		
Quercus affinis		●
Q. emoryi	●	●
Q. canbyi Trel.		●
Q. cupreata		●
Q. castanea	●	
Q. conspersa	●	
Q. crassifolia	●	
Q. crassipes	●	
Q. delgadoana	●	
Q. alpescens	●	
Q. depressa	●	
Q. flocculenta	●	
Q. fulva	●	
Q. furfuracea	●	
Q. galeanensis		●
Q. gentry		●
Q. graciliramis		●
Q. gravesii	●	
Q. hintoniorum		●
Q. hirtifolia	●	
Q. hypoleucoides	●	
Q. eduardii	●	
Q. hypoxantha		●
Q. jonesii	●	
Q. laurina	●	●
Q. mexicana		●
Q. miquihuanensis		●
Q. ocoteifolia	●	
Q. pinnativenulosa	●	
Q. runcinatifolia	●	
Q. rysophylla		●
Q. saliicifolia	●	
Q. saltillensis		●
Q. sapotifolia	●	

Table 1 contd. ...

...Table 1 contd.

	SMW-MEX	SME-MEX
Q. sartorii		•
Q. sideroxyla	•	•
Q. skinneri	•	
Q. tenuiloba	•	
Q. viminea	•	
Q. xalapensis	•	
Q. chihuahuensis	•	
Q. convallata	•	
Q. diversifolia	•	
Q. edwardsae	•	
Q. fusiformis	•	•
Q. germana	•	
Q. glaucoides		•
Q. gregii		•
Q. intrincata		•
Q. invaginata		•
Q. laceyi		•
Q. laeta	•	
Q. lancifolia	•	
Q. magnolifolia	•	
Q. microlepis.		•
Q. microphylla	•	•
Q. muehlenbergii	•	
Q. oblongifolia	•	
Q. obtusata		•
Q. oleoides		•
Q. opaca	•	
Q. pastorensis	•	
Q. pendicularis	•	
Q. polymorpha		•
Q. praeco	•	
Q. pringlei		•
Q. pungens		•
Q. rugosa	•	•
Q. sebifera	•	
Q. sinuata	•	
Q. splendens	•	
Q. striatula	•	

Table 1 contd. ...

...Table 1 contd.

	SMW-MEX	SME-MEX
Q. supranitida	•	
Q. thinkhamii	•	
Q. toxicodendrifolia	•	
Q. monterreyensis		•
Q. vaseyana		•
Q. verde	•	
Q. mohriana	•	
Q. toumeyei	•	
Q. tarahumara	•	
Q. basaseachicensis	•	
Q. chrysolepis		•
Q. arizonica	•	
Q. durifolia	•	
Q. gambelli	•	
Q. grisea	•	
Q. potosina	•	
Q. tuberculata	•	
Q. mcvaughii	•	
Q. scytophylla	•	
Q. subpathulata	•	
Q. deppressipes	•	
Q. intricata	•	
Q. tardifolia	•	
Q. clivicola		•
Q. sartorii		•
Q. agrifolia	•	
Q. dumosa	•	
Q. urbanii	•	
Q. virginiana		•
Q. coahuilensis		•
Q. carmenensis		•
Family Ericaceae		
Arbutus xalapensis var. texana		•
A. arizonica	•	
Arctostaphylos pungens	•	
A. glauca	•	
A. peninsularis	•	
A. platyphylla	•	
Total species, 149; Common species, 6	**101**	**55**

compounds which unveil them and make them detectable for very many animals. Apparently, these characteristics are evolutionary mechanisms that allow ECM fungi to survive after they have been eaten by animals. In Europe some truffle hunters follow truffle flies to their "source of food" and thus they will be able to find the black truffles! Trained dogs are used in the same way to find the truffles in the field or in truffle plantations. Maggots of many species of insects in the world use the fruiting bodies as food or a nursery to complete part of their life cycles. Beetles are very abundant the same as many species of Diptera (e.g., *Drosophila* spp.) (Schigel, 2012). These insects eat actively and are good spore dispersers but many times they are food for spiders and frogs that visit the fruiting bodies. Insects in the fruiting bodies are also sought after, by birds and many species of mice, rats and squirrels visit the fruiting bodies for food and frequently become food for snakes attracted by all of them. White-tailed deer and peccaries eat mushrooms and toadstools and become food for bobcats, wolves, coyotes, cougars, and black bear. Thus, the whole food chain is integrated in this yearly process of fruiting body production that resulted from the germination of primary spores that fuse together to form secondary mycelia. The rainy season brought the necessary water which permeates the spore cell wall and induces it to germinate in summer or autumn. The phenology of fruiting bodies production may vary from one region to another, and this is the case of both Sierras in which they are produced first in the SMW during July-August whereas at the SME they are produced mainly during September-October and the same happens in different parts of the world. Some of the main research themes of the last decades have been Global Change and Climate Change which have moved the seasons to an uncertain pattern around the world (Rilling and Allen, 2002). These changes in the weather of every region move the whole food chain habit and some species may not be fit to survive if they occur either frequently or last for long periods. In the case of ECM fungi changes in the periods of the rainy season might change the production of fruiting bodies and all the dispersers in the food chain may also have to change their feeding habits for some time. Translocation of nutrients from the soil to the forest trees via the ECM fungi might also change with climate change (Rilling and Allen, 2002). Apart from these, forest fires, cattle ranchers, logging activities, and other human activities cause changes in the forests that are difficult to overcome for the ECM fungal communities. Thus, forests have multidiverse and multisymbiotic associations, they are complex and many times they are very specialized and energy in the forest might not be unlimited. Once the ecosystems are altered by for example a forest fire the whole nutrition is gone. Multisymbiotic relationships occurring in the forests as food chains have taken millions of years to reach their current state of stability and climate change is happening every day. Ecosystems are fragile and reorganization of the species in the food chain is not as simple as establishing a new plantation with desirable edible mushrooms to replace a natural forest. The multiple associations of animals and ectomycorrhizal fungi are widespread in every terrestrial ecosystem of the world and are generally overlooked. However, most species are interconnected in the nutritional chain (i.e., fungi, hosts, and animals) and they have a high ecological significance (Elliot et al., 2022).

5.1 *Ectomycorrhizal Fungi*

Some common species from the Sierra Madre Occidental SMW, Sierra Madre Oriental SME and boletes from submontane thorn scrubs close to the SME are represented in Figures 2, 3 and 4. Table 2 compiles ectomycorrhizal fungi in Mexico, USA and Canada.

6. Data Analysis

6.1 *Qualitative Analysis*

Qualitative similarity indices are non-parametric techniques that allow the analysis of the percentage of similarity between two communities based on the number of species they share, i.e., present at both sites, as well as the total number of species present. The index used was Jaccard.

6.1.1 Jaccard's Index of Similarity

According to the similarity matrix obtained with the Jaccard's index, it can be observed that the sites with the greatest similarity of fungal species are SME-Mex/NE-USA (79%), sharing 196 species and SMW-MEX/NW-USA (72%) sharing 176

Fig. 2. Some common species from the Sierra Madre Occidental SMW: *Amanita rubescens* (A), *A. cochiseana* (B), *Boletus chippewaensis* (C), *Cortinarius magnivelatus* (D), *Suillus spraguei* (E) and *Tricholoma magnivelare* (F).

Fig. 3. Some common species from the Sierra Madre Oriental SME: *Amanita rubescens* (A), *Boletus inedulis* (B), *Strobilomyces dryinus* (C), *Laccaria laccata* (D), *Amanita vaginata var. Plúmbea* (E) and *Turbinellus floccosus* (F).

Fig. 4. Some common Boletes from submontane thorn scrubs close to the SME: A. *Phlebopus portentosus (*A), *Phlebopus brassiliensis* (B), *Phlebopus aff. Mexicanus* (C), *Boletus paulae* (D), *Boletinellus rompelli* (E) and *Hortiboletus rubellus* (F).

Table 2. Ectomycorrhizal fungi reported and their distribution in Sierras in Mexico, northwest of USA, northeast of USA and northeast Canada.

	SMW-MEX	NW-USA	SMEi-MEX	NE-USA	NE-CAN
ASCOMYCOTA					
ORDER EUROTIALES					
Family Elaphomycetaceae					
Elaphomyces granulatus			•	•	
ORDER PEZIZALES					
Family Discinaceae					
Hydnotrya cerebriformis			•	•	
Family *Pezizaceae*			•	•	
Hydnobolites cerebriformis			•		
Pachyphlodes carnea			•	•	
P. citrina	•	•	•	•	
P. marronina			•		
P. virescens			•	•	
Sarcosphaera coronaria	•	•	•		
Family *Pyronemataceae*					
Genea hispidula			•	•	
G. arenaria			•	•	
Gilkeya compacta			•	•	
Sphaerosporella brunnea	•	•			
Tricharina gilva	•				
Trichophaea hemisphaerioides	•	•			
Family *Tuberaceae*					
Tuber candidum			•	•	
T. dryophilum	•				
T. lyonii			•	•	
T. regimontanum	•				
BASIDIOMYCOTA					
ORDER AGARICALES					
Family Tricholomataceae					
Infundibulicybe gibba	•	•	•	•	
Family Amanitaceae					
Amanita abrupta	•	•	•	•	
A. amerifulva	•		•	•	
A. amerirubescens	•				
A. atkinsoniana	•				
A. basii	•	•	•		
A. bisporigera	•		•	•	•

Table 2 contd. ...

...Table 2 contd.

	SMW-MEX	NW-USA	SMEi-MEX	NE-USA	NE-CAN
A. calyptroderma	•	•			•
A. ceciliae	•		•	•	•
A. chlorinosma	•		•		•
A. citrina	•	•	•	•	•
A. cochiseana	•	•	•		
A. cokeri	•		•		
A. crocea	•		•		
A. daucipes	•			•	
A. flavoconia	•	•	•	•	•
A. flavorubescens	•	•		•	
A. frostiana	•	•	•	•	•
A. fulva	•	•			•
A. gemmata	•	•	•	•	•
A. jacksonii	•		•	•	
A. magniverrucata	•	•			
A. multisquamosa			•	•	
A. muscaria	•		•		
A. ocreata	•	•			
A. onusta	•	•	•		
A. pantherina	•	•	•		
A. peckiana	•	•			
A. perpasta	•	•			
A. phalloides	•	•		•	
A. polypyramis			•	•	
A. porphyria	•	•			
A. rhopalopus	•	•			•
A. rubescens	•	•	•	•	•
A. smithiana	•	•			
A. spreta	•	•		•	
A. strobiliformis	•	•			
A. tuza	•				
A. vaginata	•	•	•	•	•
A. variabilis	•	•			
A. verna	•		•		
Aspidella solitaria	•		•		
Zhuliangomyces illinitus			•	•	

Table 2 contd. ...

...Table 2 contd.

	SMW-MEX	NW-USA	SMEi-MEX	NE-USA	NE-CAN
Family Cortinariaceae					
Calonarius odorifer	•	•			
Cortinarius alboviolaceus	•		•		•
C. atkinsonianus			•	•	
C. brunneus	•	•			
C. camphoratus			•	•	
C. caperatus	•	•			•
C. cinnamomeus	•	•	•	•	•
C. collinitus	•	•	•	•	•
C. corrugatus	•	•			
C. elegantissimus	•	•			
C. evernius			•	•	
C. flexipes			•	•	
C. hemitrichus					
C. iodes			•	•	
C. magnivelatus	•	•	•		
C. marylandensis			•	•	
C. pinetorum					
C. purpureus	•				
C. sanguineus	•	•	•	•	•
C. semisanguineus	•	•	•	•	•
C. smithii	•				
C. traganus	•				
C. violaceus	•	•	•	•	•
Thaxterogaster corrugis	•	•			•
T. talus			•		
Family Hydnagiaceae					
Laccaria amethystina	•	•	•	•	•
L. bicolor	•	•	•	•	
L. laccata	•		•		
L. laccata	•		•		•
L. proxima	•				
L. ochropurpurea	•	•	•	•	•
Family Inocybaceae					
Inocybe confusa	•				
I. geophylla			•	•	
I. geophylla var. geophylla			•	•	

Table 2 contd. ...

...Table 2 contd.

	SMW-MEX	NW-USA	SMEi-MEX	NE-USA	NE-CAN
I. hystrix	•		•	•	
I. lacera	•	•	•	•	
I. nemorosa			•	•	
Inosperma calamistratum	•	•	•	•	
I. erubescens			•		
I. maculatum	•	•	•		
Pseudosperma rimosum	•	•	•	•	
P. sororium	•	•	•	•	
Family Tricholomataceae					
Leucopaxillus albissimus	•	•	•	•	
L. gentianeus	•	•	•	•	
L. paradoxus	•				
Tricholoma caligatum			•	•	
T. equestre	•	•	•	•	•
T. magnivelare	•	•			
T. pardinum			•	•	
T. saponaceum	•	•			
T. sejunctum	•	•	•	•	
T. sulphurescens	•	•	•	•	
T. terreum	•	•	•	•	
T. ustaloides	•	•	•		
T. vaccinum	•	•	•	•	•
T. virgatum			•	•	•
ORDER BOLETALES					
Family Boletaceae					
Aureoboletus auriporus	•		•	•	
A. betula	•	•			
A. moravicus			•	•	
A. russellii	•	•	•	•	•
A. projectellus			•	•	
A. gracilis	•	•	•	•	•
Baorangia bicolor	•	•	•	•	•
Boletellus ananas	•		•	•	
B. chrysenteroides	•	•	•	•	•
B. coccineus	•				
B. flocculosipes	•				
B. pseudochrysenteroides	•		•		
B. merulioides			•	•	•

Table 2 contd. ...

...Table 2 contd.

	SMW-MEX	NW-USA	SMEi-MEX	NE-USA	NE-CAN
B. rompelii			•	•	
Boletus aureissimus	•		•	•	
B. barrowsii	•	•	•		
B. chippewaensis	•	•			
B. edulis	•	•	•		•
B. flammans	•	•	•	•	
B. luridellus			•	•	
B. paulae			•		
B. pinophilus	•	•	•	•	
B. pseudopinophilus	•	•	•	•	
B. rubriceps	•	•			
B. sensibilis	•	•			
B. subgraveolens			•	•	
B. subluridellus	•	•	•	•	
B. variipes	•	•	•	•	
B. vermiculosus	•	•			
Buchwaldoboletus hemichrysus	•	•			
Butyriboletus brunneus			•	•	
B. floridanus	•	•	•	•	
B. frostii	•	•	•	•	•
B. peckii			•	•	
B. regius			•	•	
Caloboletus rubripes	•	•			
Chalciporus piperatus			•	•	
Cyanoboletus pulverulentus			•		
Harrya chromipes	•	•	•	•	•
Heimioporus ivoryi			•	•	
Hemileccinum rubropunctum	•		•	•	
Hortiboletus campestris			•	•	
Hortiboletus rubellus			•	•	
Imleria badia	•		•	•	
Leccinellum albellum			•	•	
L. griseum	•	•	•	•	
L. quercophilum	•	•			
L. rugosiceps	•		•		
Leccinum aeneum	•	•			
L. aurantiacum	•	•			

Table 2 contd. ...

...Table 2 contd.

	SMW-MEX	NW-USA	SMEi-MEX	NE-USA	NE-CAN
L. insigne	•	•	•	•	•
L. manzanitae	•	•			
L. versipelle	•	•			
L. vulpinum	•	•			
Neoboletus luridiformis	•	•	•	•	
Phylloporus leucomycelinus	•	•	•	•	•
P. rhodoxanthus	•	•	•	•	
P. cyaneotinctus	•	•	•	•	
P. porphyrosporus	•	•	•	•	•
Pulchroboletus rubricitrinus	•	•			
P. ravenelii	•	•	•	•	
Retiboletus griseus	•	•			•
R. ornatipes	•			•	•
Rubinoboletus caespitosus					
R. eastwoodiae	•	•			
R. pulcherrimus	•	•			
Strobilomyces confusus	•	•	•	•	
S. strobilaceus	•	•	•	•	•
S. dryophilus			•	•	
Suillellus amygdalinus	•	•			
S. luridus	•	•	•	•	
S. hypocarycinus			•	•	
S. subvelutipes	•	•	•	•	
Sutorius eximius	•	•	•	•	•
Tylopilus alboater	•	•	•	•	
T. balloui			•	•	
T. felleus	•	•	•	•	•
T. ferrugineus	•	•	•	•	
T. plumbeoviolaceus	•	•	•	•	
T. subcellulosus			•		
T. tabacinus			•	•	
T. williamsii			•	•	
Xanthoconium affine			•	•	
Xerocomellus porosporus			•	•	
X. truncatus	•	•	•	•	
Xerocomus illudens	•	•	•	•	
X. subtomentosus			•	•	

Table 2 contd. ...

...Table 2 contd.

	SMW-MEX	NW-USA	SMEi-MEX	NE-USA	NE-CAN
Family Boletinellaceae					
Phlebopus brasiliensis			•		
P. mexicanus			•		
P. portentosus			•	•	
Family Diplocystidiaceae					
Astraeus hygrometricus	•	•	•	•	
A. pteridis	•	•	•		
Family Gomphidiaceae					
Chroogomphus jamaicensis			•		
C. rutilus	•	•			•
C. vinicolor	•	•	•	•	•
Gomphidius glutinosus	•	•	•	•	•
G. subroseus	•	•			
Family Gyroporaceae					
Gyroporus castaneus	•	•	•	•	
G. subalbellus			•	•	
Family Paxillaceae					
Melanogaster variegatus					
Paragyrodon sphaerosporus	•				
Paxillus involutus	•	•	•	•	•
Family Rhizopogonaceae					
Rhizopogon luteolus	•	•			
R. occidentalis	•	•			
R. roseolus	•	•	•		
R. subcaerulescens	•	•			
Family Sclerodermataceae					
Pisolithus arhizus	•	•	•	•	
Scleroderma areolatum	•		•	•	
S. cepa	•	•	•	•	
S. citrinum			•	•	
S. polyrhizum	•	•			
S. texense			•	•	
S. verrucosum	•		•	•	
Family Suillaceae					
Suillus americanus			•	•	•
S. cothurnatus	•	•	•	•	
S. brevipes	•	•	•	•	

Table 2 contd. ...

...*Table 2 contd.*

	SMW-MEX	NW-USA	SMEi-MEX	NE-USA	NE-CAN
S. cavipes			•	•	
S. granulatus	•	•	•	•	•
S. lakei			•	•	•
S. luteus	•	•	•	•	
S. placidus			•	•	•
S. pseudobrevipes	•	•	•	•	•
S. spraguei	•	•			•
S. tomentosus	•	•	•	•	•
ORDER CANTHARELLALES					
Family Hydnaceae					
Cantharellus cibarius	•	•	•	•	•
C. cinnabarinus	•	•	•	•	•
C. ignicolor			•	•	
C. lateritius			•	•	
C. minor	•	•			•
Clavulina amethystina	•				
C. cinerea	•	•	•	•	
Craterellus cornucopioides	•	•	•	•	•
C. tubaeformis		•	•	•	•
Hydnum albidum			•	•	
H. repandum	•		•	•	
H. rufescens	•				
ORDER GOMPHALES					
Family Clavariadelphaceae					
Clavariadelphus truncatus	•	•	•		
C. pistillaris	•	•	•	•	
Family Gomphaceae					
Gautieria sp.			•	•	
Gomphus clavatus			•	•	•
Ramaria botrytis	•		•		
R. flava	•		•		
R. formosa	•	•	•	•	
R. rubricarnata			•		
R. subbotrytis	•	•	•		
Turbinellus floccosus	•		•		•

Table 2 contd. ...

...*Table 2 contd.*

	SMW-MEX	NW-USA	SMEi-MEX	NE-USA	NE-CAN
ORDER HYMENOCHAETALES					
Family Hymenochaetaceae					
Coltricia cinnamomea	•	•	•	•	•
C. focicola	•	•			
C. montagnei			•	•	•
C. perennis			•	•	•
ORDER RUSSULALES					
Family Albatrellaceae					
Albatrellopsis ellisii	•	•			
A. flettii	•	•			
Albatrellus confluens	•	•			
Family Russulaceae					
Lactarius argillaceifolius	•	•	•	•	
L. camphoratus			•	•	
L. chelidonium			•	•	
L. chrysorrheus			•	•	•
L. corrugis	•	•	•	•	
L. croceus	•	•			
L. deterrimus	•	•	•	•	•
L. indigo			•	•	•
L. lignyotus			•	•	
L. olympianus	•		•		
L. paradoxus			•	•	
L. rimosellus			•	•	
L. rufus	•	•	•	•	•
L. rubrilacteus	•	•			
L. salmoneus			•	•	
L. scrobiculatus	•	•	•	•	•
L. subpalustris	•	•			
L. torminosus	•		•	•	•
L. uvidus	•	•	•	•	
L. vellereus	•	•			
L. zonarius		•	•		
Lactifluus corrugis			•	•	
L. hygrophoroides					•
L. piperatus	•	•	•	•	•
L. volemus var. *volemus*	•	•	•	•	

Table 2 contd. ...

...*Table 2 contd.*

	SMW-MEX	NW-USA	SMEi-MEX	NE-USA	NE-CAN
Laeticutis cristata	•	•	•		
Multifurca furcata	•		•	•	
Russula acrifolia	•				
R. aeruginea	•	•	•		
R. albonigra	•	•	•	•	
R. americana	•	•			
R. atropurpurea			•	•	
R. brevipes	•	•	•	•	•
R. claroflava	•				
R. cyanoxantha	•	•	•	•	
R. decolorans			•	•	•
R. emetica	•	•	•	•	•
R. flavida	•	•	•	•	
R. foetens	•	•	•	•	
R. foetentula			•		
R. fragilis			•	•	
R. mariae	•		•		•
R. nana	•	•			
R. adusta	•		•		•
R. ochroleucoides			•	•	
R. olivacea	•		•	•	•
R. parvovirescens			•	•	
R. risigallina	•		•	•	
R. rosea	•		•		
R. sanguinea	•	•	•	•	•
R. silvicola	•				
R. subgraminicolor			•	•	
R. subalutacea	•	•			
R. virescens	•	•	•	•	
R. xerampelina	•	•	•	•	•
ORDER SEBACINALES					
Family Sebacinaceae					
Sebacina schweinitzii	•	•	•	•	
ORDER THELEPHORALES					
Family Bankeraceae					
Hydnellum scabrosum					
H. scrobiculatum			•		
Sarcodon imbricatus	•	•	•	•	

Table 2 contd. ...

...Table 2 contd.

	SMW-MEX	NW-USA	SMEi-MEX	NE-USA	NE-CAN
Family Thelephoraceae					
Polyozellus multiplex	•	•			
Phellodon niger	•		•	•	
Thelephora caryophyllea	•	•	•	•	•
T. palmata	•	•			•
T. terrestris	•	•	•	•	•
Tomentella sp.	•		•		
Total species, 350; Common species, 45	**248**	**185**	**253**	**165**	**82**

Note: The number of records covering the total regions is 350 species, of which 253 are present in the SME-MEX, 248 different species at the SMW-MEX, 185 species at NW USA, 165 in NE USA and 82 in the NE-CAN regions.

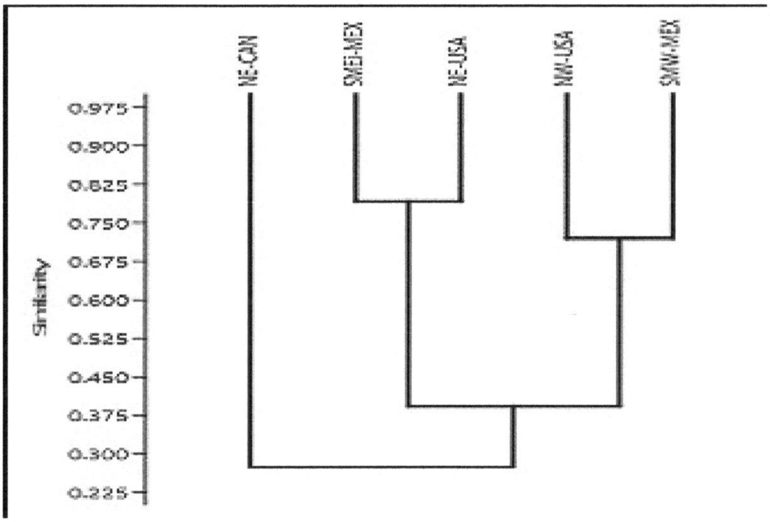

Fig. 5. Similarity analysis of ECM species.

related species, that is, present in both sites, being greater than 50% of the total species present throughout the area (n = 175). The Jaccard index ranges from 0 to 1 (0–100%). The lowest similarity values were for all NE-CAN-related areas (less than 30%) mainly because it is the area with the lowest species richness (Tables 3, 4).

6.1.2 Jaccard's Similarity Cluster

The cluster analysis indicates that two groups are formed between the study areas, the first and with greater similarity is formed by SME-MEX/NE-USA, while a second is formed by NW-USA/SMW-MEX (Fig. 5).

Table 3. Distribution of the species in different regions.

Region	Species richness	(%)
SMW-MEX	248	70.8
NW-USA	185	74
SME-MEX	253	72.2
NE-USA	165	65
NE-CAN	82	23.42
Total species richness (S)	350	100%

Table 4. Values above the diagonal indicate the percentage of similarity (%) according to the Jaccard's index (values below the diagonal indicate the number of species that share between pairs of sites).

	SMW-MEX	NW-USA	SME-MEX	NE-USA	NE-CAN
SMW-MEX		71.94	46.09	38.55	25.86
NW-USA	176		35.91	36.21	28.50
SME-MEX	151	114		79.15	26.42
NE-USA	122	99	196		28.76
NE-CAN	66	57	69	60	

7. Generalists with a Wide Distribution

A categorization of species was carried out based on their presence in the distribution areas. Generalist species that occur at the five study sites. A very frequent species is that species present in four sites of the five studied and a particular species is that present in a single record site. The results show that 52 species are reported at the five sites and can be considered generalists, i.e., with a wide distribution. Also, there are 51 very frequent species, that is, they occur in four sites and 28 are specific species, that is, they are only reported in a single site. Regarding the specific distribution species, 18 of them are reported only in the SMW-MEX area and 10 in SME-MEX.

8. Ecological Interactions of Mushrooms with Animals

Current knowledge on the diversity of macromycetes species from the north of Mexico is increasing due to the use of molecular methods that are helping to identify some species (Garza et al., 2022). Diversity of oaks and conifer species is high in both the SMW and SME and together with soil, climate and altitudinal variables they help to understand how the establishment of macromycetes species in different locations establish (Garza et al., 2019; Sridhar and Deshmukh, 2019). Oak diversity can reach more than 70 species and that of pines can be of more than 50 species (García, 2016). So far, results show that all oak and conifer species form ectomycorrhizas with a high diversity of macromycetes and they are widely dispersed in the geographic panorama in the north of Mexico (Pérez Silva and Acosta, 1986; Garcia and Garza, 2001; Quiñonez et al., 2008, 2014, 2015; García et al., 2019). The distances from one locality to another are huge and the phenology of fruiting body production varies from one place to another and from one Sierra to another, many fungal species have been found either in the Pacific or Atlantic coast of the USA, Canada and Mexico. Others have not been studied so far and molecular studies are required to carry out identification studies. Molecular studies are solving many taxonomic problems and help the understanding of fungal distribution in some geographical regions (de la Fuente et al., 2020; Garza et al., 2022). However, even when correct identification has been carried out still many other data are required to understand a bit more about how they are surviving in the different locations (Maser et al., 2008). Thus, it is well known that animal dispersion occurs and many species are involved in these procedures (Maser et al., 2008). Thus, chipmunks, squirrels, rats, and mice are actively involved in eating mushrooms, deer, peccary, black bear, and other big game species eat mushrooms or depredate mushrooms eater animal species! and, in this way, they also might contribute to dispersing spores in the forests. Thus, close relationships between animal spore dispersers and ectomycorrhizal fungi may explain to a certain extent the isolation that some species (e.g., *Pinus monophylla* in Baja California). If the animal dispersers are varied and different species occur in each of the big Sierras and their locations in the North of Mexico, this information could be very helpful to try to understand a little bit more how important these ecological relationships are for the functioning of forest ecosystems (Elliot et al., 2022). This may also explain to a certain extent how many ectomycorrhizal fungal species are associated with *Pinus cembroides* throughout its distribution in the north and central parts of Mexico as well as in some locations from Arizona. This condition also happens with many other species of ectomycorrhizal fungi that form associations with other conifers (e.g., *Pinus arizonica* or oaks *Quercus fusiformis*) in eastern and western north America and they also grow in the North of Mexico in both Sierras. Perhaps future studies of ectomycorrhizal fungi could elaborate a little more on ecological interactions of these fungal associations (e.g., mushroom dispersion by animals), local climate and soil conditions, altitudinal ranges, geologic substrates, phenology of fruiting bodies and their associated host's species (Elliot et al., 2022). The latter could help to generate more information on the multiple relationships occurring between ectomycorrhizal fungi species in the forests. Both small and big animals (e.g., insects, mites, earthworms, slugs, snails, nematodes, spiders, frogs and toads, birds, snakes, deer, peccary, bear, rats, mice, squirrels, chipmunks and other species) play important roles either individually or together in the nourishment, recycling, and dispersion of ectomycorrhizal fungal species in the forests (Figs. 6–10) (Maser et al., 2008; Elliot et al., 2022). Mushroom dispersion is an active process, and it has been like that for millions of years up to the present day. Thus, studies considering a more complete analysis could show a different panorama as the mutualistic symbiosis is in fact part of multiple organisms' symbioses based on nutrition and occurring in the forest ecosystems of the world. There are several rare and edible species of mushrooms in Mexico (Figs. 11, 12). However, animal grazing, tree felling and forest fire have severe threat to the biodiversity of mushrooms in Mexico (Fig. 13).

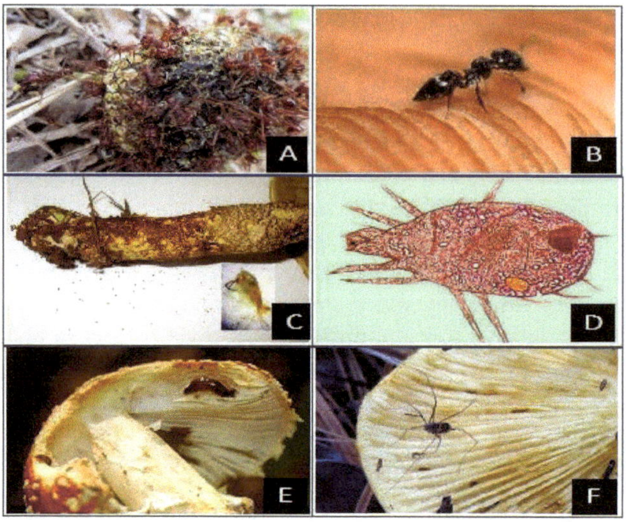

Fig. 6. Interactions of animals and ectomycorrhizal fungi at both Sierras: *Pisolithus arhizus* and *Atta Mexicana* (A); *Agaricus campestris* gills with *Crematogaster* sp. Ant (B); *Leccinum insigne* with springtails *Lepidocyrtus* sp. (C); Mite (D), *Amanita muscaria* var. *flavivolvata* and *Milax gigates* slugs (E); *Lactarius* sp. with Staphilinidae insects in the gills and the spider *Leiobunum* sp. (F).

Fig. 7. Interactions of animals and ectomycorrhizal fungi at both Sierras: Hammer flat worm *Bipallium adventium* on *Hypomyces lateritius* (A); Bluish slug *Milax gagates* on *Ramaria flava* (B); *Helix* sp. feeds on many mushroom species (C); *Rumina decollata* eating *Fuligo septica* (D); Insect feeding of *Lentinus arcularius* (E); *Rhinella marina* toad (F).

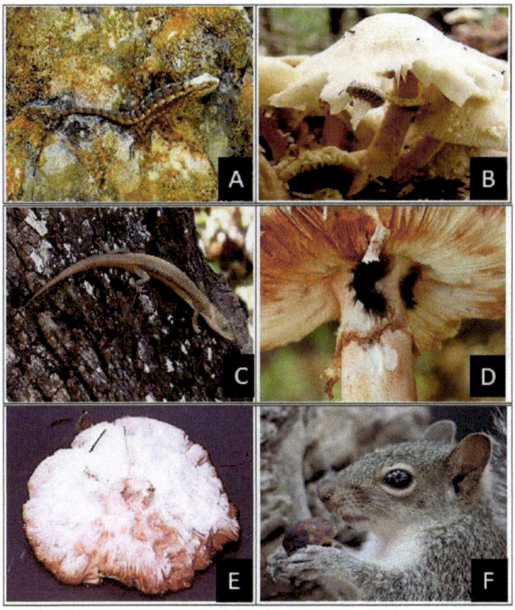

Fig. 8. Interactions of animals and ectomycorrhizal fungi at both Sierras: *Sceloporus olivaceus lizard* (A); *Armadillidum vulgare* (B); *Forest caiman lizard Gerrhonotus ophiurus* (C); *Wooly bear carterpillar Pyrrhactia isabella* feeding on *Amanita novinupta* (D); *Amanita rubescens* gills eaten by rodents (E); Forest grey squirrel *Sciurus caroliniensis* (F).

Fig. 9. Interactions of animals and ectomycorrhizal fungi at both Sierras: Mouse *Peromyscus maniculatus* (A); *Russula* sp., with Staphilinidae insects (B); *Lynx rufus* (C); Coyote *Cannis latrans* (D); *Accipiter cooperi* hawk (E); Forest blue bird *Apelocoma wollweberi* (F).

Fig. 10. Interactions of animals and ectomycorrhizal fungi at both Sierras: *Cyanocorax yncas* Chara (A); *Psilorhinus morio* chara papán (B); Wild pig *Pecari tajacu* (C); White tailed deer *Odocoileus virginianus* (D); North American *Bison bison* (E); Black bear *Ursus americanus* (F).

Fig. 11. Rare species found at both Sierras (SMW and SME) in the north of México: *Sclerogaster* sp. (A); *Rhizopogon* sp. (B); *Leccinum aeneus* (C); *Xerocomus* sp. (D); *Xerocomellus carmenae* a new species recently described (E) (Garza et al., 2022); *Paxillus* sp. (F).

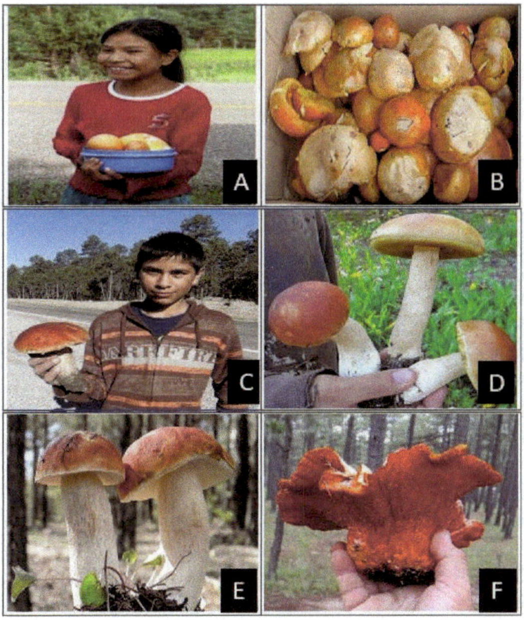

Fig. 12. Edible species and people at the Sierra Madre Occidental (SMW): Girl selling *Amanita cochiseana* by the road (A); *A. cochiseana* (B); Boy showing *Boletus rubriceps* (C); Different *Boletus* species of the edulis complex collected for food at SMW (E); *Boletus rubriceps*; *Hypomyces lactifluroum* (F) (Quiñones et al., 2014).

Fig. 13. Some of the activities that threat biodiversity in forests: Cows introduced at forests (A); *Lactarius deliciosus* (B); Lumber activities (C); *Leccinum versipele* (D); Forest fire (E); *Boletinellus merulioides* (F).

9. Discussion

Current knowledge on the diversity of ectomycorrhizal fungi from the north of Mexico is increasing and molecular methods will help to solve some taxonomic problems with some species (Garza et al., 2022). Diversity of oaks and conifer species is high in both the SMW and SME and together with soil, climate, and altitudinal variables they help to understand how the establishment of macromycetes species in different locations is going on (Garza et al., 2019; Sridhar and Deshmukh, 2019). Oak diversity can reach more than 70 species and that of pines can be of more than 50 species (García, 2016; Frank et al., 2007; Olimpia et al., 2017). So far results show that all oak and conifer species form ectomycorrhizas with a high diversity of macromycetes and they are widely dispersed in the geographic panorama in the north of Mexico (García and Garza, 2001; García et al., 2014; de la Fuente et al., 2020). The distances from one locality to another are huge and the phenology of fruiting bodies production varies from one place to another and from one Sierra to another, many fungal species have been found either in the pacific or Atlantic coast of the USA, Canada, and Mexico (Barron, 1999; Cripps et al., 2016; McKenny and Stuntz, 1987). Others have not been studied so far (e.g., truffles) and molecular studies are required to carry out identification studies. Phylogenetic studies are solving many taxonomic problems and help the understanding of fungal distribution in some geographical regions (de la Fuente et al., 2020). However, even when correct identification has been carried out still many other data (e.g., mycophagy) are required to understand a bit more about how they are surviving in different locations (Maser et al., 2008; Elliot et al., 2018, 2022). Thus, it is well known that animal dispersion (e.g., mycophagy) occurs and many species are involved in these interesting ecological interactions. Thus, chipmunks, squirrels, rats, and mice are actively involved in eating mushrooms, deer, peccary, bear, and other big game species eat mushrooms or eat mushrooms eater species! and, in this way, they also contribute to dispersing spores in the forests (Maser et al., 2022). Thus, close relationships between animal spore dispersers and ectomycorrhizal fungi may explain to a certain extent the survival in an almost isolated condition that some species might have (e.g., *Pinus monophylla*) in Baja California. If the animal dispersers are varied and different species occur in each of the big Sierras and their locations in the North of Mexico, this information could be very helpful to try to understand a little bit more how important these ecological relationships are for the functioning of forest ecosystems (Elliot et al., 2022). This may also explain to a certain extent how, many ectomycorrhizal fungal species are associated with *Pinus cembroides* throughout its distribution in the north and central parts of Mexico as well as in some locations from Arizona. This condition also happens with many other species of ectomycorrhizal fungi that form associations with other conifers (e.g., *Pinus arizonica* or oaks *Quercus fusiformis*) in eastern and western North America and they also grow in the north of Mexico in both Sierras. Perhaps future studies of ectomycorrhizal fungi could elaborate a little more on ecological interactions of these fungal associations (e.g., mushrooms dispersion by animal species), local climate and soil conditions, altitudinal ranges, geologic substrates, phenology of fruiting bodies and their associated hosts' species. The latter could help to generate more information on the multiple relationships and interactions occurring between ectomycorrhizal fungi species in the forests (Elliot et al., 2022). Both small and big animals (e.g., insects, mites, earthworms, slugs, snails, nematodes, spiders, birds, snakes, deer, peccary, bear, rats, mice, squirrels, and chipmunks) and several species play important roles either individually or together in the nourishment, recycling, and dispersion of ectomycorrhizal fungal species in the forests (Maser et al., 2008; Pacheco et al., 2014; Santos and Gatica, 2014; Schigel, 2012). Mushroom dispersion is an active process, and it has been like that for millions of years up to the present day. Thus, studies considering a more complete analysis could show a different panorama as the mutualistic symbiosis is in fact part of multiple symbioses occurring in the forest ecosystems of the world. The loss of ectomycorrhizal diversity in the forests can be translated into a degenerated ecosystem that has very different from the original condition and its functionality will be affected and, this condition will bring new challenges for all interacting species (e.g., from the fruit body to the squirrel and up to the wolf) to stay alive.

Conclusion

Temperate forests in the north of Mexico have a great diversity of ectomycorrhizal associations, pines, oak, and other hosts are abundant and many species of animals are nutritionally linked and associated through the mycophagia interactions during the rainy season. These associations are generally overlooked but they play important roles in the distribution of species in the forests. Soil, hosts, climate conditions, altitude and slope also play an important part in the distribution of the different species. Human-related activities (e.g., forest fires, illegal logging, cattle ranchers and forest fragmentation) together with climate change are threatening the temperate forests and all their associated species. Some species are common in both Sierras as well as in the south of the USA and Canada and they might have shared a common origin in the past.

Acknowledgements

The authors would like to thank Jonas Delgadillo for the photographs used in Fig. 8(A, C, D and F) and Fig. 10(C and E); We also thank Mariela González for the photographs used in Fig. 8(F); We are grateful to Mario Alberto García Aranda for the photograph used in Fig. 10(D).

References

Arora, D. (1986). Mushrooms Demystified. Ten Speed Press Berkeley. Second Edition, p. 936.

Baroni, T. (2017). Mushrooms of the Northeastern United States and Eastern Canada. Timber Press Field Guide, p. 600.

Barron, G. (1999). Mushrooms of Ontario and Eastern Canada. Long Pine Field Guide, p. 330.

Bessette, A.E., Bessette, A.R. and Fischer, D.W. (1997). Mushrooms of Eastern North America. Syracuse University Press, p. 563.

Bessette, A.E., Roody, W.C. and Bessette, A.R. (2016). Boletes of Eastern North America. Syracuse University Press, p. 471.

Bessette, A.E., Bessette, A.R. and Lewis, D. (2019). Mushrooms of the Gulf Coast States. A Field Guide to Texas Louisiana, Alabama, and Florida. University of Texas Press, p. 601.
Beug, M.W., Bessette, A.E. and Bessette, A.R. (2014). Ascomycete Fungi of North America. A Mushroom Reference Guide, University of Texas Press, p. 477.
Binder, M. and Bresinsky, A. (2002). Derivation of a polymorphic lineage of gasteromycetes from boletoid ancestors. Mycologia, 94: 85–98.
Binder, M. and Hibbett, D.S. (2006). Molecular systematics and biological diversification of Boletales. Mycologia, 98: 971–981.
Cripps, C.L. Evenson, V.S. and Kuo, M. (2016). The Essential Guide to Rocky Mountain Mushrooms by Habitat. Board of Trustees, University of Illinois, p. 259.
de la Fuente, J.I., Garcia, J., Lopez, C.Y., Oros, I., Vela, R.Y. et al. (2020). An annotated checklist of the macrofungi (Ascomycota, Basidiomycota, and Glomeromycota) from Quintana Roo, Mexico. Check List, 16(3): 627–648. https://doi.org/10.15560/16.3.627.
Elliot, T.F. and Stephenson, S.L. (2018). Mushrooms of the Southwest. Timber Press Field Guide, p. 390.
Elliott, T.F., Truong, C., Jackson, S., Zuniga, C.L., Trappe, J.M. and Vernes, K. (2022). Mammalian mycophagy: A global review of ecosystem interactions between mammals and fungi. Fungal Syst. Evol., 9: 99–159. doi: 10.3114/fuse.2022.09.07.
Flores, E., Carrillo, A., Wehenkel, C., Garza, F. and Hernández, J.C. (2018). Diversidad de macromicetos en bosques de pino en el municipio Madera. Chihuahua Revista Mexicana de Ciencias Forestales, 9(50): 342–360. https://doi.org/10.29298/rmcf.v9i50.240.
Frank, J., Barry, S., Madden, J. and Southwork, D. (2007). Oaks belowground: Mycorrhizas, truffles, and small mammals. USDA, Forest Service, Tech. Rep. PSW-GTR, p. 217.
García, G., Valenzuela, R., Raymundo, T., García, L., Guevara, L. et al. (2014). Macrohongos asociados a encinares (*Quercus* spp.) en algunas localidades del estado de Tamaulipas, México. Capitulo En Biodiversidad Tamaulipeca 2, SEP., Instituto Tecnológico de Ciudad Victoria, Tam, 5(1): 103–140.
García, J. and Garza, F. (2001). Conocimiento de los hongos de la Familia Boletaceae de México. Revista Ciencia UANL Volume IV, # 3 Julio–Septiembre 2001: 336–343.
García, L.J. (2016). Taxonomy and phytogeography of the Fagaceae family (Magnoliophyta: Fagales) in Tamaulipas and Nuevo Leon, Mexico. Tesis Doctorado. Universidad Autónoma de Nuevo León, Facultad de Ciencias Forestales, p. 210.
Garza, F. (1986). Hongos ectomicorrícicos en el estado de Nuevo León. Rev. Mex. Mic., 2: 197–205.
Garza, F., García, J. and Castillo, J. (1985). Macromicetos asociados al bosque de *Quercus rysophylla* en algunas localidades del centro del estado de Nuevo León. Rev. Mex. Mic., 1: 423–437.
Garza, F., García, J., Estrada, E. and Villalón, H. (2002). Macromicetos, ectomicorrizas y cultivos de *Pinus culminicola* en Nuevo León. Revista Ciencia UANL Volume V, # 2, Abril-Junio, 204–210.
Garza, F., García, J., Guevara, G., Martínez, C.R., Ayala, O. and de la Fuente, J.I. (2022). *Xerocomellus carmeniae* (Boletales, Basidiomycota), a new fungus from northeastern Mexico. Acta Bot. Mex., 129: e2039. https://doi.org/10.21829/ abm129.2022.2039.
Garza, F., Quiñónez, M., Garza, L., Carrillo, A., Villalón, H. et al. (2019). Some edible, toxic and medicinal mushrooms from temperate forests in the north of Mexico. pp.144–198. *In*: Sridhar, K.R. and Deshmukh, S.K. (eds.). Advances in Macrofungi. Diversity, Ecology, and Biotechnology. CRS Press, Boca Raton, USA.
González, S., González, M., Tena, J.A., Raucho, I. and López, I. (2012). Vegetación de la Sierra Madre Occidental, México: una síntesis. Acta Bot. Mex., 100: 351–403.
Guevara, G., Cázares, E., Bonito, G., Healy, R., Stielow, B. et al. (2014). Hongos hipogeos de Tamaulipas, México. Capitulo 4. En Biodiversidad Tamaulipeca 2, SEP, Instituto Tecnológico de Ciudad Victoria, Tam., 1: 87–101.
Hall, I., De la Mare, P., Bosselman G., Persley, C. and Wang, Y. (2019). Commercial inoculation of *Pseudotsuga* with an ectomycorrhizal fungus and its consequences. pp. 305–322. *In*: Sridhar, K.R. and Deshmukh, S.K. (eds.). Advances in Macrofungi. Diversity, Ecology, and Biotechnology. CRS Press, Boca Raton, USA.
Halling, R.E., Osmundson, T.W. and Neves, M.A. (2008). Pacific Boletes: Implications for biogeographical relationships. Mycol. Res., 112: 437–447.

Heads, S., Miller, A., Crane, J., Thomas, M., Ruffatto, D. et al. (2017). The oldest fossil mushroom. Plos One, 12: e0178327.

Laferriere, E.J. and Gilbertson, R.L. (1992). Fungi of Nabogame, Chihuahua, Mexico. Mycotaxon, 44: 73–87.

Li, Y.-N. and Yang, Z.L. (2021). The Boletes of China: *Tylopilus* s.l. Science Press, Springer Nature Singapore Pte Ltd., p. 415. https://doi.org/10.1007/978-981-16-2986-0.

Maser, C., Claridge, A. and Trappe, J.M. (2008). Trees Truflles and Beasts: How Forest Function. Rutgers University Press, New Brunswick, New Jersey, and London, p. 267.

McKenny, M. and Stuntz, D. (1987). The New Savory Wild Mushrooms. The University of Washington Press Seattle and London p. 242.

Mirov, P. (1967). The Genus *Pinus*. The Ronal Press Co. New York, p. 602.

North, M.P. (2002). Seasonality and abundance of truffles from Oak woodlands to red Fir Forests. USDA Forest Service Tech. Rep. PSW-GTR-183.

Nuhn, M.E., Binder, M., Taylor, A.F.S., Halling, R.E. and Hibbett, D., (2013). Phylogenetic overview of the Boletineae. Fungal Biol., 117: 479–511.

Olimpia, M., García, O., Garibay, R., Hernández, E., Arellano, E. and Oyama, K. (2017). Word-wide meta-analysis of *Quercus* forests ectomycorrhizal fungal diversity reveals southwestern Mexico as a hotspot. Mycorrhiza, 27: 811–822.

Pacheco, J., List, R. and Ceballos, G. (2014). Mamíferos. La biodiversidad en Chihuahua: Estudio de estado, CONABIO, México, pp. 320–325.

Pérez-Silva, E. and Aguirre-Acosta, E. (1986). Flora Micológica del estado de Chihuahua, México I. Inst. Biol., 57: 17–32.

Quiñonez, M. and Garza, F. (2015). Hongos silvestres comestibles de la Sierra Tarahumara de Chihuahua, Mexico. Universidad Autónoma de Chihuahua, Mexico, p. 238.

Quiñonez, M., Garza, F., Mendoza, J.R., García, J., Saenz, J. and Bolaños, H.R. (1999). Guía de Hongos de la región de Bosque Modelo. Universidad Autónoma de Chihuahua, Facultad de Zootecnia, Gobierno del Estado de Chihuahua, Mexico, p. 118.

Quiñónez, M., Garza, F., Sosa, M., Lebgue, T., Lavin, P. and Bernal, S. (2008). Índices de diversidad y similitud de hongos ectomicorrizógenos en bosques de Bocoyna, Chihuahua, México. Revista Ciencia Forestal en México, 33(103): 59–78.

Quiñónez, M., Ruan, F., Aguilar, I.E., Garza, F., Toutcha, T. et al. (2014). Knowledge and use of edible mushrooms in two municipalities of the Sierra Tarahumara, Chihuahua, México. J. Ethnobiol. Ethnomed., 10: 67. 10.1186/1746-4269-10-67.

Retallack, G. and Landing, E. (2014). Affinities and architecture of Devonian trunks of *Prototaxites loganii*. Mycologia, 106: 1143–1158.

Reyes, V.M. and Nuñez López, D. (2014). Geografía y fisiografía. La biodiversidad en Chihuahua: Estudio de estado, CONABIO, México, pp. 23–25.

Rilling, M.C. and Allen M.S. (2002). Global change and mycorrhizal fungi Ecological Studies. pp. 157. *In*: van der Heijden, M.G.A. and Sanders, I. (eds.). Mycorrhizal Ecology, Springer-Verlag, Berlin and Heidelberg.

Salinas, M.M., Estrada, A.E. and Villarreal, J.A. (2017). Endemic vascular plants of the Sierra Madre Oriental, México. Phytotaxa, 328: 1–52.

Sánchez, N., Soria, I., Romero, L., López, M., Rico, R. and Portillo, A. (2015). Los hongos Agaricales de las áreas de encino del estado de Baja California, México. Estudio de Diversidad Volumen I. University of Nebraska, Lincoln, pp. 215–226.

Santos Barrera, G. and Gatica Colima, A. (2014). Anfibios y reptiles: Pastizal. La biodiversidad en Chihuahua: Estudio de estado, CONABIO, México, pp. 300–306.

Schigel, D.S. (2012). Fungivory and host associations of Coleoptera: A bibliography and review of research approaches. Mycology, 3(4): 258–272.

Siegel, N. and Schwarz, C. (2017). Mushrooms of the Red Coast a Comprehensive Guide of the Fungi of Coastal Northern California. Ten Speed Press, Berkeley, p. 588.

Sridhar, K.R. and Deshmukh, S.K. (2019). Advances in Macrofungi. Diversity, Ecology, and Biotechnology. CRS Press, Boca Raton, USA, p. 365.

Villarreal Ruiz, L. and Nery Luna, C. (2019). Biodiversity and ecology of Boreal Pine woodlands: ectomycorrhizal fungi in the Anthropocene. pp. 40–69. *In*: Sridhar, K.R. and Deshmukh, S.K. (eds.). Advances in Macrofungi. Diversity, Ecology, and Biotechnology. CRS Press, Boca Raton, USA.
Wu, G., Feng, B., Xu, J., Zhu, X.T., Li, Y.C. et al. (2014). Molecular phylogenetic analyses redefine seven major clades and reveal 22 new generic clades in the fungal family Boletaceae. Fungal Divers., 69: 93–115.
Wu, G., Li, Y.C. and Zhu, X.T. (2016). One hundred noteworthy boletes from China. Fungal Divers., 81: 25–188.

7

On the *Amanita* in Southwest India

Kandikere R Sridhar,[1,*] *Shivannegowda Mahadevakumar*[2,3] and *Namera C Karun*[4]

1. Introduction

The cosmopolitan Amanitaceae is an important agaric family that encompass morphologically, ecologically and economically valued *Amanita* species (Yang, 1997, 2015). The global estimation of *Amanita* is up to 1000 species with acceptance of about 700 accepted species (Bas, 2000; Yang, 2000; Tulloss, 2005; Cui et al., 2018). Among these, only half of them are described with 100 as poisonous species and 50 as edible species. For the last two decades new or cryptic species of *Amanita* brought to the limelight (Yang, 2011; Kumar et al., 2021). *Amanita* has a wider perspective on ecosystem functioning owing to their broad perspectives on ectomycorrhizal, edible, medicinal and poisonous features. The global record reveals up to 613 ectomycorrhizal *Amanita* species known to colonize more than 10 plant families (Cui et al., 2018; Tulloss and Yang, 2020). Based on morphology and molecular biology the genus *Amanita* has been divided into three subgenera (*Amanita*, *Amanitina* and *Lepidella*) comprising eleven sections (e.g., *Amanita*, *Amarrendiae*, *Amidella*, *Arenariae*, *Caesareae*, *Lepidella*, *Phalloideae*, *Roanokensis*,

[1] Department of Biosciences, Mangalore University, Mangalagangotri, Mangalore, Karnataka, India.
[2] Forest Pathology Department, Division of Forest Protection, KSCSTE - Kerala Forest Research Institute, Peechi, Thrissur, Kerala, India.
[3] Botanical Survey of India, Andaman and Nicobar Regional Centre, Haddo 744102, Port Blair, Andaman, India.
[4] Western Ghats Macrofungal Research Foundation, Bittangala, Virajpet, Kodagu, Karnataka, India.
* Corresponding author: kandikere@gmail.com

Strobiliformes, *Vaginatae* and *Validae*) (Cui et al., 2018). The poisonous species have been included in the section *Phalloideae*, while edible species in *Caesareae*. Owing to a lack of fossil records, the origin of the genus *Amanita* is still not recognized. The lethal amanitas (*Phalloideae*) seem to be originated during Paleocene and dispersed into Eurasia and North America via Beeringian Land Bridge with an extension to Central America in Oligocene to Miocene period (Cai et al., 2014). The edible aminatas (*Caesareae*) have been predicted origin during Palaeocene and Eocene in the Pantropical regions (Sánchez-Ramírez et al., 2015). The application of molecular markers revealed several cryptic new species in *Amanita*, which clarified the origin of *Amanita* up to a greater extent (Yang, 2011; Cui et al., 2018).

Forests in tropical regions consist of ecologically unique ectomycorrhizal fungal associations with a variety of tree species (Carrales et al., 2018, 2022). Studies on the Indian Amanitaceae were initiated by Butler and Bisby (1931). Recently, from the Indian subcontinent Kumar et al. (2021) validated up to 66 species of *Amanita*. Most of the contributions to *Amanita* come from the temperate coniferous and broad-leaved forests of the northwestern Himalayas of the Indian subcontinent (Kumar et al., 2021). However, the Western Ghats and southwest India provided suitable ecological and climatic conditions to support a variety of *Amanita* species with several host tree species. About 30 species of *Amanita* are described from different locations in southwest India and many of them are associated with tree species such as ectomycorrhizas. There is ample scope for describing many more species of *Amanita* from the deep forest forays of the Western Ghats through morphological and molecular tools. *Amanita* species have been described and assessed from the states of Maharashtra, Karnataka, Kerala and Tamil Nadu. This chapter intends to consolidate the ecology and host tree species of *Amanita* occurring in the southwestern part of India with emphasis on the edible, ectomycorrhizal and bioactive potential of *Amanita konajensis*.

2. *Amanita* in Southwestern India

Amanita is known for up to 30 species from different ecosystems of the Western Ghats and southwest coastal region of India (Maharashtra, Karnataka, Kerala and Tamil Nadu) (Table 1). Kerala state is represented by as many as 18 species followed by Karnataka (11 spp.), Maharashtra (4 spp.) and Tamil Nadu (1 spp.). Among them, *Amanita farinosa*, *A. muscaria* and *A. pantherina* have wide global distribution, while currently *A. albofloccosa*, *A. bharatensis*, *A. konkanensis*, *A. konajensis* and *A. sampajensis* have narrow distribution in the Indian subcontinent. Among the *Amanita* in southwest India, three species are known for their edibility (*A. hemibapha*, *A. konajensis* and *A. vaginata*). Three species (*Amanita anguistilamellata*, *A. hemibapha* and *A. muscaria*) were repeatedly reported from southwest India, whereas 14 species (*A. ceciliae*, *A. crenulata*, *A. eriophora*, *A. farinosa*, *A. flavoflocossa*, *A. franzii*, *A. griseofusca*, *A. lignitincta*, *A. oberwinkleriana*, *A. ovalispora*, *A. pantherina*, *A. sampajensis*, *A. vaginata* and *A. verna*) were reported by one or a few studies.

Table 1. Revised list of *Amanita* species reported from southwest India (*, non-lethal; **, edible).

	Geographic location	Remarks	Distribution	Reference
Amanita alblofloccosa A.V. Sathe & S.D. Deshp.	Maharashtra	From Pune	India	Sathe et al. (1980) and Kumar et al. (1990)
Amanita angustilamellata (Höhn.) Boedijn	Karnataka	Scrub jungles; poisonous	India, Sri Lanka, China, Indonesia, Malaysia and Singapore	Vrinda et al. (1997); Pradeep and Vrinda (2007, 2010); Mohanan (2011); Karun and Sridhar (2014); Greeshma et al. (2016); Pavithra et al. (2016) and Bijeesh et al. (2022)
	Kerala	Associated with trees: *Hopea parviflora, Hopea ponga, Myristica malabarica, Terminalia paniculata, Terminalia bellarica* and *Vateria indica*; poisonous		
Amanita aureofloccosa Bas	Kerala	Associated with trees: *Acacia auriculiformis, Hopea ponga* and *Vateria indica*	India, Cango, Democratic Republic and South America	Pradeep and Vrinda (2010); Mohanan (2011) and Karun and Sridhar (2014)
**Amanita ballerina* Raspe, Thongbai & K.D. Hyde	Karnataka	Moist deciduous forest; non-lethal	India and Thailand	Kantharaja and Krishnappa (2022)
Amanita bharatensis Sathe & J.T. Daniel	Kerala	From Munnar	India	Sathe and Daniel (1980) and Florence (2004)
Amanita bisporigera G.F. Atk.	Karnataka	Moist deciduous forest; poisonous	India, Canada, Mexico, Costa Rica, Mexico, United States and South America	Mohanan (2011), Bijeesh et al. (2022) and Kantharaja and Krishnappa (2022)
	Kerala	Associated with trees: *Diospyros malabarica, Hopea parviflora* and *Hopea ponga*; poisonous		
Amanita ceciliae (Berk. & Broome) Bas	Kerala	Semi-evergreen and wet evergreen forests	India, Europe and England	Mohanan (2011)
Amanita crenulata Peck	Kerala	Forests; poisonous	India, Canada and United States	Bijeesh et al. (2022)

Table 1 contd. ...

...Table 1 contd.

	Geographic location	**Remarks**	**Distribution**	**Reference**
Amanita elata (Massee) Corner & Bas	Kerala	Associated with trees: *Calophyllum calaba, Hopea parviflora, Hopea ponga, Tectona grandis, Terminalia paniculata* and *Vateria indica*	India, Sri Lanka, China, Malaysia and Singapore	Pradeep and Vrinda (2010) and Mohanan (2011)
Amanita eriophora (Bark.) E.J. Gilbert	Karnataka	Moist deciduous forest	India, Singapore and Cambodia	Kantharaja and Krishnappa (2022)
Amanita farinosa Schwein	Kerala	Associated with trees: *Canophyllum inophyllum* and *Terminalia paniculata*; Poisonous	India, China, Korea, Tibet, Portugal, Russia, Canada, United States, Costa Rica, Mexico, United States and South America	Bijeesh et al. (2022)
Amanita flavofloccosa Nagas. & Hongo	Maharashtra	Associated with trees: *Dalbergia* sp. and *Gliricidia* sp.	India and Japan	Senthilarasu (2014)
**Amanita franzii* Zhu L. Yang, Y.Y. Cui & Q. Cai	Karnataka	Moist deciduous forest; non-lethal	India and China	Kantharaja and Krishnappa (2022)
Amanita griseofarinosa Hongo	Kerala	Associated with trees: *Garcinia morella, Hopea parviflora, Hopea racophloea* and *Vateria indica*; poisonous	India, China, Korea and Japan	Pradeep and Vrinda (2010) and Mohanan (2011)
Amanita griseofusca J. Khan & M. Kiran	Karnataka	Semi-evergreen forest	India and Pakistan,	Kantharaja and Krishnappa (2022)
***Amanita hemibapha* (Berk. & Broome) Sacc.	Karnataka	Associated with tree: *Vateria indica*; edible	India, China, South Korea, Japan, Pakistan, Thailand, Sri Lanka and United States	Natarajan et al (2005a); Vrinda et al. (2005); Natarajan et al. (2005a, b); Pradeep and Vrinda (2007, 2010); Mohanan (2011) and Ravikrishnan et al. (2017)
	Kerala	Associated with trees: *Diospyros malabarica, Hopea parviflora, Myristica fragrans* and *Vateria indica*; edible		

***Amanita konajensis* K.R. Sridhr, Mahadevak., B.R. Nutan & N.C. Karun	Karnataka	Associated with trees: *Acacia auriculiformis, Acacia mangium, Anacardium occidentale, Hopea ponga* and *Terminalia paniculata*; edible at tender stage	India	Karun and Sridhar (2014) and Ghate and Sridhar (2016)
Amanita konkanensis A.V. Sathe & M.S. Kulk.	Maharashtra	From Sawantwadi	India	Kulkarni (1992) and Bhatt et al. (2003)
Amanita lignitincta Zhu L. Yang ex Y.Y. Cui, Q. Cai & Zhu L. Yang	Karnataka	Semi-evergreen forest	India and China	Kantharaja and Krishnappa (2022)
Amanita magniverrucata Thiers & Ammirati	Kerala	Associated with trees: *Hopea parviflora, Mesua ferrea* and *Xanthophllus arnottianum*; poisonous	India, Mexico and United States	Pradeep and Vrinda (2010), Mohanan (2011) and Bijeesh et al. (2022)
Amanita muscaria (L.) Lam.	Kerala	Associated with tree: *Pinus roxburghii*; poisonous	India, Africa, China, Indonesia, Japan, Tibet, Australia, Europe, Russia, Hawaii, New Zealand, United States and South America	Natarajan (1977); Sasthe and Sasangan (1977); Natarajan and Kannan (1982); Kannan and Natarajan (1988); Pradeep and Vrinda (2007); Mohanan (2011) and Bijeesh et al. (2022)
	Tamil Nadu	Associated with tree: *Acacia mearnsii, Cupressus macrocarpa, Eucalyptus globulus* and *Pinus patula*; poisonous		
Amanita oberwinkleriana Zhu L. Yang & Yoshim.	Kerala	Associated with tree: *Hopea parviflora*; poisonous	India, China, Japan and Korea	Bijeesh et al. (2022)
Amanita ovalispora Boedijin	Karnataka	Semi-evergreen forest	India, China and Indonesia	Kantharaja and Krishnappa (2022)
Amanita pantherina (DC.) Krombh.	Kerala	Associated with trees: *Hopea parviflora* and *Xanthophyllum flavescens*; poisonous	India, Africa, Isreal, Japan, Nepal, Sumatra, Vietnam, Europe, Turkey, Canada, United States and Mexico	Bijeesh et al. (2022)

Table 1 contd. ...

...Table 1 contd.

	Geographic location	Remarks	Distribution	Reference
Amanita phelloides (Vaill. Ex Fr.) Link	Kerala	Associated with trees: *Acacia auriculiformis* and *Acacia mangium*; poisonous	India, Africa, Iran, Europe, Turkey, United States, South America, Australia and New Zealand	Vrinda et al. (2005b) and Bijeesh et al. (2022)
Amanita pseudoporphyria Hongo	Kerala	Associated with trees: *Diospyros malabarica, Hopea parviflora, Terminalia paniculata,* and *Vateria indica*; poisonous	India, Nepal, China, Japan, Korea and Thailand	Mohanan (2011) and Bijeesh et al. (2022)
Amanita sampajensis A.V. Sathe & S.M. Kulk	Maharashtra	Sampaje forests	India	Sathe and Kulkarni (1980)
***Amanita vaginata* (Bull.: Fr.) Lam.	Karnataka	Associated with tree: *Dipterocarpus indicus*; edible	India, Europe, Scotland, Iran, United States and Australia	Natarajan et al. (2005a)
Aminata verna (Bull.) Lam.	Kerala	Associated with tree: *Vateria indica*; poisonous	India, Europe, Korea, Vietnam, Thailand, Turkey and United States	Bijeesh et al. (2022)
Amanita volvata (Peck) Lloyd	Kerala	Associated with tree: *Hopea parviflora*; poisonous	India, Indonesia, Japan, Canada and United States	Pradeep and Vrinda (2010) and Bijeesh et al. (2022)

2.1 *Mycorrhizal Amanita*

Amanita are well known for their association with a wide variety of tree species over 10 families of plants (e.g., Caesalpiniaceae, Casuarinaceae, Dipterocarpaceae, Fagaceae, Myrtaceae, Nothofagaceae, Pinaceae and others) (Cui et al., 2018; Sridhar and Karun, 2019) indicates their importance in ecosystem services in forests. In southwest India also many *Amanita* species are associated with the trees belonging to the Dipterocarpaceae (Table 1). Trees belonging to the family Dipterocarpaceae are a major ectomycorrhizal family in Southeast Asia (Carrales et al., 2018). Mustaffa et al. (2012) have reported that 56% of dipterocarps are critically endangered in Malaysia. *Amanita angustilamellata, A. elata, A. hemibapha, A. muscaria* and *A. pseudoporphyria* have an association with a wide variety of tree species. Up to 27 tree species (18 genera) have been colonized by *Amanita* in southwest India. Among 27 tree species, a highest of 11 species of *Amanita* associated with *Hopea parviflora* followed by eight species with *Vateria indica*, and six species with *Hopea ponga*, three species each with *Acacia auriculiformis* and *Diospyros malabarica*, while the rest 22 tree species associated with one or two *Amanita* species. Among the tree species, the five tree species belonging to Dipterocarpaceae (*Dipterocarpus indicus, Hopea parviflora, H. ponga, H. racophloea* and *Vateria indica*) supported as many as 15 species of *Amanita*. In an earlier study, 10 species of *Amanita* were known to be associated with eight tree species belonging to Dipterocarpaceae (Sridhar and Karun, 2019). Some of the exotic tree species (e.g., *Acacia, Eucalyptus* and *Pinus*) were also associated with five species of *Amanita* in southwest India (range, 1–3 spp.). Similarly, 15 exotic tree species were associated with as many as 29 ectomycorrhizal fungi (including *Amanita*) in southwest India (range, 1–13 spp.) (Sridhar and Karun, 2019).

2.2 *Medicinal and Poisonous Amanita*

Amanita belongs to the section *Phalloideae* are usually poisonous and result in fatal worldwide. *Amanita* species reported from southwest India composed of deadly poisonous (*A. phalloides*), poisonous, non-toxic and edible species (see Table 1). However, some *Amanita* spp. belong to the section *Phalloideae* also known as non-lethal or lacking toxins such as amatoxins (α-amanitin and β-amanitin) and phallotoxins (e.g., *A. ballerina, A. chuformis, A. franzii, A. levistriata* and *A. pseduogemmata*) (Codjia et al., 2022). Among these, two species are known from southwest India (*A. ballerina* and *A. franzii*). However, many species that are known from southwest India are yet to be evaluated to follow their toxicity, edibility and medicinal value. Although many *Amanita* is poisonous or lethal, never the less they have medicinal value at a sub-lethal concentration (hormetic doses). Thus, there is a need to assess the poisonous *Amanita* to produce toxins and other metabolites that are of interest in medicine (Sevindik, 2020).

2.3 *Non-Toxic and Edible Amanita*

Many *Amanita* spp. are also non-lethal owing to a lack of toxins (amatoxins and phallotoxins) (e.g., *A. ballerina, A. chuformis, A. franzii, A. levistriata* and

A. pseduogemmata) (Codjia et al., 2022). Among the *Amanita* found in southwest India, three species are edible (*A. hemibapha, A. konajensis* and *A. vaginata*). It is worth investigating these *Amanita* spp. should be assessed for its nutritional and nutraceutical value.

3. Nutraceutical, Bioactive and Functional Potential

3.1 *Amanita hemibapha*

Edible *Amanita hemibapha* is associated with five tree species in southwest India (*Diospyros malabarica, Hopea parviflora, Myristica fragrans* and *Vateria indica*). The uncooked edible wild *A. hemibapha* isolated from the foothills of the Western Ghats is known for proteins, dietary fiber, carbohydrates and calorific value (Ravikrishnan et al., 2017). Major minerals present in uncooked *A. hemibapha* include calcium, potassium, magnesium, iron and phosphorus. It also possesses major quantities of essential amino acids like valine, lysine, leucine, isoleucine and threonine. Major quantities of palmitic, stearic, oleic and linoleic acids were present in uncooked *A. hemibapha.* In addition, it also possesses many essential fatty acids like linoleic, eicosadenoic, eicosapentaenoic and docosahexaenoic acids.

Uncooked *Amanita hemibapha* consists of moderate quantities of total phenolics, tannins, phytic acid, flavonoids, vitamin C, β-carotene and lycopene, while the trypsin inhibition activity was very low (Ravikrishnan et al., 2021a). It also consists of 31 active principles valuable in cancer prevention, health promotion, pharmaceutical significance, nutritional value, cosmetic value, environmental and industrial applications. *Amanita hemibapha* consists of several soluble sugars, organic acids (aqueous and methanol extracts) and polyphenols (methanol extract) Ravikrishnan et al., 2021b). It has considerable inhibitory activity against Gram-positive bacteria (*Bacillus cereus, B. subtilis, Staphylococcus aureus* and *Streptococcus pneumonia*), Gram-negative bacteria (*Enterobacter aerogenes, Excherichia coli, Haemophilus influenza, Proteus vulgaris* and *Pseudomonas aeruginosa*) and yeast (*Candida albicans*) (Ravikrishnan and Sridhar, 2023). *Amanita hemibapha* also showed a high level of ferrous ion-chelating capacity and total antioxidant activity, while a moderate level of DPPH radical-scavenging activity as well as reducing power.

3.2 *Amanita konajensis*

Amanita konajensis is one of the ethnic delicacies in the scrub jungles of southwest India consumed by tribals and native dwellers (Karun and Sridhar, 2016). Our group was constantly collecting details pertaining to this mushroom over a decade and collected details pertaining to its occurrence, edibility, and other aspects, yet the identity was not established. The identity of this *Amanita* species has recently been established based on detailed morphological and molecular sequence data as *A. konajensis* (Crous et al., 2022). Traditionally it is called in the vernacular language Kannada 'Motte-Anabe' (meaning, 'egg mushroom'). It commonly occurs in lateritic soil with pebbles. This mushroom will be collected and consumed at the egg or a bit advanced stage (dumb-bell shape) usually underneath *Acacia* and cashew

(*Anacardium occidentale*) trees. Uncooked and cooked tender sporocarps possess a significantly higher quantity of total lipids as well as calorific value, while the opposite for the crude protein (Greeshma et al., 2018a). No significant difference was seen in crude fiber and carbohydrates among uncooked and cooked samples. Potassium and iron contents were high in uncooked as well as cooked samples. The ratio Na/K was favourable (< 1) (combat the blood pressure), while the Ca/P ratio was not favourable (< 1) (unable to control the loss of calcium). Cooking increased many essential amino acids (e.g., isoleucine, methionine, cystine, phenylalanine, tyrosine, threonine and valine), while the in vitro protein digestibility was high in uncooked samples with high protein efficiency ratios qualified the nutrient benefits of this mushroom.

Several bioactive compounds were almost significantly higher in uncooked compared to cooked samples (contents of total phenolics, tannins, flavonoids, vitamin C, phytic acid, lycopene and β-carotene), while trypsin-inhibition activity was significantly lower and hemagglutinin activity was substantially decreased in cooked samples (Greeshma et al., 2018b). Antioxidant activities (total antioxidant, ferrous ion-chelation, reducing power and DPPH radical-scavenging and ABTS radical-scavenging activities) were also higher in cooked compared to uncooked samples. Improved bioactive components and bioactive potential qualify tender *A. konajensis* as suitable ethnic food to support the nutrition and health of local consumers. Tender cooked samples of *A. konajensis* consist of soluble sugars (glucose, fructose and trehalose), organic acids (tartaric acid, and malic acids) and polyphenols (gallic acid, myricetin, ethyl catechol and p-coumaric acid) reveals the possibilities to use in the production of health-promoting diets (Greeshma et al., 2021).

The protein solubility of tender *A. konajensis* at pH 2–8 was significantly higher in uncooked compared to cooked samples (Greeshma et al., 2018c). Cooking has not changed the least gelation concentration, while water-absorption as well as oil-absorption capacities, were higher in cooked than uncooked samples. The emulsion properties (emulsion activity and stability), and foam properties (foam capacity and foam stability) were significantly higher in cooked compared to uncooked samples. Assessment by the Principal Component Analysis revealed that the protein solubility, emulsion stability and foam capacity in uncooked samples were controlled by the crude protein, total lipids and crude fiber, while almost all the functional properties except for protein solubility were influenced by the crude fiber and carbohydrates. The high emulsion properties and foam capacity in cooked tender *A. konajensis* will be advantageous in the formulation of nutraceutical products.

4. *Amanita konajensis*

4.1 *Brief Description*

Developmental stages, mature fruit body and sections of tender and near mature fruit body of *A. konajensis* are given in Fig. 1. Developing basidia with sterigmata attached to basidiospores and released basidiospores of *A. konajensis* are given in Fig. 2. The immature sporocarp is oval to dumbbell-shaped. The mature fruit body measures 5.5–10.6 cm. The pileus was light grey to greyish-brown, hemispherical

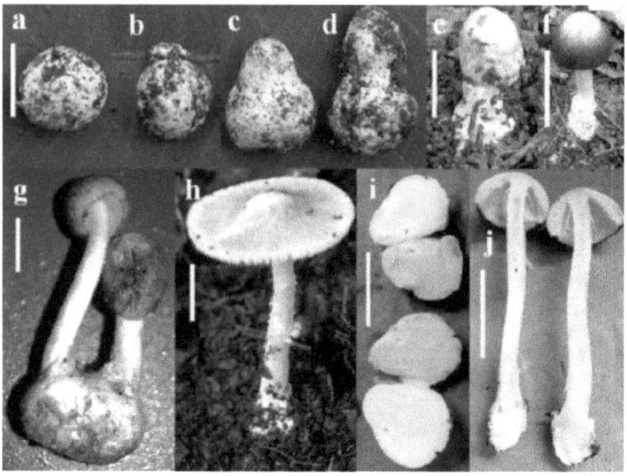

Fig. 1. Developmental stages of *Amanita konajensis* (A-F), fused volva of developing fruit bodies (consumed at a-e stages) (g), mature fruit body (h), longitudinal sections of tender fruit bodies (i) and longitudinal section of maturing fruit body (j) (Scale bar, 20 mm).

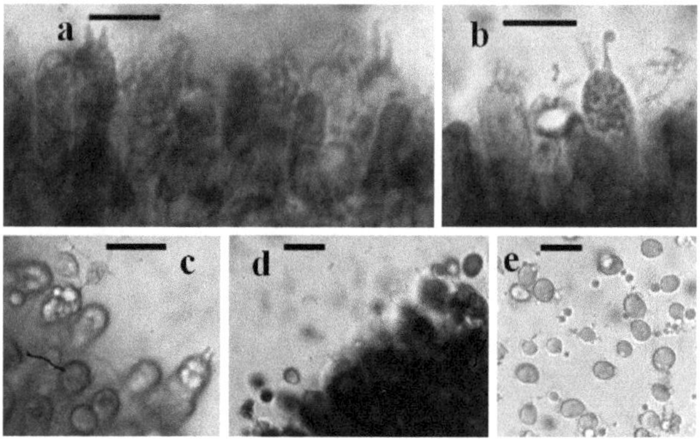

Fig. 2. Developing basidia with sterigmata of *Amanita konajensis* (a, b), immature basidia with sterigmata and basidiospores (c, d) and mature basidiospores (e) (Scale bar, 20 μm).

to convex on maturity, smooth, viscid, non-striated and measures 1.5–6.1 cm. The lamellae were white, free, narrow to inflated, crowded, regular and short gills of 3–4 lengths. Stipes were white to smokey-white with age, fibrillose, cylindrical stuffed, equal or slightly tapering towards the apex and measuring 3.7–8.0 × 0.5–1.2 cm. The annulus was white, membranous, superior, skirt-like, persistent, striate on the upper surface, smooth to the silky inner surface, emerging from the apex region and measuring 1.6–2.8 × 1.4–2.4 cm. Basidia were hyaline, long, cylindrical, thin-walled and club-shaped with 2–4 sterigmata with each basidiospore measuring 18.2–28.6 × 7.8–11.7 μm. Basidiospores hyaline, smooth, thin-walled, sub-spherical to broadly ellipsoidal and measuring 9.1–11.7 × 6.5–9.1 μm. Cheilocystidia are hyaline, short and broadly clavate. It is consumed in its immature stage but not in its mature state. Morphological features of *A. konajensis* have been compared to the nearest resembling *Amanita marmorata* and *A. marmorata* subsp. *myrtacearum* (Table 2) (Gilbert, 1941; Miller et al., 1996; Crous et al., 2022).

Table 2. Comparison of *Amanita marmorata* and *Amanita konajensis*.

	Amanita marmorata **Cleland & Gilbert (Gilbert, 1941)**	***Amanita marmorata*** **subsp.** ***myrtacearum*** **O.K. Miller, D. Hemmes & G. Wong (Miller et al., 1996)**	***Amanita konajensis*** **K.R. Sridhar, Mahadevak., B.R. Nuthan & N.C. Karun (Crous et al., 2022)**
Distribution	Neutral bay, Sidney and New south Wales	Islands of Kauai, Lanai, Oahu, Maui, Molokai and Hawaii	Scrub jungles, Konaje, Mangalore, India
Substratum	Particolous	Particolous	Particolous
Immature fruit body (cm) Oval shaped Dumb-bell shaped (U=upper end × L=lower end)	 – –	 – –	 0.5(0.6–1.9)2.1 × 0.4(1.2–1.6)2.0 (n = 11) 2.5(2.6-3.5)3.6 × 0.9(1.1–1.6)1.7 U (n = 11) × 1.1(1.2–2.0)2.2 L (n = 11)
Mature fruit body (cm)	–	–	5.4(5.5–10.6)11.3 (n = 28)
Pileus (cm)	1.7-–2.1	3.5(4.2–5.5)9.5	1.2(1.5–6.1)6.3 (n = 28)
Pileus	Dried dull brown, broadly convex and no signs of universal veil	Light grey to smoke grey occasionally brownish grey, convex to broadly convex at age, glabrous, subviscid and non-striate	Light grey to greyish brown hemi-spherical to convex on maturity, smooth, viscid and non-striate
lamellae	Dull brown, narrowly attached, subdistant and broad	White, narrowly free, close, alternate with one another lamellulae	White, free, narrow to in-flated, crowded, regular and short gills of 3–4 lengths
Stipes (cm)	4.3 cm long × 1.0–1.2 mm wide	4(8)10 long × 0.3(0.5–1.0)l.5 wide	3.5(3.7–8)8.4 long × 0.4(0.5–1.2)1.4 wide (n = 28)
Stipe	Cylindrical and wide at apex	White, equal or enlarging to form a clavate base and slightly raised loose fibrils	White to smoky white with age, fibrillose, cylindrical, stuffed and equal or slightly tapering towards the apex
Annulus (cm)	–	–	0.6(0.7–1.8)1.9 (n = 28)
Annulus		White, skirt like, superior and often almost at the lamellae	White, membranous, superior, skirt-like, flaring, persistent, striate on upper surface, smooth to silky inner surface and emerging out from the apex region
Volva (cm)	1.4 × 1.1	–	1.4(1.6–2.9)3.3 × 1.3(1.4–2.4)2.6 (n = 28)

Table 2 contd. ...

...Table 1 contd.

	Amanita marmorata **Cleland & Gilbert (Gilbert, 1941)**	***Amanita marmorata*** **subsp.** ***myrtacearum*** **O.K. Miller, D. Hemmes & G. Wong (Miller et al., 1996)**	***Amanita konajensis*** **K.R. Sridhar, Mahadevak., B.R. Nuthan & N.C. Karun (Crous et al., 2022)**
Volva	Dull brown, saccate and not separable from stipe	White, thick, saccate and persistent	White, saccate, membranous, lobed, smooth and spongy/puffy
Basidiospore (µm)	7–10 × 6(7)9	6.7(7.6–9.0)12.0 × 5.5(6.0–8.0)10.5	7.8(9.1–11.7)13 × 5.2(6.5–9.1)10.4 (n = 50)
Basidiospore	Amyloid, globose to subglobose, smooth and thin-walled	Hyaline, thin walled and subglobose to less frequently globose	hyaline, smooth, thin-walled and sub-spherical to broadly ellipsoidal
Basidia (µm)	39–43 × 8.5–11.5	(32–43) × (8–11)	15.6(18.2–28.6)31.2 × 5.2(7.8–11.7)13 (n = 25)
Basidia	Hyaline, clavate, thin walled, four spored and simple septate.	Hyaline, thin walled, clavate and 4-spored	hyaline, long, cylindrical, thin walled and club-shaped with 4 sterigmata containing a spore each
Cheilocystidia (µm)	13–22 × 6.5–8.0	(20–44) × 10(24)38	–
Cheilocystidia	Clavate to pyriform, simple septate, thin walled and creating a sterile layer on some lamellar edges.	Hyaline, clavate, thin walled and forming a sterile layer over the lamellae	Hyaline, short and broadly clavate
Ectomycorrhizal host tree species		*Auracaria columnaris* (G. Forst.) Hook., *Casuarina equisetifolia* L. ex J.R. & Forst., *Eucalyptus robusta* Sm., *E. saligna* Sm., *Eucalyptus* sp., *Melaleuca quinquenervia* (Cav.) S.T. Blake (imported from eastern states of Australia)	*Anacardium occidentale* L. *Acacia auriculiformis* A. Cunn. ex Benth. *Acacia mangium* Willd
Fresh weight (g) Immature fruit body Mature fruit body	 – –	 – –	 2.99 (n = 22) 7.14 (n = 23)
Edibility Immature phase Mature phase	 ? Inedible	 ? Inedible	 Edible Inedible

4.2 *Molecular Studies*

Amanita konajensis was sampled every year during the months of May and June and specimens were preserved for further investigation (morphological and molecular studies). Once, our morphological details suggested that it was new and we subjected it to molecular identification. Briefly, genomic DNA was isolated following the CTAB method (Zhang et al., 1998, Mahadevakumar et al., 2018). The ITS-rDNA region was amplified using the ITS1-ITS4 primer pair (White et al., 1990). PCR amplified products are visualized and subjected to sequencing by a Sanger Sequencer. Sequences were examined and curated and consensus sequences were subjected to nBLAST analysis for comparison. Based on a megablast search of NCBI's GenBank nucleotide database, the closest hits using the ITS sequence of KRS4, KRS6, KRS11 and MKS12 isolates shared 100 sequence similarity to KRS1 (Holotype UOM2021-10) *Amanita konajensis* followed by *Amanita marmorata* (voucher RET 685-9), *Amanita eucalypti* (voucher PERTH 8809828) and the type of *Amanita gardneri* (voucher PERTH 08776121). Further, the evolutionary history was inferred using the Neighbor-Joining method (Saitou and Nei, 1987). The percentage of replicate trees in which the associated taxa clustered together in the bootstrap test (1000 replicates) is shown next to the branches (Felsenstein, 1985). The tree is drawn to scale, with branch lengths in the same units as those of the evolutionary distances used to infer the phylogenetic tree. The evolutionary distances were computed using the Kimura 2-parameter method (Kimura, 1980) and are in the units of the number of base substitutions per site. This analysis involved 27 nucleotide sequences (Table 3). All ambiguous positions were removed for each sequence pair. There were a total of 813 positions in the final dataset. Evolutionary analyses were conducted in MEGA X (Kumar et al., 2018). Morphological examination, molecular sequence data, along with phylogenetic studies, suggested that the species we collected over the years is *Amanita konajensis* (Fig. 3).

4.3 *Ecology*

Amanita konajensis was recorded in lateritic habitats (e.g., arboretum and *Areca* plantation) during the initiation of monsoon season (June third week to end of July: about six weeks). When cyclone Tauktae hit the southwest India during March-April 2021, a pre-monsoon shower was initiated in early April. During such storm rains, *A. konajensis* started fruiting in mid-April and ended by the end of May (6 weeks), which was more unusual than the normal conditions. Usually, this mushroom erupts during early monsoon with thunderstorms on the lateritic stony terrain as ectomycorrhizal with tree species like *Acacia auriculiformis, A. mangium, Hopea ponga* and *Terminalia paniculata*). In addition, *A. konajensis* is also associated with cashew trees (*Anacardium occidentale*), it seems this mushroom is an important ectomycorrhizal fungus that protects the cashew trees from pathogens and supplies nutrients. A variety of mushrooms are capable to form fairy rings, which indicates depletion of soil potassium (Rodriguez et al., 2021). Such rings are known to shape

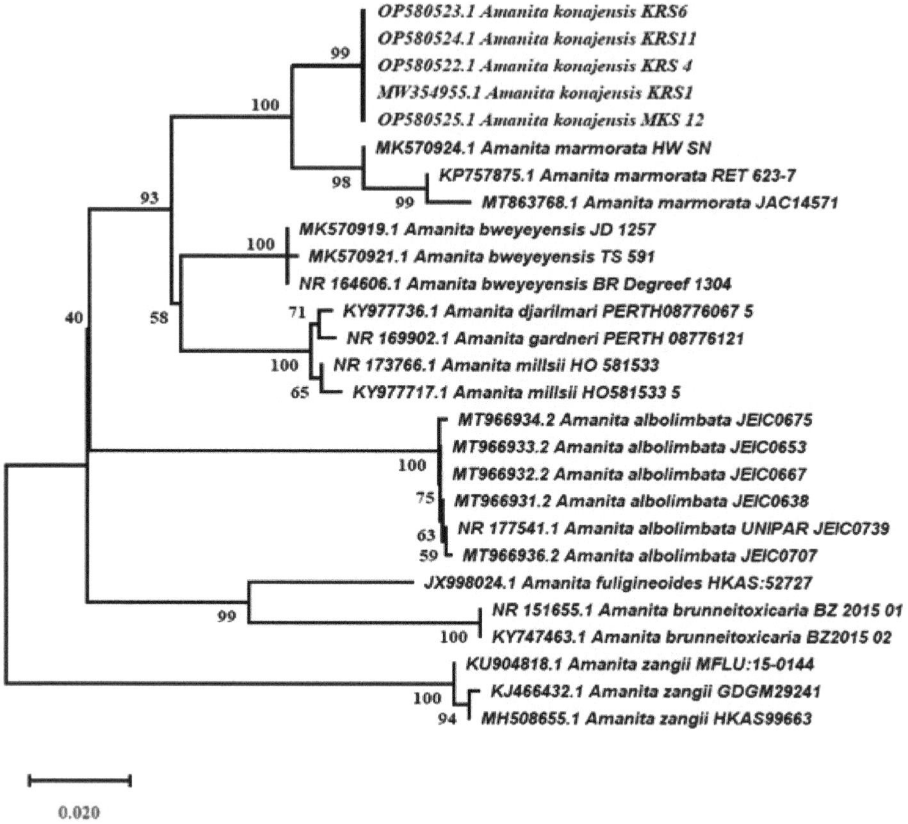

Fig. 3. Phylogenetic tree of *Amanita* species constructed by Neighbour-Joining method showing *Amanita konajensis* forming a distinct clade with reference sequence (type species *A. konajensis* KRS1). The scale bar represents the expected number of nucleotide changes per site. The tree is rooted to *Amanita zangii* (MFLU:15-0144; GDGM29241; HKAS99663).

the soil microbial communities through altering soil conditions. In cashew and *Acacia* groves, *A. konajensis* formed fairy rings of about 1.5–3 m diameters. Such a ring of *A. konajensis* indicates as favourable soil or subterranean network with desired tree species.

Amanita konajensis has been found among the top 12 species of mushrooms found in low and moderately disturbed coastal dunes on the southwest coast of India (Ghate and Sridhar, 2016; Ghate et al. 2016). It is likely associated with some of the coastal dune tree species. From our experience with the occurrence of macrofungi for the last 8–10 years, *A. konajensis* seems to be a narrow disjunct distribution especially in lateritic scrub jungles and coastal dunes.

Table 3. List of nucleotide sequences data retrieved from GenBank database and used in phylogenetic analysis of *Amanita konajensis**.

	Taxon	Voucher specimen	GenBank Accession Number**
1.	*Amanita bweyeyensis*	BR Degreef 1304	NR_164606.1
2.	*Amanita bweyeyensis*	JD 1257	MK570919.1
3.	*Amanita bweyeyensis*	TS 591	MK570921.1
4.	*Amanita marmorata*	HW SN	MK570924.1
5.	*Amanita marmorata*	RET 623-7	KP757875.1
6.	*Amanita marmorata*	JAC14571	MT863768.1
7.	*Amanita konajensis*	KRS1*	MW354955.1
8.	***Amanita konajensis***	**KRS4**	**OP580522.1**
9.	***Amanita konajensis***	**KRS6**	**OP580523.1**
10.	***Amanita konajensis***	**KRS11**	**OP580524.1**
11.	***Amanita konajensis***	**MKS12**	**OP580525.1**
12.	*Amanita djarilmari*	PERTH08776067_5	KY977736.1
13.	*Amanita millsii*	HO 581533	NR_173766.1
14.	*Amanita millsii*	HO581533_5	KY977717.1
15.	*Amanita gardneri*	PERTH 08776121	NR_169902.1
16.	*Amanita albolimbata*	UNIPAR JEIC0739	NR_177541.1
17.	*Amanita albolimbata*	JEIC0707	MT966936.2
18.	*Amanita albolimbata*	JEIC0675	MT966934.2
19.	*Amanita albolimbata*	JEIC0653	MT966933.2
20.	*Amanita albolimbata*	JEIC0667	MT966932.2
21.	*Amanita albolimbata*	JEIC0638	MT966931.2
22.	*Amanita brunneitoxicaria*	BZ 2015_01	NR_151655.1
23.	*Amanita brunneitoxicaria*	BZ2015_02	KY747463.1
24.	*Amanita fuligineoides*	HKAS:52727	JX998024.1
25.	*Amanita zangii*	MFLU:15-0144	KU904818.1
26.	*Amanita zangii*	GDGM29241	KJ466432.1
27.	*Amanita zangii*	HKAS99663	MH508655.1

*, Holotype specimen of *Amanita konajensis* (Crous et al., 2022).
**, Sequence in Bold are from the present investigation.

Conclusion

Documenting *Amanita* from wide geographic regions will be an important step to understanding their ecological, nutraceutical and bioactive significance for conservation. The Himalayas and the Western Ghats are the potential heritage geographic hotspots of a variety of *Amanita*. The southwest region of the Indian subcontinent offers ecologically and climatically suitable geographic conditions to support a variety of *Amanita* species (the Western Ghats, foothills, coastal plains and maritime habitats). Dipterocarpaceae members are logged for quality timber in spite of their association with many ectomycorrhizal fungi (including *Amanita*) of ecological and economic significance. So far, up to 30 species of *Amanita* consisting of ectomycorrhizal, poisonous, medicinal, non-toxic, edible and nutraceutical properties have been reported. Some of the *Amanita* is traditionally consumed by the tribals in southwest India possibly by their ethnic knowledge (e.g., *Amanita hemibapha* and *A. konajensis*). *Amanita* is one of the genera that attracted continuous attention from several investigators worldwide for revision and reappraisal. Identification of poisonous and non-poisonous *Amanita* has primary significance to avoid fatal deaths by consumption. Nevertheless, many *Amanita* is ectomycorrhizal and have a major role in ecosystem services in the forest ecosystems. Their nutraceutical and medicinal significance have not been explored extensively owing to their poisonous nature. There are several unexplored habitats and ecosystems in southwest India worth exploring *Amanita* for future environmental, industrial and nutraceutical applications.

Acknowledgements

The authors are grateful to the Department of Biosciences, Mangalore University for extending the facilities to carryout studies on *Amanita* of southwest India.

References

Bas, C. (2000). A broader view on *Amanita*. Bollettino del Gruppo Micologico g Bresadola, 43: 9–12.

Bhatt, R.P., Tulloss, R.E., Semwal, K.C., Bhatt, V.K., Moncalvo, J.M. and Stephenson, S.L. (2003). The Amanitaceae of India. A critically annotated checklist. Mycotaxon, 88: 249–270.

Bijeesh, C., Pradeep, C.K. and Vrinda, K.B. (2022). Poisonous Mushrooms of Kerala, India: An Illustrated Handbook. Jawaharlal Nehru Tropical Botanic Garden and Research Institute, Trivandrum, Kerala, India, p. 187.

Butler, E.J. and Bisby, G.R. (1931). Fungi of India. Imperial Council of Agricultural Research, India, Science Monograph 1–XVIII; p. 237.

Cameron, E.K., Martins, I.S., Lavelle, P., Mathieu, J., Tedersoo, L. et al. (2018). Global gaps in soil biodiversity data. Nat. Ecol. Evol., 2: 1042–1043.

Codjia, J.E.I., Wang, P.M., Ryberg, M., Yorou, N.S. and Yang, Z.L. (2022). *Amanita* sect. Phalloideae: Two interesting non-lethal species from West Africa. Mycol Prog., 21: 39. 10.1007/s11557-022-01778-0.

Corrales, A., Koch, R.A., Vasco-Palacios, A.M., Smith, M.E., Ge, Z.-W. and Henkel, T.W. (2022). Diversity and distribution of tropical ectomycorrhizal fungi. Mycologia, 10.1080/00275514.2022.2115284.

Crous, P.W., Boers, J., Holdom, D., Osieck, E.R., Steinrucken, T.V. et al. 2022. Fungal planet description sheets 1383–1435. Persoonia, 48: 261–371.

Cui, Y.-Y., Cai, Q., Tang, L.-P., Liu, J.-W. and Yang, Z.L. (2018). The family Amanitaceae: Molecular phylogeny, higher-rank taxonomy and the species in China. Fungal Diversity, 91: 5–230.

Felsenstein, J. (1985). Confidence limits on phylogenies: An approach using the bootstrap. Evolution, 39: 783–791.

Ghate, S.D. and Sridhar, K.R. (2016). Spatiotemporal diversity of macrofungi in the coastal sand dunes of southwestern India. Mycosphere, 7: 458–472.

Ghate, S.D., Sridhar, K.R. and Karun, N.C. (2014). Macrofungi on the coastal sand dunes of south-western India. Mycosphere, 5: 144–151.

Gilbert, E.-J. (1941). pp. 1–336. *In*: Bresadola. Mycol., 27 (Suppl. 1).

Greeshma, A.A., Anu-Appaiah, K.A., Pavithra, M. and Sridhar, K.R. (2021). Biochemical profile of two ectomycorrhizal edible mushrooms of the Western Ghats. Fungal Biotec, 1: 39–49.

Greeshma, A.A., Sridhar, K.R. and Pavithra, M. (2018a). Nutritional perspectives of an ectomycorrhizal edible mushroom *Amanita* of the southwestern India. Curr. Res. Environ. Appl. Mycol., 8: 54–68.

Greeshma, A.A., Sridhar, K.R. and Pavithra, M. (2018c). Functional attributes of ethnically edible ectomycorrhizal wild mushroom *Amanita* in India. Microb. Biosys., 3: 34–44.

Greeshma, A.A., Sridhar, K.R., Pavithra, M. and Ghate, S.D. (2016). Impact of fire on the macrofungal diversity of scrub jungles of Southwest India. Mycology, 7: 15–28.

Greeshma, A.A., Sridhar, K.R., Pavithra, M. and Tomita-Yokotani, K. (2018b). Bioactive potential of nonconventional edible wild mushroom *Amanita*. pp. 719–738. *In*: Gehlot, P. and Singh, J. (eds.). Fungi and their Role in Sustainable Development: Current Perspectives, Springer Nature, Singapore.

Kannan, K. and Natarajan, K. (1988). *In vitro* synthesis of ectoendomycorrhizae of *Pinus patula* with *Amanita muscaria*. Curr. Sci., 57: 338–340.

Kantharaja, R. and Krishnappa, M. (2022). Amanitaceous fungi of central Western Ghats: Taxonomy, phylogeny and six new reports to Indian mycota. J. Threat. Taxa, 14: 20890–20902.

Karun, N.C. and Sridhar, K.R. (2014). A preliminary study on macrofungal diversity in an arboretum and three plantations of the southwest coast of India. Curr. Res. Environ. Appl. Mycol., 4: 173–187.

Kimura, M. (1980). A simple method for estimating evolutionary rate of base substitutions through comparative studies of nucleotide sequences. J. Mol. Evol., 16: 111–120.

Kulkarni, S.M. (1992). *Amanita konkanensis*: A new species of Agaricales. Biovigyanam, 18: 56–58.

Kumar, A., Bhatt, R.P. and Lakhanpal, T.N. (1990). The Amanitaceae of India. Bishen Singh Mahendra Pal Singh Publication, Dehradun, India, p. 160.

Kumar, A., Mehmood, T., Atri, N.S. and Sharma, Y.P. (2021). Revised and an updated checklist of Amanitaceae from India with its specific distribution in Indian States. Nova Hedw., 112: 223–240.

Kumar, S., Stecher, G., Li, M., Knyaz, G. and Tamura, K. (2018). MEGA X: Molecular evolutionary genetics analysis across computing platforms. Mol. Biol. Evol., 35: 1547–1549.

Mahadevakumar, S., Chandana, C., Deepika, Y.S., Sumashri, K.S. and Janardhana, G.R. (2018). Pathological studies on the southern blight of China aster (*Callistephus chinensis*) caused by *Sclerotium rolfsii*. Eur. J. Pl. Pathol., 151: 1081–1087. 10.1007/s10658-017-1415-2.

Miller, O.K., Hemmes, D.E. and Wong, G. (1996). A new subspecies of Amanita section Phalloideae from Hawaii. Mycologia, 88: 140–145.

Mohanan, C. (2011). Macrofungi of Kerala. KFRI Handbook No. 27, Kerala Forest Research Institute, Peechi, Kerala, India, p. 597.

Mustaffa, M.F.J., Suratman, M.N. and Isa, N.N.M. (2012). Dipterocarpaceae: Survival amidst degradation: A brief compilation for conservative prescription. IEEE Symposium on Business, Engineering and Industrial Applications. Bandung (Indonesia), 149–153.

Natarajan, K. (1977). South Indian Agaricales III. KAVAKA - Trans. Mycol. Soc. India, 5: 35–39.

Natarajan, K. and Kannan, K. (1982). Cellulose production by *Aminita muscaria*. Curr. Sci., 51: 559–561.

Natarajan, K., Kumaresan, V. and Narayanan, K. (2005b). A checklist of Indian agarics and boletes (1984–2002). KAVAKA - Trans. Mycol. Soc. India, 33: 61–128.

Natarajan, K., Senthilarasun, G., Kumaresan, V. and Riviere, T. (2005a). Diversity in ectomycorrhizal fungi of a dipterocarp forest in Western Ghats. Curr. Sci., 88: 1893–1895.

Pavithra, M., Sridhar, K.R., Greeshma, A.A. and Karun, N.C. (2016). Spatial and temporal heterogeneity of macrofungi in the protected forests of Southwestern India. Int. J. Agric. Technol., 12: 105–124.

Pradeep, C.K. and Vrinda, K.B. (2007). Some noteworthy agarics from Western Ghats of Kerala. J. Mycopathol. Res., 45: 1–14.

Pradeep, C.K. and Vrinda, K.B. (2010). Ectomycorrhizal fungal diversity in three types and their association with endemic, indigenous and exotic species in the Western Ghats forests of Thiruvanthapuram district Kerala. J. Mycopathol. Res., 48: 279–289.

Ravikrishnan, V. and Sridhar, K.R. (2023). Biological activity of six wild edible mushrooms of the Western Ghats. *In*: Semwal, K., Stephenson, S.L. and Husen, A. (eds.). Promising Medicinal Mushrooms. CRC Press, Boca Raton, USA (in press).

Ravikrishnan, V. and Sridhar, K.R. and Rajashekhar, M. (2021). Bioactive attributes of edible wild mushrooms of the Western Ghats. pp. 20–38. *In*: Sridhar, K.R. and Deshmukh, S.K. (eds.). Advances in Macrofungi: Pharmaceuticals and Cosmeceuticals. CRC Press, Taylor & Francis Group, Boa Raton, USA.

Ravikrishnan, V., Ganesh, S. and Rajashekhar, M. (2017). Compositional and nutritional studies on two wild mushrooms from Western Ghat forests of Karnataka, India. Int. Food Res. J., 24: 679–684.

Ravikrishnan, V., Sridhar, K.R. and Rajashekhar, M. (2021a). Biochemical profile of six edible wild mushrooms of the Western Ghats. pp. 142–162. *In*: Sridhar, K.R. and Deshmukh, S.K. (eds.). Advances in Macrofungi: Industrial Avenues and Prospects. CRC Press, Taylor & Francis, Boca Raton, USA.

Rodriguez, A., Iáñbez, M., Bol, R., Brűggemann, N., Lobo, A. et al. (2022). Fairy ring-induced soil potassium depletion gradients reshape microbial community composition in a montane grassland. European J. Soil Sci., 73: e13239. 10.1111/ejss.13239.

Saitou, N. and Nei, M. (1987). The neighbor-joining method: A new method for reconstructing phylogenetic trees. Mol. Biol. Evol., 4: 406–425.

Sánchez-Ramírez, S., Tulloss, R.E., Amalfi, M. and Moncalvo, J.M. (2015). Palaeotropical origins, boreotropical distribution and increased rates of diversification in a clade of edible ectomycorrhizal mushrooms (*Amanita* section Caesareae). J. Biogeogr., 42: 351–363.

Sathe, A.V. and Daniel, J. (1980). Agaricales (Mushroom) of Kerala state. pp. 9–42. *In*: Sathe, A.V., Deshpande, S., Kulkarni, S.M. and Daniel, J. (eds.). Agaricales of South West India. Maharashtra Association for the Cultivation of Science, Pune, India.

Sathe, A.V. and Kulkarni, S.M. (1980). Agaricales (Mushroom) of Karnataka State. pp. 43–73. *In*: Sathe, A.V., Deshpande, S., Kulkarni, S.M. and Daniel, J. (eds.). Agaricales of South West India. Maharashtra Association for the Cultivation of Science, Pune, India.

Sathe, A.V. and Sasangan, K.C. (1977). Agaricales from South West India - III. Biovigyanam, 3: 119–121.

Senthilarasu, G. (2014). Diversity of agarics (gilled mushrooms) of Maharashtra, India. Curr. Res. Env. & Appl. Mycol., 4: 58–78.

Sevindik, M. (2020). Poisonous mushroom (nonedible) as an antioxidant source. pp. 1–25. *In*: Ekiert, H.M. and Ramawat, K.G. (eds.). Plant Antioxidants and Health. Springer Nature Switzerland.

Sridhar, K.R. and Karun, N.C. (2019). Diversity and ecology of ectomycorrhizal fungi in the Western Ghats. pp. 479–517. *In*: Singh, D.P., Gupta, V.K. and Prabha, R. (eds.). Microbial Interventions in Agriculture and Environment, Volume 1, Research Trends, Priorities and Prospects. Springer Nature Singapore.

Tulloss, R.E. (2005). *Amanita* - distribution in the Americas, with comparison to eastern and southern Asia and notes on spore character variation with latitude and ecology. Mycotaxon, 93: 189–231.

Vrinda, K.B., Pradeep, C.K. and Kumar, S.S. (2005a). Occurrence of a lesser known edible *Amanita* in the Western Ghats of Kerala. Mushr. Res., 14: 5–8.

Vrinda, K.B., Pradeep, C.K. and Kumar, S.S. (2005b). Occurrence of the deadly *Amanita phalloides* in the Western Ghats of Kerala. J. Mycopathol. Res. 43: 259–262.

Vrinda, K.B., Pradeep, C.K., Pradeep, N.S. and Abraham T.K. (1997). Agaricales from Western Ghats - II. Ind. J. Forest., 20: 314–318.

White, T., Bruns, T., Lee, S. and Taylor, J. (1990). pp. 315–322. *In*: Innis, M.A., Gelfand, D.H., Sninsky, J.J and White, T.J. (eds.). Amplification and Direct Sequencing of Fungal Ribosomal RNA Genes for Phylogenetics. PCR Protocols, a Guide to Methods and Applications, Academic Press, New York.

Yang, Z.L. (1997). Die *Amanita*-Arten von Südwestchina. Bibl. Mycol., 170: 1–240.

Yang, Z.L. (2000). Species diversity of the genus *Amanita* (Basidiomycetes) in China. Acta Botanica Yunnanica, 22: 135–142.

Yang, Z.L. (2011). Molecular techniques revolutionize knowledge of basidiomycete evolution. Fungal Divers., 50: 47–58.

Yang, Z.L. (2015). Atlas of the Chinese species of Amanitaceae. Science Press, Beijing, China, p. 213.

Zhang, Y.P., Uyemoto, J.K. and Kirkpatrick, B.C. (1998). A small scale procedure for extracting nucleic acids from woody plants infected with various phytopathogens for PCR assay. J. Virol. Meth., 71: 45–50.

Wood-Inhabiting Macrofungi

8

Substrate Ecology of Wood-Inhabiting Basidiomycetes

Ivan V Zmitrovich,[1,*] *Stanislav P Arefiev*,[2] *Vladimir I Kapitonov*,[3] *Anton G Shiryaev*,[4] *Kiran R Ranadive*[5] and *Margarita A Bondartseva*[1]

1. Introduction

Woody vegetation prevails in the humid and cold climatic belts of the planet, in the arid zone it is scarce and sparse, while in the humid tropical forests it is rich and diverse. This vegetation type is the main demping component of terrestrial habitats and exhibits the maximum biodiversity of zonal biomes. The maintenance of this type of vegetation in the terrestrial environment is closely related to the interaction of woody plants with fungi, namely mycorrhizal symbiosis, decomposition of wood and phylloplane epiphytism or endophytism.

[1] V.L. Komarov Botanical Institute, Russian Academy of Sciences, 2 Professor Popov Street, Saint Petersburg, 197376, Russia.

[2] The Institute for the Development of the North – A Structural Subdivision of the Federal State Budgetary Institution of Science, The Federal Research Center Tyumen Scientific Center of the Siberian Branch of the Russian Academy of Sciences, Russia.

[3] Tobolsk complex scientific station Ural Branch of the Russian Academy of Sciences, 15 Academician Osipov, Tobolsk, 626152, Russia.

[4] Institute of Plant and Animal Ecology, Ural Branch of Russian Academy of Sciences, 202, 8 Marta Street, 620144 Ekaterinburg, Russia.

[5] Annasaheb Magar Mahavidyalaya, Hadapsar, Maharashtra, India.

* Corresponding author: iv_zmitrovich@mail.ru

The tree's life history is somehow linked with a diversity of wood-inhabiting fungi, among which Basidiomycota (basidiomycetous fungi, basidiomycetes) represent a group of fungal organisms most adapted to deep xylolysis. Various groups of basidiomycetes are spatially associated with wood at its different stages (from living trees to litter) and most of their representatives are capable of wood decomposing and utilizing. However, some of the wood-inhabiting basidiomycetes are mainly mycorrhizal or mycoparasitic.

Enzymes of wood-destroying fungi are adapted to the decomposition of various wood components. Therefore, three main groups of wood-decomposers can be distinguished: "ancestral soft rot" fungi, white-rot fungi and brown-rot fungi. All these groups are capable of oxidizing the C-C components of lignocellulose molecular composites. White-rot fungi target the lignin molecules, while brown-rot fungi attack the cellulose and hemicelluloses. The "ancestral soft-rot" fungi are capable of weak lignin degradation but are characterized by high levels of hydrolytic activity directed at plant cell wall polysaccharides.

The most adapted wood destroyers are represented by some groups of Basidiomycota, those belonging to the order Polyporales (class Agaricomycetes). They are characterized by excellent parallel morphological adaptations for sporulation on various substrates, like small twigs, trunks, stumps, fallen logs as well as amorphous woody debris in the forest litter.

It is difficult to overestimate the role of wood-inhabiting basidiomycetes in forest ecosystems. Under humid taiga conditions, these fungi are the key mediators for the enrichment of forest soils and humus. Humic acids released during the decay of wood as well as litter in such forests are a powerful factor in the formation of relief in taiga landscapes. In mesophilic nemoral and tropical rainforests, wood-inhabiting basidiomycetes are important mediators for maintaining a balance of biomass, while their pathogenic role is important in arid ecosystems.

2. The Group Delimitation

The term "wood-inhabiting basidiomycetes" is generally applied to a topical group that includes Basidiomycota whose sporulation is associated with wood in various conditions. The vegetative mycelium is associated with wood and the nutrition of fungi will be augmented by the degradation of wood, then such a fungus can be called a "wood-destroying basidiomycete".

In a number of cases, the mycelium of basidiomycetes spreads along the forest litter into the humus-soil horizon and is able to form ectomycorrhizae, while the fruiting bodies are obligately or facultatively associated with woody substrates. Although these fungi coincide with the topical group of wood-inhabiting basidiomycetes, their trophic affiliation is differing. The situation is analogous to some obligate parasitic wood-destroying basidiomycetes that develop their sporulations or/fruit bodies on woody substrates.

Some lichenized Basidiomycota dependence on algae as facultative and whose mycelium is able to penetrate and destroy woody substrates, can also be attributed to representatives of the group of wood-inhabiting basidiomycetes.

3. Biodiversity

From a historical perspective, the wood was colonized by basidiomycetes heterochronously and polyphyletically. As a result, all the main phylogenetic radiations of Basidiomycota possess elements adapted to wood (Table 1). There are an estimated up to 30,000 species of these fungi worldwide.

Table 1. The taxonomical details of wood-inhabiting Basidiomycota.

Classes	**Orders**	**Key genera of wood-inhabiting fungi**
Pucciniomycetes	Platygloeales	**Platygloea*
Tremellomycetes	Tremellales	*Tremella*
Dacrymycetes	Dacrymycetales	<u>*Calocera*</u>, *Cerinomyces*, <u>*Dacrymyces*</u> and *Femsjonia*
Agaricomycetes	Agaricales	*Armillaria*, *Arrhenia*, *Campanella*, *Caripia*, *Cellypha*, *Chaetocalathus*, *Cheimonophyllum*, *Chondrostereum*, *Chromocyphella*, *Claudopus*, *Clitopilus*, *Collybia*, *Coprinus*, *Coronicium*, *Cotylidia*, *Crepidotus*, *Crinipellis*, *Cylindrobasidium*, *Deflexula*, *Episphaeria*, *Favolaschia*, *Filoboletus*, *Fistulina*, *Flagelloscypha*, *Flammulina*, *Gymnopilus*, *Hemipholiota*, *Henningsomyces*, *Hohenbuehelia*, *Hypholoma*, *Hypsizygus*, *Kuehneromyces* *Lachnella*, *Lampteromyces*, *Lentinula*, *Lycoperdon*, *Merismodes*, *Mucronella*, *Mycena*, *Nothopanus*, *Omphalina*, *Ossicaulis*, *Panellus*, *Pellidiscus*, *Phaeosolenia*, *Pholiota*, *Pleurocybella*, *Pleurotus*, *Pluteus*, *Podoscypha*, *Porotheleum*, *Radulomyces*., *Rectipilus*, *Resupinatus*, *Rhodotus*, *Rimbachia*, *Schizophyllum*, *Stigmatolemma*, *Strobilurus*, *Stromatocyphella*, *Stropharia*, *Tricholomopsis*, *Trogia*, *Typhula* and *Volvariella*
	Atheliales	<u>*Amylocorticium*</u>, <u>*Amylocorticiellum*</u>, **Athelia*, *Ceraceomyces*, **Piloderma*, <u>*Serpulomyces*</u> and *Tylospora*
	Auriculariales	*Auricularia*, *Basidiodendron*, *Bourdotia*, *Craterocolla*, *Ductifera*, *Eichleriella*, *Exidia*, *Exidiopsis*, *Heterochaete*, *Oliveonia*, *Pseudohydnum*, *Stypella* and *Tremellostereum*
	Boletales	<u>*Bondarcevomyces*</u>, <u>*Coniophora*</u>, <u>*Gyrodontium*</u>, <u>*Jaapia*</u>, <u>*Leucogyrophana*</u>, <u>*Meiorganum*</u>, *Paxillus*, <u>*Serpula*</u>, <u>*Tapinella*</u> and *Tylopilus*
	Cantharellales	**Botryobasidium*, **Ceratobasidium*, *Clavulicium*, **Suillosporium* Pouzar, **Thanatephorus* and *Tulasnella*
	Corticiales	*Corticium*, *Cytidia*, *Punctularia* and *Vuilleminia*
	Gloeophyllales	<u>*Boreostereum*</u>, <u>*Gloeophyllum*</u>, <u>*Neolentinus*</u> and <u>*Veluticeps*</u>
	Gomphales	*Hydnocristella*, *Kavinia*, *Lentaria* and *Ramaricium*
	Hymenochaetales	*Asterodon*, *Basidioradulum*, *Cyclomyces*, *Fibricium*, *Hydnochaete*, *Hymenochaete*, *Hyphodontia*, *Inonotopsis*, *Inonotus*, *Leucophellinus*, *Oxyporus*, *Phellinidium*, *Phellinus*, *Phylloporia*, *Pyrrhoderma*, *Repetobasidium*, *Resinicium*, *Schizopora*, *Sidera*, *Stipitochaete*, *Subulicium*, *Trichaptum* and <u>*Tubulicrinis*</u>

Table 1 contd. ...

...Table 1 contd.

Classes	Orders	Key genera of wood-inhabiting fungi
	Polyporales	*Abortiporus, Amauroderma, Amylocystis, Anomoporia, Antrodia, Antrodiella, Aurantiporus, Auriculariopsis, Auriporia, Bjerkandera, Bulbillomyces, Byssomerulius, Ceriporia, Cerrena* and *Climacocystis, Climacodon, Cryptoporus, Dacryobolus, Daedalea, Daedaleopsis, Datronia, Dichomitus, Donkioporia, Flaviporus, Fomes, Fomitopsis, Ganoderma, Gloeoporus, Grammothele, Grifola, Haddowia, Hapalopilus, Haploporus, Hexagonia, Humphreya, Hydnophlebia, Hymenogramme, Hyphoderma, Hyphodermella, Hypochnicium, Intextomyces, Ischnoderma, Irpex, Jahnoporus, Junghuhnia, Laetiporus, Lentinus, Leptoporus, Lopharia, Loweomyces, Megasporoporia, Meripilus, Microporus, Mycorrhaphium, Nigroporus, Oligoporus, Pachykytospora, Panus, Parmastomyces, Perenniporia, Phaeolus, Phanerochaete, Phlebia, Phlebiella, Phlebiopsis, Polyporus, Porogramme, Postia, Pycnoporellus, Pyrofomes, Scopuloides, Skeletocutis, Sparassis, Spongipellis, Steccherinum, Trametes, Theleporus, Tyromyces, Wolfiporia* and *Xenasma*
	Russulales	*Acanthobasidium, Acanthophysellum, Aleurodiscus, Amylonotus, Amylosporomyces, Asterostroma, Auriscalpium, Bondarzewia, Clavicorona, Conferticium, Dentipellis, Dentipratulum, Dichopleuropus, Dichostereum, Echinodontium, Gloeocystidiellum, Gloeodontia, Gloiodon, Gloiothele, Hericium, Heterobasidion, Laurilia, Laxitextum, Lentinellus, Megalocystidium, Pseudoxenasma, Scytinostroma, Scytinostromella, Stecchericium, Stereofomes, Stereum, Vararia, Wrightoporia* and *Xylobolus*
	Sebacinales	**Sebacina*
	Thelephorales	**Amaurodon*, **Pseudotomentella*, **Thelephora*, **Tomentella*, **Tomentellopsis* and **Tomentellago*
	Trechisporales	*Litschauerella, Subulicystidium* and *Trechispora*

Note: Underlined, the genera associated with a brown-rot; *, Genera associated with white-rot or "ancestral soft-rot".

The most dominant order of wood-inhabiting Basidiomycota is the Polyporales (class Agaricomycetes), which is consist of almost xylotrophic taxa. Other diverse and rich taxa include Agaricales, Hymenochaetales and Russulales. Nodes of brown-rot producers within Basidiomycota are not numerous and have a species affiliation are well taxonomically delimited (order Dacrymycetales in Dacrymycetes; orders: Gloeophyllales and Boletales in Agaricomycetes; families Dacryobolaceae, Fomitopsidaceae, Laetiporaceae and Sparassidaceae in Polyporales). Obviously, the colonization of litter, woody debris and woody surfaces was carried out at many levels of organization of basidiomycetes, these adaptations and certain convergence or unification, were superimposed on various types of organization and biochemical pathways.

4. Wood Decay Fundamentals

Wood decomposition is the most common style of life for wood-dwelling basidiomycetes. Historically, this regime dates back to the Devonian transformation of plant ecomorphs from plagiotropic to orthotropic via the processes of shoot lignification as a result of transformations in the secondary metabolism of plants (Gensel et al., 2020). Close relatives of Devonian plants with associated symbiotic fungi did not exclude horizontal gene transfer between plants and fungi, especially expressed by genes associated with the construction and destruction of lignin molecular composites (Ragan and Chapman, 1978).

The main sign of wood organization is the presence of vessels with secondarily thickened walls. Cellulose and hemicellulose fibrils make up the fibrillar core, while the linocellulose complex makes up their amorphous matrix. Basidiomycetes are adapted to the destruction and development of such a specific substrate. This process occurs in two stages: (1) the development of an enzymatic system for the hydrolysis of polysaccharides and (2) the development of an enzymatic system for the oxidation of lignin and polysaccharides. The principal polysaccharides of the cell walls of woody plants and their corresponding hydrolytic enzymes hydrolyze internal glycosidic bonds (Table 2). As a result of hydrolysis, the disruption of the microfibrillar structure takes place in the wood.

In basidiomycetes, the ability to hydrolyze polysaccharides is linked with the ability to oxidize C–C bonds in polysaccharides as well as lignin-containing composites (= multi-component substances: cellulose, lignin, pectins, etc.). Hence, depending on the oxidation targets (polysaccharides or lignocelluloses), there are three types of wood decomposition: "ancestral soft rot", white-rot and brown-rot. The major targets of wood-inhabiting brown-rot fungi are cellulose and hemicelluloses; lignin is slightly modified by demethylation (Eriksson et al., 1980, 1990; Wright,

Table 2. Details of hydrolytic biodegradation of the wood with specific examples.

Fractions of polysaccharides of wood cell wall	Hydrolytic enzymes of wood-inhabiting fungi	Products of degradation	Example species, authors
Cellulose	1,4-β-D-glucan cellobiohydrolases	cellobiose, cellooligomers, D-glucose	*Postia placenta*, *Hypocrea jecorina* and *Phanerochaete chrysosporium* (Cowling, 1961; Aro et al., 2005)
Pectins	rhamnogalactouronan hydrolase, rhamnogalactouronan lyase; endo-β-1,6-galactanase, exogalactanase; L-arabinofuranosidase	D-galactouronic acid; D-galactose L-arabinose	*Ceratobasidium* spp. (González García et al., 2006); *Tremella aurantialba* (Jing et al., 2007)

1985). The wood loses its fibrillar structure, becomes brittle and cracks into a red-brown mass owing to the modification of lignin.

Koenigs (1972) established that such wood decomposers are capable to produce huge amounts of H_2O_2. This substance is produced by extracellular enzymes (peroxidases). The attack mechanism of C–C bonds of crystalline cellulose and similar composites is known as the Fenton reaction: $H_2O_2 + Fe^{2+} + H^+ \rightarrow H_2O + Fe^{3+} + HO^{\cdot}$; $RCH\ (OH) + 2HO^{\cdot} \rightarrow RHO + CO_2 + H_2O$. To overcome the destruction of the hyphae wall and attack the lignified parts of the secondary cell wall, OH radicals must be produced at a distance from the hyphae and fungal reducing agents must be stable enough to diffuse before they enter into the reduction reaction: Fe (III) and oxygen to Fe (II) and peroxide. The OH radical production occurs in several ways in various systems, including the secretion of hydroquinones, cellobiose dehydrogenases, low molecular weight glycopeptides including phenolate chelators (Gamauf et al., 2007).

Brown-rot fungi are adapted to rapid xylolysis as well as capable to cause heart-rot (in stumps and logs of conifers), which are characterized by highly resistant lignocellulosic complexes. The wood cracks into polygons or chips and this lignin-rich material are a humus precursor in forest soils. Under xerophilic conditions in bare and decorticated wood, brown-rot fungi show rather weak activity, causing the so-called "dry-rot". Representatives of the genera *Dacrymyces* and *Gloeophyllum* are typical producers of dry-rot. However, when the process of evaporation of water from the wood is hindered by forest shading, soil conditions, or an abundance of dead wood, rot remains active and in several cases is accompanied by self-wetting of the wood owing to metabolic processes. Such active wet brown-rot is characteristic of some fungi (*Serpula*, *Tapinella* and many species of *Antrodia* and *Fomitopsis*).

White-rot fungi are fine-tuned to deep degradation of lignin and partial decomposition of polysaccharides. Lignin is a highly inert, high molecular weight biopolymer that is composed of many multilayer polyphenolic compounds. White-rot fungi degrade this stable biopolymer through multiple oxidative steps involving peroxidases and laccases (phenol oxidases), which act non-specifically to form lignin free radicals, followed by spontaneous cleavage reactions (Rabinovich et al., 2001; Gamauf et al., 2007). Laccases are represented in tree fungi by lignin peroxidases (LiPs), manganese peroxidases (MnPs) and universal peroxidases (VP), have a high level of redox potential and, thus, are able to oxidize polyphenolic and other polycyclic aromatic compounds.

There are two main types of white-rot:

(1) Non-selective delignification largely affects hardwood and simultaneously destroys cellulose, lignin and hemicellulose. The walls of the vessels gradually degrade from the lumen to the middle plate. Wood residues are represented by discolored lignin derivatives and broken straw-white cellulose filaments. This pattern is typical for most producers of white-rot by ascomycetes and basidiomycetes.

(2) Selective delignification affects hard as well as soft wood. In this case, lignin and hemicellulose-cellulose are attacked first and then the cellulose. Wood residues

in many cases retain their regular crystalline structure or are permeated with regular pockets. The color change occurs in several cases to a white or straw color, but in some cases may be changed to an unusual red or green color due to polyphenol residues. This pattern is typical only for some basidiomycetes, such as *Ceriporiopsis subvermispora*, *Phanerochaete* spp., *Pleurotus* spp. and such white-rot fungi can be a good source for biotechnological applications (Schmidt, 2006).

White-rot fungi are consisting of a huge pool of wood destroyers that inhabit a wide range of ecological niches from live trees and shrubs to forest litter.

The concept of "ancestral soft-rot" was formulated in the course of research by the Nagy group (Nagy et al., 2015). Prior to this, inactive and visually white decay caused by ascomycetes and early divergent basidiomycetes was also classified as white-rot. However, studies have shown that families of oxidative enzymes involved in lignin degradation have undergone a burst of diversification in the taxa of Agaricomycetes after the divergence of Auriculariales. It turned out that early divergent basidiomycetes from Cantharellales (*Botryobasidum* and *Tulasnella*) and Sebacinales (*Sebacina*) lack important families of lignin degradation enzymes (for example, peroxidases of class II), but are characterized by a diverse repertoire of families of enzymes involved in cellulose degradation. It is these types of basidiomycetes that produce "ancestral soft-rot", affecting mainly cellulose, but having a certain laccase activity, i.e., capable of lignin modification.

The white-rot fungi are a derivative and homogeneous group of xylotrophic fungi that have evolved from brown rot producers through a complex system of oxidative enzymes (Nobles, 1958). In contrast, Gilbertson (1980) argues that brown-rot fungi are phylogenetically younger than white rot fungi and have multiple and independent origins. The remains of precursors of gymnosperms (*Callyxylon* spp.) have characteristics comparable to white-rot (Stubblefield et al., 1985). Gilbertson also confirms this by identifying silent laccase genes in the brown-rot genome (D'Souza et al., 1996). The oxidative nature of cellulose degradation by brown-rot can be affected as a secondary strategy for the rapid colonization of woody substrates.

5. Trophic Differentiation

The major terminological revision of fungal feeding modes presented by Cooke and Whipps (1980), there are five major feeding modes for plant fungi (facultative biotrophs, obligate biotrophs, facultative necrotrophs, obligate saprotrophs and obligate necrotrophs). Biotrophs and necrotrophs encompass the old undefined category of "parasites". The latter represents a lifestyle, not a way of nutrition; thus, it should be excluded from the trophic category.

The true biotrophs live in association with grasses and leaves of woody plants. Their mycelium has adapted to the weak exploitation of the host protoplast due to the development of the so-called "interaction zone", appressoria and haustoria. The wood-inhabiting basidiomycetes in most cases lack these specialized structures since cells of xylem are equipped with secondarily thickened walls and usually do not have a protoplast. However, in the composition of Tremellales and some Pucciniomycetes, there are fungi that combine xylosaprotrophic ability with biotrophic (mycoparasitic). For instance, *Tremella* representatives are able to colonize only fungal hymenium cells (the so-called "intrahymenial parasites"), either to transform the fruiting of the entire host or to colonize the wood affected by xylotrophs and the vegetative mycelium of xylosaprotrophs, penetrating into them with haustoria or similar structures (Olive, 1946; Bandoni, 1961, 1987; Zugmaier et al., 1994; Torkelsen, 1997). Some wood-inhabiting basidiomycetes, such as *Septobasidium* (Pucciniomycetes), parasitize the bark of scale insects. According to the Cooke and Whipps' classification, the predominant mass of basidiomycetes living in wood ranges from obligate necrotrophs to obligate saprotrophs.

5.1 Obligate Necrotrophs

Unlike facultative necrotrophs, these fungi are unable to assimilate necrotic tissues and are therefore adapted to the destruction of cells and protoplasts. The wood-decomposing ability is weak. Such a group can be considered as a specialized derivative of the previous one. The function of obligate necrotrophs leads to the death of seedlings. Taxonomically, obligate woody necrotrophs belong to the anamorphic genera of Ascomycota (*Camarosporium*, *Cucurbitaria*, *Cytospora*, *Ottia*, *Phoma* and *Phomopsis*), but some genera of basidiomycetes (for example, *Cylindrobasidium*, *Chondrostereum*) sometimes show a tendency to this kind of activity (Davydkina, 1980).

5.2 Obligate Saprotrophs

This previous group of wood-inhabiting basidiomycetes has the ability to saprotrophically utilize the cell wall skeleton and the capability to interact with living plant cells. However, the predominant wood-inhabiting basidiomycetes belong to obligate saprotrophs. This group includes both brown rot fungi and white rot fungi. Taxonomically, the main pool is wood-dwelling Basidiomycota belonging to the orders Polyporales, Agaricales, Hymenochaetales, Cantharellales, Auriculariales and Dacrymycetales. In connection with the wider range of xylosaprotrophs, a finer organization of this group is required.

5.2.1 Pathogenic Saprotrophs

These fungi colonize the heartwood of living trees and shrubs or young branches that are rich in water, mineral nutrients and oxygen. The source of nutrients for the fungi is the carbohydrates or lignin of the xylem vessels, together with the proper water level. On the death of the tree, this type of fungi ceases its activity (*Chondrostereum*, *Climacodon*, *Fomes*, *Fomitopsis* spp., *Ganoderma* spp., *Inonotus* spp., *Laetiporus*, *Lentinus*, *Oxyporus* spp., *Phaeolus*, *Phellinus*, *Pholiota* spp., *Piptoporus*, *Pleurotus* spp. and *Vuilleminia* spp.). Some of these genera (for example, *Inonotus* spp., *Phellinus* spp.) die simultaneously with the death of their host tree and some of them (e.g., *Chondrostereum*, *Ganoderma* spp., *Laetiporus*, *Piptoporus* and others) are able to continue their activities after the death of the host tree as non-pathogenic saprotrophs.

5.2.2 *Non-Pathogenic Saprotrophs*

These fungi form a major pool of saprotrophs. They colonize standing dead trees, fallen trees, branches, stumps and wood remains of lost structures in the forest floor. Thus, these fungi are associated with wood at all stages of its decay. Some species occupy spatially and temporally localized niches. The initial stages and niches are associated with the same set of "terminal pathogenic saprotroph" taxa. The fallen logs are infected with many decorators (such as *Ceriporiopsis*, *Cylindrobasidium*, *Junghuhnia*, *Oxyporus*, *Stereum* and others), surface and bulk wood decomposers (*Antrodia*, *Fomitopsis*, *Gloeophyllum*, *Lentinus*, *Ossicaulis*, *Postia*, *Skeletocutis* and *Tyromyces*), followed by decomposers of wood residues (*Anomoporia*, *Athelia*, *Botryobasidium*, *Coprinus* spp., *Lycoperdon* spp., *Serpulomyces* and *Stropharia* spp.). Stumps and deadwood are inhabited by some characteristic ("leaders") species, such as *Cerrena*, *Clavicorona*, *Fistulina*, *Fomitopsis*, *Ganoderma*, *Hypholoma*, *Lentinus* and *Tapinella*.

5.2.3 *Facultative Humus Saprotrophs/Mycorrhiza-Formers*

The woody litter is connected with the humus horizon of the soil by a continuous series of lignocellulosic composites (Table 3).

The mycelium of some of the wood-inhabiting basidiomycetes penetrates into these horizons. In some instances, they form ectomycorrhizal covers. The ability for ectomycorrhizal formation is known for such genera of litter decomposers as *Amphinema*, *Byssocorticium*, *Coltricia*, *Piloderma*, *Thelephora*, *Tomentella* and *Tylospora* (Erland, Taylor, 1999; Rinaldi et al., 2008; Zmitrovich, 2008). According to Chen et al. (2001), the genes coding ligninolytic enzymes commonly associated with white-rot fungi are prevalent in a wide taxonomic range of ectomycorrhizal fungi (*Amanita*, *Bankera*, *Cortinarius*, *Hydnellum*, *Paxillus*, *Ramaria*, *Rozites*, *Tricholoma*, *Tylopilus*, *Xerocomus* and others). However, these ectomycorrhiza-forming organisms do not belong to the group of wood-inhabiting basidiomycetes. These fungi can be classified as facultative mycorrhiza-formers (Bon, 1991). Their origin is associated with lignotrophic basidiomycetes of many phylogenetic lineages inhabiting the litter (Hibbett, Donoghue, 1995; Binder, Hibbett, 2006; James et al., 2006).

Table 3. The features and rates of litter decomposition in boreal-nemoral forests (Heal and Dighton, 1985).

	Mosses	Herbaceous plants	Angiosperm leaves	Coniferous needles	Wood
Cellulose (%)	16–35	20–37	6–22	20–31	36–63
Lignin (%)	7–36	3–30	9–42	20–58	17–35
C:N ratio	13–50	29–160	21–71	63–327	294–327
Decay (% year^{-1})	20	30–70	40–60	3–50	1–90

6. Topical Groupings

Topical groupings are important to discuss here for two main reasons: (1) some ectomycorrhizal fungi and mycoparasites are capable of fruiting on wood and thus live in wood, (2) xylotrophic fungi show the same patterns.

Topical aspects of wood-inhabiting basidiomycetes have been discussed by Isikov and Konoplya (2004) as well as Arefiev (2010). The primary arrangement of topical groups can be based on the topology of shoots (I, II branching orders, stem) (Isikov and Konoplya, 2004). Arefiev (2010) emphasized the relationship of the fungus with tree bark and distinguished transcortical and cortical species. Based on the wood debris and its structural decomposition through time, the general topical groupings of wood-inhabiting basidiomycetes may be classified as under:

6.1 Wood Debris-Inhabiting Fungi

This group is associated with buried wood, rotten fallen branches included in the bedding and rotten stumps. The fruiting bodies of these fungi are, as a rule, near-stem or negatively geotropic. Many characteristic representatives are *Calocera*, *Hypholoma*, *Macrotyphula* and *Stropharia*.

6.2 Corticolous Fungi

The nutrient source of these fungi is outside the bark substrate. As a rule, they are associated with ectomycorrhizas and decomposition of the litter (*Paxillus involutus* and *Tylopilus felleus*), either with surface deposits of lignocellulose (*Coprinellus disseminatus*), or with epiphytic algae/moss protonema (*Mycena pseudocorticola*). Some fungi also parasitize bark scale insects (*Septobasidium*), but conventionally these groups are not mentioned when discussing wood-dwelling fungi. Their fruiting bodies are usually negatively geotropic.

6.3 Transcortical Fungi

These necrotrophic as well as xylosaprotrophic basidiomycetes infect shoots through natural holes in the bark. Further development of the mycelium is associated with the undergrowth of the bark, while sporulation develops mainly in the places of penetration. Depending on the sequence of shoots, two main subgroups can be distinguished here: (1) crown fungi (*Vuilleminia*); (2) stem fungi (*Basidioradulum*, *Cylindrobasidium*, *Datronia*, *Hyphoderma*, *Oxyporus*, *Radulomyces*).

6.4 Decortical Fungi

These fungi attack wood after significant damage to the bark (frost cracks, keels and ulcers, insects or fungal exfoliation). Mycelium infests the total mass of wood in principle, this group includes all active tinder fungi: *Fomitopsis*, *Antrodia*, *Gloeophyllum*, *Bjerkandera* and many genera of corticoid fungi.

7. Substrate Groupings

A substrate is "a material on which an organism grows or to which it attaches itself" (Kirk et al., 2001). Thus, it is possible to isolate the nutrient substrate and the binding substrate (Yurchenko, 2006). In most cases, both substrates are intermerged, but in some instances, sporulation and nutrient uptake are spatially incongruent.

The key substrates are inhabited by the most important wood-inhabiting basidiomycetes (Table 4). It should be noted that many wood-destroying fungi

Table 4. Overview of key species of wood-destroying fungi on most common substrates of basic world biomes (Zmitrovich et al., 2015).

Key substrata	Key fungal pathogens/decomposers	Literature
Trees and shrubs (arranged by their distribution patterns from north across the equator and south)		
Betula sect. *Albae*	*Inonotus obliquus*, *Fomes fomentarius*, *Fomitopsis betulina* (Figs. 1, 1), *Phellinus laevigatus*, *Trichaptum biforme*, *Antrodiella faginea*,*Gloeoporus dichrous* and *Cerrena unicolor*	Arefiev (2010)
Populus tremula	*Phellinus tremulae* (Figs. 1, 2), *Inocutis rheades* (Figs. 1, 5), *Peniophora rufa*, *Punctularia strigosozonata* and *Gloeoporus pannocinctus*	Ershov and Ezhov (2009)
Alnus incana	*Vuilleminia comedens*, *Phellinus alni* (Figs. 1, 3), *Ph. conchatus*, *Fomitiporia punctata*, *Stereum rugosum*, *S. subtomentosum* and *Lyomyces crustosus*	Strid (1975); Zmitrovich (2012)
Salix spp.	*Trametes suaveolens* (Figs. 1, 4), *Haploporus odorus*, *Phellinus conchatus*, *Ph. punctatus*, *Neofavolus suavissimus* and *Daedaleopsis confragosa*	Andersson et al. (2009)
Picea abies/obovata	*Porodaedalea chrysoloma*, *Rhodofomes roseus* (Figs. 2, 3), *Pycnoporellus fulgens* (Figs. 2, 1) and *Trichaptum abietinum*	Andersson et al. (2009)
Abies sibirica	*Fomitiporia hartigii* (Figs. 2, 4) and *Hymenochaete cruenta*	Bondarceva and Parmasto (1986)
Pinus sylvestris	*Porodaedalea pini* (Figs. 2, 2), *Fomitopsis pinicola* (Figs. 2, 5), *Phlebiopsis gigantea*, *Trichaptum fuscoviolaceum* and *Diplomitoporus flavescens*	Sinadsky (1983)
Larix sibirica	*Fomitopsis officinalis*, *Laetiporus montanus*, *Phaeolus schweinitzii*, *Porodaedalea niemelaei*, *Rhodofomes cajanderi* and *Trichaptum laricinum*	Ezhov et al. (2011); Spirin et al. (2006)
Juniperus communis	*Amylostereum laevigatum*	Davydkina (1980)
Quercus robur	*Laetiporus sulphureus* (Figs. 3, 1), *Fomitiporia robusta*, *Inocutis dryophila* and *Xylobolus frustulatus*	Chamuris (1988); Larsen and Cobb-Poulle (1990); Spirin (2002); Ghobad-Nejhad and Kotiranta (2008)
Tilia cordata	*Cerioporus squamosus* (Figs. 3, 2), *Spongipellis spumea* and *Neolentinus schaefferi*	Malysheva and Malysheva (2008)

Table 4 contd. ...

...Table 4 contd.

Key substrata	Key fungal pathogens/decomposers	Literature
Juglans regia	*Inonotus plorans*	Ghobad-Nejhad and Kotiranta (2008)
Ceratonia siliqua	*Ganoderma australe* and *Fuscoporia torulosa*	Ţura et al. (2011)
Robinia pseudoacacia	*Fulvifomes robiniae*	Larsen and Cobb-Poulle (1990)
Quercus virginiana	*Phellinus coffeatoporus* and *Ph. grenadensis*	Larsen and Cobb-Poulle (1990)
Eucalyptus camadulensis	*Laetiporus gilbertsonii* and *Ganoderma australe*	Burdsall and Banik (2001); Ţura et al. (2011)
Tamarix aphylla	*Inocutis tamaricis*, *Fuscoporia torulosa*, *Ganoderma austral* and *Peniophora tamaricicola*	Ghobad-Nejhad and Kotiranta (2008); Ţura et al. (2011).
Cupressus sempervirens	*Fuscoporia torulosa*	Ţura et al. (2011)
Olea dioica	*Inonotus cuticularis* (Figs. 3, 3)	Ranadive et al. (2015)
Casuarina cunninghamiana	*Inonotus ochroporus* and *I. patouillardii*	Gottlieb et al. (2002)
Ficus religiosa	*Ganoderma lucidum* (Figs. 3, 4)	Ranadive et al. (2015)
Gliricidia sepium	*Ganoderma chalceum* (Figs. 3, 5)	Ranadive et al. (2015)
Tectona grandis	*Earliella scabrosa* (Figs. 4, 1)	Ranadive et al. (2015)
Dalbergia melanoxylon	*Favolus grammocephalus*	Ranadive et al. (2015)
Terminalia bellerica	*Pseudofavolus tenuis* (Figs. 4, 3)	Ranadive et al. (2015)
Eugenia jambolana	*Funalia leonina* (Figs. 4, 4)	Ranadive et al. (2015)
Mangifera indica	*Ganoderma philippii* (Figs. 4, 5).	Ranadive et al. (2015)
Sapthodia campanulata	*Inonotus rickii*	Ranadive et al. (2015)
Azadirachta indica	*Cubamyces flavidus* (Figs. 4, 6)	Ranadive et al. (2015)
Ficus benghalensis	*Podoscypha petaloides*	Ranadive et al. (2015)
Gnetum ula	*Lentinus tricholoma* (Figs. 4, 2)	Ranadive et al. (2015)
Brugiera gymnorhiza	*Trametes cingulata* and *Cubamyces flavidus*	Gilbert et al. (2008)
Rhizophora apiculata	*Funalia sanguinaria* and *Trametes nivosa*	Gilbert et al. (2008)
Sonneratia alba	*Phellinus fastuosus*, *Inonotus luteoumbrinus* and *Trametes cingulata*	Gilbert et al. (2008)
Avicennia germinans	*Phellinus swieteniae* and *Trichaptum biforme*	Gilbert and Sousa (2002)
Nothofagus dombei	*Pseudoinonotus crustosus* and *Nothophellinus andinopatagonicus*	Larsen and Cobb-Poulle (1990); Gottlieb et al. (2002)

Table 4 contd. ...

...Table 4 contd.

Key substrata	Key fungal pathogens/decomposers	Literature
Small schrubs and semi-arboreous plants		
Calluna vulgaris	*Acanthobasidium norvegicum*, *Acanthophysium apricans*, *Corticium macrosporopsis*, *Hastodontia hastata*, *Phanerochaete ericina*, *Ph. martelliana* and *Sistotrema dennisii*	Domański (1988, 1991, 1992); Yurchenko (2006)
Rubus idaeus	*Ceratobasidium cornigerum*, *Peniophora cinerea*, *Acanthobasidium norvegicum*, *Vuilleminia macrospora* and *Efibula tuberculata*	Domański (1988, 1991, 1992); Yurchenko (2006)
Actinidia spp.	*Peniophora sphaerocystidiata*	Yurchenko (2006)
Lignified herbaceous and succulent plants		
Chamaerion angustifolium	*Rhizoctonia pseudocornigera*, *Peniophora cinerea* and *Sistotrema octosporum*	Yurchenko (2006)
Humulus lupulus	*Aleurodiscus cerussatus*	Yurchenko (2006)
Juncus sp.	*Tomentella juncicola*	Domański (1992)
Carnegiea spp.	*Hyphoderma fouquieriae*, *Peniophora tamaricicola* and *Phanerochaete omnivorum*	Nakasone and Gilbertson (1978)
Opuntia spp.	*Crustoderma opuntiae* and *Uncobasidium calongei*	Nakasone and Gilbertson (1978); Yurchenko (2006)
Woody lianas		
Vitis spp.	*Fomitiporia mediterranea* and *Phylloporia ampelina*	Fischer and Gonzalez Garcia (2015)
Actinidia kolomikta	*Peniophora incarnata*, *Schizopora flavipora* and *Xylodon sambuci*	Elena and Paplomatas (2002); Shiryaev et al. (2021)
Non-lignified epiphytes		
Bryophyta	*Rhizoctonia bicornis*, *Sistotrema muscicola*, *Lindtneria leucobryophila*, *Athelia epiphylla*, *Tomentella sublilacina*, *Amphimena byssoides* and *Tubulicrinis subulatus*	Eriksson and Ryvarden (1973); Eriksson et al. (1984); Domański (1988, 1991); Yurchenko (2001, 2006)
Chlorophycophyta (epiphytic)	*Athelia epiphylla*, *A. phycophila*, *A. andina*, *Resinicium bicolor*, *Hyphoderma* spp., *Sistotrema sernanderi*, *Sistotremastrum suecicum*, *Sidera lenis* and *Xylodon rimosissimus*	Jülich (1972); Eriksson et al. (1981, 1984); Eriksson and Ryvarden (1975, 1976); Yurchenko and Golubkov (2003); Yurchenko (2006); Zmitrovich (2008); Miettinen and Larsson (2011)

Table 4 contd. ...

...Table 4 contd.

Key substrata	Key fungal pathogens/decomposers	Literature
Lichenized Ascomycota	*Athelia arachnoidea, A . epiphylla, A . salicum, Botryobasidium candicans, Peniophora cinerea* and *Sistotrema brinkmannii*	Eriksson et al. (1978); Parmasto (1998); Yurchenko and Golubkov (2003); Zmitrovich (2008)
Chitinous substrata	*Hyphoderma setigerum, Peniophora cinerea, Antrodiella pallescens, Sistotrema brinkmannii, Peniophora incarnata* and *Phanerochaete laevis*	Yurchenko and Zmitrovich (2001); Miettinen et al. (2006); Yurchenko (2006)
Humus soil horizon	*Piloderma croceum, Tylospora fibrillosa, Amphinema byssoides, Byssocorticium* spp., *Tomentella* spp., *Tomentellastrum* spp., *Tomentellopsis* spp., *Conohypha terricola, Echinotrema clanculare, Sistotrema hypogaeum, Waitea circinate* and *Dacryobasidium coprophilum*	Jülich (1984); Yurchenko (2006); Zmitrovich (2008)
Anthropogenic composites	*Coniophora marmorata, Serpula lacrymans* and *Leucogyrophana olivascens*	Bondartsev (1956); Jülich (1984); Yurchenko (2006)

are able to form secondary colonization of non-lignified substrates, such as moss protonemata or algae cells. Such phenomena are exhaustively presented by Yurchenko (2001, 2006). A greater amount of fungal mycelium infests the core lignin-containing substrate and surface-associated hyphae form appressoria in the zone of interaction with green epiphytic cells. In Some species like *Athelia*, these associations of green cells and mycelium are quite stable (Zmitrovich, 2008) as well as in some representatives of the Rickenellaceae family (Larsson et al., 2006).

The other type of secondary phenomenon is related to the colonization of herbaceous plants by wood-rot fungi. These plants possess H-lignin (hydrophenilous lignin) (Manskaya and Kodina, 1975; Zmitrovich, 2010) and are likely derived from woody predecessors (Church, 1919; Chadefaud, 1950; Takhtajan, 1950). The most predominant lignifying component of these plants is in xerophylizied forms, where parenchymatous live tissues are reduced. Those fungi colonizing such substrates are seem to be omnivorous and xerotolerant.

It is obvious that the capacity of a wide range of biopolymer decomposition capabilities is present in several taxa, while a real substrate specialization has ecological control and has been correlated to the insolation niche occupied by the fungi and their biomorphic status. Representative wood-inhabiting basidiomycetes in different habitats are presented in Figs. 1–4.

8. Woody Lianas – A Poorly Studied Niche

Woody lianas (vines) are characterized by completely lignified perennial shoots, but at the same time, they are unable to maintain an orthotropic growth system. For

Fig. 1. Key species of wood-inhabiting basidiomycetes on deciduous trees in boreal belt of the planet: 1 – *Fomitopsis betulina* (Polyporales) on *Betula pubescens*; 2 – *Phellinus tremulae* (Hymenochaetales) on *Populus tremula*; 3 – *Phellinus alni* (Hymenochaetales) on *Alnus incana*; 4 – *Trametes suaveolens* (Polyporales) on *Salix fragilis*; 5 – *Inocutis rheades* (Hymenochaetales) on *Populus tremula* (Photo credit: V.I. Kapitonov).

wood-inhabiting basidiomycetes, the latter circumstance is not very important, but the small size and strength of the stems and the presence of small woody residues lingering inside the multi-stemmed structure leave certain nuances in the species composition of the basidiomycetes inhabiting this unusual life form.

We have studied the alien vines of Ekaterinurg city (Urals), the town lying in the south taiga vegetation subzone. The species *Typhula micans* was recorded as predominate (16 host plant species), then follow *T. setipes* (14 host plant species), *Lyomyces sambuci* (13 host plant spp.), *Typhula crassipes* (13 host plant spp.), *T. culmigena* (12 host plant spp.), *T. juncea* (10 host plant spp.), *Pterulicium gracile* (11 host plant spp.), *Bjerkandera adusta* (9 host plant spp.), *Irpex lacteus* (9 host

Fig. 2. Key species of wood-inhabiting basidiomycetes on coniferous trees in boreal belt of the planet: 1 – *Pycnoporellus fulgens* (Hymenochaetales) on *Picea obovata*; 2 – *Porodaedalea pini* (Hymenochaetales) on *Pinus sylvestris*; 3 – *Rhodofomes roseus* (Polyporales) on *Picea obovata*; 4 – *Fomitiporia hartigii* (Hymenochaetales) on *Abies sibirica*; 5 – *Fomitopsis pinicola* (Polyporales) on *Pinus sylvestris* (Photo credit: V.I. Kapitonov).

plant spp.), *Peniophora cinerea* (9 host plant spp.), *Cylindrobasidium laeve* (8 host plant spp.), *Schizophyllum commune* (7 host plant spp.). All these plant species are native and common species in the Urals and Siberia.

An average of 24 basidiomycete species have been recorded on East Asian and North American grape varieties. The mean species richness of fungi among the 17 most fungi-rich grape species is significantly higher for East Asian and North American grape species (24 fungal species), whose natural ranges are the most distant from the city of Ekaterinburg, than for geographically nearer European and native North Eurasian grape species, the average species the richness of which is 9 and 8 species, respectively.

Fig. 3. Key species of wood-inhabiting basidiomycetes in warm temperate and subtropical forest belt: 1 – *Laetiporus sulphureus* (Polyporales) on *Quercus robur*; 2 – *Cerioporus squamosus* (Polyporales) on *Tilia cordata*; 3 – *Inonotus cuticularis* (Hymenochaetales) on *Olea dioica*; 4 – *Ganoderma lucidum* (Polyporales) on *Ficus religiosa*; 5 – *Ganoderma chalceum* (Polyporales) on *Gliricidia sepium* (Photo credit: I.V. Zmitrovich and K.R. Ranadive).

The species revealed are mainly common inhabitants of native taiga plant species, however, some exotic tropical and subtropical East Asian fungal species (never recorded on other substrates in the natural forests of the Urals and Siberia) are also recorded here. For example, *Hydnophlebia chrysorhiza* was found in East and South Asia (Japan, South Korea and India), Africa (Cameroon), America (USA, Canada, Brazil and Venezuela). Another interesting find is *Tomentella olivascens*, found in broad-leaved forests of Europe (France, Ukraine and Caucasus), East Asia (Kuriles and Primorsky Krai) and North America (USA). The species *Radulomyces rickii* is native to East Asia (Japan), Africa (Ethiopia), Europe (Great Britain, France, the Netherlands, Italy, southern Scandinavia and Western Russia), America

Fig. 4. Key species of wood-inhabiting basidiomycetes in tropical belt of the planet: 1 – *Earliella scabrosa* (Polyporales) on *Tectona grandis*; 2 – *Lentinus tricholoma* (Polyporales) on *Gnetum ula*; 3 – *Pseudofavolus tenuis* (Polyporales) on *Terminalia bellerica*; 4 – *Funalia leonina* (Polyporales) on *Eugenia jambolana*; 5 – *Ganoderma philippii* (Polyporales) on *Mangifera indica*; 6 – *Cubamyces flavidus* (Polyporales) on *Azadirachta indica* (Photo credit: K.R. Ranadive).

(USA), Australia and New Zealand. The species *Steccherinum bourdotii* is known in the mixed forests of Asia (India), Europe (from southern Scandinavia to the Alps, western Russia), Africa (Uganda) and North America (USA). On the other hand, the discovery of an arcto-alpine species of Eurasian polypore, Cerioporus scutellatus – traditionally known on dead branches of the shrub *Duschekia fruticosa*, which develops in the Arctic or the alpine belt of mountains in the permafrost zone, is interesting.

Among the studied vines, the mycobiota of the Vitaceae family is the richest. For 100 years, 63 species of fungi associated with vineyards have been identified (15 species of macro- and 48 microfungi). At the same time, all types of macrofungi are local species, while only 20% of microfungi develop under the natural conditions of the Middle Urals (Shiryaev et al., 2022).

9. Trophic Plasticity and Width of Trophic Niche

The association of wood-inhabiting basidiomycetes with a certain tree species is one of the leading factors in their evolution and distribution trends (Bondartsev, 1971). Many significant features of the substrate ecology of wood-inhabiting basidiomycetes in Northern Eurasia are demonstrated by studies carried out in Western Siberia (Mukhin, 1993; Arefiev, 2010). According to the results of work carried out over the entire latitudinal-zonal spectrum of the forest zone of this macrogegion – from the forest-steppe to the forest-tundra – Prof. Mukhin (1993) identifies three main substrate groups out of 345 species of xylotrophic basidiomycetes he noted.

Euritrophs of the first level with almost the same frequency develop on the wood of most coniferous and deciduous trees and confidently classifies only two species as *Fomitopsis pinicola* and *Hydnoporia tabacina*. Euritrophs of the second level develops on either different coniferous species (105 species, for example, *Rhodofomes roseus*, *Gloeophyllum protractum*) or on different hardwood species (136 species, for example, *Fomes fomentarius*, *Trametes versicolor*).

Finally, stenotrophic species are characteristic of only one tree species (98 species). The range of stenotrophic fungal species can practically coincide with the range of the host plant (for example, the ranges of the *Abies* and its stem pathogen *Fomitiporia hartigii*, or occupy the part of the range of the host plant. For example, *Fomitopsis betulina*, common in most of the range of white birch *Betula alba* s.l., is absent at its northern limit in Western Siberia and Greenland (Knudsen et al., 1993).

There is no sharp boundary between these substrate groupings of wood-inhabiting basidiomycetes. So, in exceptional cases, the second-level euritrophs, characteristic of conifers, can be found on deciduous ones and vice versa. Moreover, between extreme eurytrophs and stenotrophs there is rather extensive ecological continuum of species with more or less pronounced substrate preferences. For example, in Western Siberia, *Picea obovata* is a fairly pronounced preference for *Rhodofomes roseus*, whereas *Betula alba* s.l. for *Fomes fomentarius*. The geographical variability of the substrate ecology of wood-inhabiting basidiomycetes is of great importance. Thus, in Western Siberia, *Punctularia strigosozonata* occurs almost exclusively on the aspen *Populus tremula*, but in the peculiar taiga-steppe conditions of Transbaikalia, this fungus is quite common on birch in burnt areas (Shiryaev et al., 2013).

10. The State of Woody Substrate and Features of the Species Composition

The tendencies of substrate differentiation of wood-inhabiting basidiomycetes of the *Betula* complex are highly demonstrative. The birch bark, impervious to water and gases and resistant to decomposition, creates special, most autonomous conditions for the decomposition of the wood enclosed in it after the death of the trees. Therefore, depending on the thickness of the bark and its integrity, birch substrates differ especially in their quality, which is reflected in the composition of associated wood-inhabiting basidiomycetes. Many species of fungi withstand strong changes in temperature and humidity and, for example, the mycelium of *Fomes fomentarius* develops in more stable conditions inside the trunk. These substrate differences in birch are more pronounced than in other species, the bark of which quickly loses its protective qualities after the death of the tree. Partial or complete violation of the properties of birch bark occurs when it is mechanically damaged as well as under the influence of fire. On dead young shoots of birch, with not yet formed birch bark as well as on burnt trunks of birch the fungal species of shrunken shrubs, the conditions for the decomposition of wood which are unstable and strongly depend on fluctuations in the external environment are characteristic.

The thickness of the substrate is also essential for creating a stable autonomous environment inside the woody substrate. In addition, thick trunks retain their living tissues longer than thin trunks in the event of wind breaking or felling, which also determines the different species composition of the fungi inhabiting them. Fungi that populate a still-living broken trunk must have a somewhat virulence and wounded species can be classical saprotrophs. With the same trunk size, trees on the vine dry out longer than those cut-down or broken by the wind. With an increase in the size of the substrate, the diversity of its parts and individual loci increases. So, large birch trees broken by the wind can combine the butt part burnt as a result of a ground fire; exposed timber fault; an intact part of the trunk, enclosed in birch bark with slowly dying living cells; the top of the trunk with poorly developed bark and branches partially damaged by falling. These parts of the trunk are populated by different types of fungi at different times. At the same time, in the central intravital rot, the development of which caused the trunk to break, its causative agent (for example, *Phellinus igniarius* s.l.) and its successors can persist for a long time. On such substrates, up to 20 species of basidiomycetes are recorded, populating the corresponding parts and loci in the order of succession. When assigning such a substrate to one particular category (in this case, the wound substrate is a deadwood of mechanical origin), this heterogeneity of its quality is not taken into account, but it can be traced in basidiomycete species composition.

The calculation of the Czekanovsky coefficient (Table 5) shows that the similarity of the composition of fungi on whole-bark dried-out and mechanically damaged cut or broken birch substrates in Tyumen is only 40%, which is lower than that of all other hardwoods. In such xeromorphic shrubs as *Crataegus sanguinea*, *C. altaica*, *C. pentagina* and *Caragana arborescens*, the similarity of the composition of aphyllophoroid fungi on these types of substrates reaches about 80%, however, there is still some difference associated with the factors considered above. Even more dramatic is the difference in the composition of *Betula alba*-associated fungi (on both broken, felled, or shrunken vine trees), in this case, the similarity is only 2–3%. In another stand-forming species, *Populus tremula*, this similarity is already higher (5–17%), in the *Caragana arborescens* it reaches 68–90%. This is due to the fact that growing birch is affected only by specialized stem parasites (which persist for some time after the death of the tree) even after reaching a considerable age (about 50 years). The *Populus tremula* is affected much earlier and its shrunken trunks are already colonized by the pathogenic basidiomycete *Phellinus tremulae*. The *Caragana arborescens* and other shrubs didn't have highly specialized pathogens in the Tyumen environments.

Table 5. The similarity of species composition of basidiomycetes associated with different substrate categories for a number of tree species (Tyumen, Western Siberia).

Tree species	Czekanovski coefficient, %		
	mechanically damaged – shrunken	**mechanically damaged – growing**	**shrunken – growing**
Native species			
Betula alba s.l.	40	2	3
Alnus incana	43	–	–
Populus tremula	46	5	17
*Acer negundo**	48	20	17
*Malus baccata**	49	23	18
Tilia cordata	56	53	38
Salix caprea	68	36	35
Padus avium	62	–	–
Populus nigra	72	–	–
Caragana arborescens	82	–	–
Alien species			
Ulmus glabra	39	20	52
Corylus avellana	55	25	62
Cerasus vulgaris × *Padus avium*	63	44	45
Syringa joskiaea	65	54	64
Phellodendron amurense	71	36	33
Quercus robur	74	57	60
Fraxinus pennsylvanica	78	72	50
Caragana sp.	79	68	90

*The most naturalized tree invaders.

In alien tree species least adapted to Siberian environment (e.g., *Quercus robur*, *Syringa vulgaris*, *S. josikaea*, *Phellodendron amurense* and *Fraxinus pennsylvanica*), the similarity of fungal species composition on different types of substrates is usually higher than in a native tree species. In alien tree species naturalized in Tyumen (e.g., *Acer negundo* and *Malus baccata*), such a similarity decreases. It is significant that in alien trees a similar set of basidiomycete species carries out not only different saprotrophic pathways of wood decomposition (wounded and dried substrates), but also pathogenic ones (growing trunks). It is obvious that such a huge differentiation of the consortium complex of wood-destroying fungi by substrate categories, in particular, the presence of specialized parasitic species in it, is an indicator of longevity and high stability of the consortium determinant.

Rather weak specialization of basidiomycetous fungi in relation to the vital state of the substrate is typical for most shrub species, including native ones. Due to their smaller size, the number of ecological niches in the waste they form is minimal, the autonomy of the internal hydrothermal regime in decaying stems is not expressed or is weakly expressed, therefore, the development of fungi is determined here mostly by external conditions and some substrate specificity. Due to the relatively short life cycle of many shrubs, which does not require some strong fungistatic mechanisms, even growing shrubs infect mainly by saprotrophic species common to them, passing to the pathogenic mode of life. On the willows (*Salix* spp.) confined to floodplains, developing under specific conditions of flooding, on the contrary, the substrates of any vitality are characterized by the stem pathogen *Phellinus igniarius*, which transformed into facultative saprotroph. In fact, a similar situation is observed in the forest tundra of Western Siberia, where xylosaprotrophic basidiomycetes are poorly represented due to hard climatic conditions and the decomposition of tree waste is carried out by stem pathogens that began their development in an autonomous hydrothermal regime inside the trunk of growing trees.

11. Alien Trees and Substrate Spectrum of Wood-Inhabiting Basidiomycetes

It is widely known that the quality of the substrate is one of the leading factors determining the distribution patterns of wood-inhabiting basidiomycetes (Bondartseva, 1965, 1992; Gordienko, 1986; Vakhmistrova, 1987). The nature of the distribution of wood-inhabiting basidiomycetes on various tree species present in urban landscapes of Tyumen, a large city located in the southern part of Western Siberia in the south taiga vegetation subzone, is highly characteristic. In total, 80 basidiomycetes were noted within the city, developing on 33 native and alien species of trees and shrubs. The volume of a quantitative study of the substrates inhabited by different basidiomycetes species amounted to 13,114 accounting entities (Arefiev, 1997, 2002).

Judging by the results obtained, each tree species has its own to one degree or another specific composition of associated basidiomycete species, whereas the most specific is the ratio of the number of fungal species on different tree substrates according to the quantitative study. The greatest basidiomycete species diversity was noted on native tree species that form plantations and have a high abundance, primarily on *Betula alba* s.l. (46 spp.), followed by *Populus tremula* (41 spp.), *Salix caprea* (27 spp.), *Pinus sylvestris* (22 spp.), *Padus avium* (22 spp.), *Alnus incana* (21 spp.). Within the alien trees, the most fungal-rich were *Acer negundo* (18 spp.) and *Malus baccata* (14 spp.), which are numerous and almost naturalized in the Tyumen region, as well as *Quercus robur* (17 spp.), which is not so adapted to local conditions and is characterized by high basidiomycete within its natural range. Within the alien shrubs, *Cotoneaster* species are the most fungal-rich (10 spp.), very common in urban plantings *C. lucidus* and rarely found *C. integerrimus*. The *Tilia cordata*, which is typical for it in natural plantations in the south of the region, was very poor in basidiomycetes (5 spp.).

At representatives of *Betula alba*-complex, the trees with which the greatest variety of aphyllophoroid fungi of the city is associated, the polypore *Fomes fomentarius*, which grows on almost every dead trunk, predominates, whereas such wound ruderal polypores as *Trametes versicolor*, *T. ochracea* and *Bjerkandera adusta* were numerous. A number of birch-associated polypores rare in Western Siberia was noted in the Tyumen city as *Laetiporus sulphureus*, *Pilatotrama ljubarskyi*, *Oxyporus populinus* and *Hericium erinaceus*.

It is interesting that the trees and shrub's introduction used in the landscaping of the city practically did not invade any new region species of wood-inhabiting basidiomycetes. On all alien and cultivated trees, mostly native fungal species develop, sometimes close (from the same genus) to fungi characteristic of the substrates within this natural range. For example, on a number of shrubs, instead of boreal *Phellinus* representatives, the cosmopolitan *Fomitiporia punctata* settles; in the central rots of *Acer negundo*, instead of the nemoral *Oxyporus populinus*, the boreonemoral *O. corticola*, widely distributed within zonal vegetation, is being settled. It is obvious that not even as isolated position of urban host population is an important issue in this case, but the urban environment itself.

The ruderal-wound fungal species are distinguished by the lowest substrate selectivity. Thus, such species as *Trametes versicolor* and closely related *T. ochracea* were found on 21 tree species. In addition to them, on treated wood, stumps, broken and felled hardwood trees, such species as *Bjerkandera adusta* was common (on a total of 18 hardwood and occasionally coniferous species), also such species as *Irpex lacteus* (16 spp.), *Cylindrobasidium laeve* (16 species), *Schizophyllum commune* (15 spp.), *Stereum hirsutum* (12 spp.), *Datronia mollis* (10 spp.), *Phlebia tremellosa* (9 species), *Cerrena unicolor* (6 species), *Trametes gibbosa* (5 spp.), *T. trogii* (5 spp.), *Lenzites betulina* (4 spp.) were found. On the trunks of large deciduous trees that died from mechanical damage in undamaged areas such species as *Fomes fomentarius* (9 trees species), *Ganoderma applanatum* (9 species), *Fomitopsis pinicola* (6 both deciduous and coniferous species) are rather common. The substrate spectrum of ruderal-wound fungi typical of coniferous species represented in the city, mainly by *Pinus sylvestris* is much narrower: *Stereum sanguinolentum* on only 3 species, *Dichomitus squalens* (1 sp.), *Diplomitoporus flavescens* (1 sp.), *Gloeophyllum sepiarium* (1 sp.), the latter species can be attributed to the group of local stenotrophs. Some other basidiomycetes, rare in the city, were found on any tree species belonging to this group (*Hapalopilus rutilans*, *Postia balsamea* and *Cerioporia purpurea*).

Some stenotrophic basidiomycete species are also common in the Tyumen city, e.g., *Fomitopsis betulina*, *Trichaptum biforme* on dry and dead birch, *T. fuscoviolaceum* on pine. For dead trunks of *Pinus sylvestris*, both in the city and within zonal taiga forests, the combination of *Fomitopsis pinicola* (in heartwood) and *Trichaptum fuscoviolaceum* (in bogwood and bark) is typical, for dark coniferous species (*Picea obovata*, *P. pungens*, *Abies sibirica*, *Pinus sibirica*) a similar combination of *Fomitopsis pinicola* and *Trichaptum abietinum*, for *Betula alba* s.l. – *Fomes fomentarius* (infested the main volume of the trunk, often in association with entomochoric *Fomitopsis pinicola*) and *Trichaptum biforme* (in peripheral wood and bark).

Sufficiently more widely than in their natural habitats, the basidiomycetes optionally causing rot diseases (central rot) of trees and shrubs are represented in urban areas. This is due to abundant damage to plantings by harvesting equipment, pruning of crowns and other anthropogenic impacts as well as a significant spread in the landscaping of alien tree species that have low resistance to pathogens in the natural and climatic environments of the Tyumen city. The gradual destruction by the rot of the central part of the trunk, which performs a mechanical function, leads to the breaking of the tree by the wind and the possible generalization of rot leads to the drying of the tree.

It is significant that pathogenic properties in the city are manifested by fungi, which in nature usually develop on dead wood residues. In particular, a mass infestation of *Syringa vulgaris* by basidiomycetes *Steccherinum ochraceum* and *Stereum hirsutum*, whereas Acer negundo by *Oxyporus corticola* or less often by *Ganoderma applanatum*, *Trametes gibbosa* and *T. trogii* was noted. The 50-year-old plantings of *Quercus rubra* in the forest park of Tyumen are especially strongly affected, where such common ruderal species as *Stereum hirsutum*, *Bjerkandera adusta*, *Trametes versicolor* and *Phlebia tremellosa* are noted as causative agents of central rot. Most of these species first populate damaged or shrunken branches, dry streams and then penetrate into the central part of the trunk.

Usually, the wound fungi, having settled on a weakened tree, cause mixed generalized rot, covering not only the central part of the trunk but also the vital conductive sapwood, which quickly leads the tree to death. This is typical for basidiomycetes *Cerrena unicolor*, *Bjerkandera adusta*, *Schizophyllum commune*, developing on *Acer negundo*, *Betula alba* s.l., *Tilia cordata* and other hardwoods. After felling a tree, the development of these fungi can proceed simultaneously with the growth of stumps and not interfere with the formation of coppice plantations.

Widely distributed in nature obligate pathogens of the central tree rot in urban areas are relatively rare. Among them, *Inonotus obliquus* (often with its successor *Gloeoporus dichrius* was noted on birch, alder and apple trees, *Phellinus igniarius* s.l. on birch, apple, willow and maple, *Ph. tremulae* and *Inocutis rheades* on aspen, *Porodaedalea pini* on pine, *Laetiporus sulphureus* and *Trametes suaveolens* on willow (the latter occasionally occurs on damaged aspen and birch). It is obvious that the relative low abundance of these specialized stem pathogens in a specific urban environment is associated with their adaptation to infection, first of all, of old trees, which are few in the city.

12. Successions of Species Groupings

Natural groupings of wood-inhabiting fungi those colonizing dried and fallen wood are temporally localized. The decay of wood in forest ecosystems passes through many phases (Kotiranta and Niemelä, 1993; Renvall, 1995; Lindgren, 2001; Spirin, 2002). As given in Table 6, in boreal forests, where soil water evaporation is non-intensive, the process of wood humification ranges between 15 and 20 years (in arid climates as well as rainy tropical forests with intensive evaporation, the decay rate is little).

A large number of works have been devoted to the successional aspect of substratum colonization by xylotrophic basidiomycetes (Ripachek, 1967; Runge, 1969, 1975; Arefiev, 1986; Lange, 1986; Renvall, Niemelä, 1992; Stavishenko,

Table 6. Humification of spruce wood in boreal forests with microsuccessions of wood-inhabiting fungi (Spirin, 2002).

Stage	Description
Fallen log (0–2 years)	Fresh wood material has intact branches and bark and their mechanical properties as in living trees. Predominate pathogenic saprotrophs: *Fomitopsis pinicola* and *Heterobasidion* spp.
Origin of decomposition (2–10 years)	The bark partially falls off and the skeletal branches break up. The wood located at the bottom side of the log changes its mechanical properties. Pathogenic saprotrophs continue the growth. The core saprotrophs species occur: *Phellinidium ferrugineofuscum* and *Rhodofomes roseus*
Intensive decomposition (10–15 years)	In this stage only the bark located above remains, whereas the log merges into the ground. The wood strongly changes its mechanical properties (becames friable and stratified); on contact with the ground the humification process starts. The pioneer species complexes are substituted as follows: *Fomitopsis rosea* → *Skeletocutis odora* and *Phlebia centrifuga*; *Heterobasidion* spp. → *Junghuhnia collabens*, *Dichostereum boreale*; *Fomitopsis pinicola* → *Pycnoporellus fulgens* and *Phellinidium sulphurascens*
Full decomposition (15–20 years)	The bark falls or is humified. The wood softens completrely and changes into a red-brown color. Saprotrophic fungi groups are represented by many ephemerous hygrophilic species as *Postia* spp., *Leptoporus mollis*, *Physisporinus* spp. and *Asterodon ferruginosus*

2000). Different authors distinguish between 3 and 7 stages (phases) of wood decomposition, which are characterized by certain types of fungi, although such a sequence of species is not entirely clear. More definite series line up when the development of basidiomata of successor fungi is fixed on the basidiomata of their predecessors and stable successional pairs of species are found (Renwall and Niemelä, 1992), for example, in Western Siberia in the central rots of broken birches *Inonotus obliquus – Gloeoporus dichrous*. It is believed that in the course of succession, increasingly weak wood-destroyers are included in it, capable of developing not only on heavily rotted blue wood, but also on tree bark, in forest litter and even the humus soil horizon. In the West Siberian taiga, the causative agent of central butt rot of Siberian pine (*Pinus sibirica*), *Heteribasidion annosum* s.l. even during the lifetime of the tree, it is replaced by *Postia sericeomollis* or, on more acidic soils, by *Serpula himantioides*, passing to facultative parasitism inside the trunk from the forest floor and bark at the base of the trunk (Arefiev, 1991, 1993). At the same time, *Fomes fomentarius*, characteristic already for the first stages of wood decomposition in the subtaiga zone of Western Siberia, persists on birch stumps up to 18 years, whereas *Ganoderma lucidum* in the middle taiga zone of the region develops in the stumps of *Pinus sibirica* up to 30 years (Arefiev, 2020).

There are also "pioneering" types of successions of xylotrophic fungi (in particular, *Chondrostereum purpureum*, *Cylindrobasidium laeve*), which appear soon after damage or death of a tree and rarely persist for more than two years using for nutrition, mainly the protoplast of wood and bark cells (Davydkina, 1974). The studies of the *Betula*-complex of aphyllophoroid macromycetes carried out in the southern part of the forest zone of the West Siberian Plain (12,740 woody substrates inhabited by fungi) (Arefiev, 1986, 1997, 2010) made it possible to objectively assess the specialization of the most important basidiomycetes species in relation to the state and size of the substrate.

According to Table 7, such species as *Trametes versicolor* and *T. ochracea* in 93% of cases were noted on a variety of mechanically damaged substrates, mainly on felling stumps (74%), that is, on relatively large-sized stem wood. At the same time, they are quite common and, on smaller damaged substrates, up to branches, occasionally occur on damaged trees (< 1%) as wound parasites or, rather, epiphytes, since in such cases they have a very limited localization. They are also able to develop on shrunken trees (7%), usually penetrating into them along the passages of the birch beetle or other xylobiont insects. *Irpex lacteus* (80%) is noticeably less common on mechanically damaged substrates; moreover, it is typically not for stumps, but for branches at the cut crown branches (49%).

In comparison with the previous species, most of the occurrences of *Fomitopsis betulina* and *Steccherinum ochraceum* occur on trees with whole bark that have already dried up on the vine (respectively, 96 and 83%). At the same time, the first type is characteristic, first of all, for dead wood (72%) and the second is for deadwood derived from dead wood (64%), which is formed as a result of the decomposition of wood and the breaking of a trunk that has lost its strength. Both of these species, especially *Fomitopsis betulina*, are sometimes found on dying trees that are still alive or on dead branches of growing trees but do not spread from them to living tissues if the tree maintains or restores active growth. They are also found on mechanically damaged substrates, more often in the crowns of large windbreak trees, the drying of which occurs slowly due to the inertia of the autonomous nutrient resources of the trunk. The branches of such substrates, as if "rooted" in the trunk, are close to dry wood in their substrate properties. Most often, dead birch wood gradually breaks up into separate fragments over a number of years, starting from the top, or is crushed when falling if the trunk breaks at the base. Losing its integrity, in the deadwood phase, it becomes to a certain extent similar to initially mechanically damaged substrates.

Table 7. Occurrence of some wood-inhabiting basidiomycetes on different categories of birch substrates* in the southern part of the forest zone (%).

Fungal species	**Shrunken at the root**			**Mechanically damaged (wounded)**				**All shrunken**	**All wounded**	**Living**
	1	2	3	4	5	6	7			
Trametes ochracea + T. versicolor	–	3.9	3.4	73.6	7.7	10.9	0.6	7.2	92.8	–
Irpex lacteus	–	10.2	10.2	18.4	49.0	12.2	–	20.4	79.6	–
Steccherinum ochraceum	1.2	18.0	63.9	3.7	3.6	9.2	0.6	83.2	16.8	–
Fomitopsis betulina	2.8	71.8	21.9	0.6	0.5	2.3	0.1	96.4	3.6	–
Inonotus obliquus	–	25.9	25.2	4.8	0.7	21.1	–	51.1	26.7	22.2

Note: *1 – dead crown trees, 2 – dead trees, 3 – fallen logs, derivatives of dead trees, 4 – cutting stumps, 5 – cutting crown, 6 – windbreak, 7 – damaged standing trees.

Therefore, fungi characteristic of this phase, in particular *Steccherinum ochraceum*, are more common on initially wounded substrates (17%) than *Fomitopsis betulina*, which is confined to fresh whole-bark dead wood (3.6%).

The stem pathogen *Inonotus obliquus* is initially able to develop only in living trees, causing a central rot and forming barren growths on them. However, this phase of the development of the fungus accounts for only 22% of its encounters. But more frequently it can be found after the death of a tree, on deadwood or windbreak trunks broken due to the development of rot, where it forms true basidiomata during the first years (there have been cases of renewal of such basidiomata on the substratum within 4 years). Judging by the fact that the species is more often found on substrates that have shrunk at the root (51% of cases) than on those that have been broken during life (27%), it can be assumed that it is close in origin to fungi that destroy entire bark substrates, primarily dried branches of living trees through which rot can penetrate into the trunk.

The study of basidiomata localization on a number of species on 300 birch substrates, carried out in the forest park of Tyumen city (Table 8), showed that species characteristic of integrated substrates (*Fomitopsis betulina*, *Fomes fomentarius* and others), even developing on mechanically damaged substrates, occupy their whole-bark parts. Species characteristic of damaged substrates (*Trametes versicolor* and *Lenzites betulina*) develops mainly on bare wood and in the area closest to it. Some of them (*Bjerkandera adusta* and *Stereum hirsutum*), penetrating through damaged branches, are able to spread along their vascular system to intact parts of the cut trunk, forming basidiomas in the traces of long-shrunken branches along a significant length of the trunk. Due to this ability, in oak plantings and some other

Table 8. Localization of basidiomata of wood-inhabiting basidiomycetes on birch substrates in the Tyumen vicinity.

Fungal species	**Localization***					
	Bare wood	**Branches**	**Branch traces**	**Smooth bark**	**Fissured bark**	**Burnt bark**
Bjerkandera adusta	21	12	15	21	28	3
Cerrena unicolor	17	–	–	7	76	–
Daedaleopsis confragosa s. l.	–	2	3	85	10	–
Datronia mollis	29	18	18	35	–	–
Fomes fomentarius	–	–	2	69	29	–
Fomitopsis betulina	–	–	–	95	5	–
Irpex lacteus	7	7	4	56	15	11
Lenzites betulina	78	3	1	7	11	–
Stereum hirsutum	37	7	21	21	7	7
Stereum subtomentosum	20	5	20	25	5	35
Trametes versicolor	56	7	3	11	23	–
Trichaptum biforme	4	4	24	37	28	2

*Percentage of substrates with a given localization of the total number of substrates with species basidiomata.

alien species, these species become real stem parasites in the Tyumen environment. Species characteristic of both mechanically damaged and whole-bark birch substrates (e.g., *Irpex lacteus*) usually inhabit thin shrunken branches, burnt parts of the trunk, or trunks that have lost their integrity during decomposition. They are most often found in open habitats and digressed forests.

Table 9 shows the results ordination of wood-inhabiting basidiomycetes in terms of the weighted proportion of the species on mechanically damaged substrates. Since it is determined by the preservation of the barrier function of the cortex, it can be said that the species are distributed along the gradient of the barrier-substrate factor. The constructed series evenly covers almost the entire range of the indicator (from 5 to 98%). Having conditionally distinguished three gradations on it, it is possible to correlate with them three barrier-substrate groups (guilds) of species.

Species with an indicator value varying from 0 to 33% (*Fomes fomentarius*, *Fomitopsis betulina*, *Hapalopilus rutilans*, *Trichaptum biforme*, *Daedaleopsis confragosa* s.l., *Xanthoporia radiata*, *Phellinus laevigatus*, *Fomitopsis pinicola*, *Steccherinum murashkinskyi*, *S. ochraceum*, *Oxyporus corticola*) are confined to dried whole-bark substrates (dry wood and derived dead wood). Species with an indicator value from 67 to 100% (*Lenzites betulina*, *Trametes trogii*, *T. versicolor*, *T. ochracea*, *T. hirsuta*, *T. gibbosa*, *Bjerkandera adusta*, *Cerrena unicolor*, *Chondrostereum purpureum* and others), as a rule, occupy mechanically damaged (cutting, windbreak) substrates.

Table 9. Weighted proportion (%) of occurrences of wood-inhabiting basidiomycetes on mechanically damaged birch substrates.

Fungal species	**%**	**Fungal species**	**%**
Fomitopsis betulina	5	*Tyromyces chioneus*	59
Xanthoporia radiata	6	*Irpex lacteus*	60
Hapalopilus rutilans	8	*Datronia mollis*	65
Daedaleopsis confragosa s. l.	13	*Trametopsis cervina*	67
Steccherinum murashkinskyi	13	*Trametes pubescens*	70
Phellinus laevigatus	14	*Ganoderma applanatum*	74
Ph. igniarius s. l.	18	*Schizophyllum commune*	79
Gloeoporus dichrous	22	*Stereum hirsutum*	80
Oxyporus corticola	22	*Cerrena unicolor*	83
Steccherinum ochraceum	23	*Cylindrobasidium laeve*	84
Plicatura crispa	25	*Bjerkandera adusta*	86
Fomitopsis pinicola	29	*Chondrostereum purpureum*	90
Fomes fomentarius	32	*Trametes gibbosa*	94
Trichaptum biforme	31	*T. versicolor* + *T. ochracea*	95
Stereum subtomentosum	40	*T. hirsuta*	97
Laxitextum bicolor	45	*Trametes trogii*	98
Inonotus obliquus	44	*Lenzites betulina*	98
Phlebia tremellosa	55		

The intermediate group with an indicator from 34 to 66% includes species that occur in a wide substrate range, but relatively more often than others develop on branches or on charred trunks (*Laxitextum bicolor*, *Stereum subtomentosum*, *Irpex lacteus* and *Datronia mollis*). Under competitive conditions, they gain an advantage when colonizing wood through weakened or specific barriers: through thin, underdeveloped or fire-damaged bark, through charred wood. This third group of fungi can be characterized as a zone of a gradual transition between the two main substrate groups of fungi within a single barrier-substrate continuum. It should be noted that the most typical birch stem parasites fell into the first group (*Phellinus igniarius* s. l. – 18%) or into the intermediate group (*Inonotus obliquus* – 44%) according to this indicator.

The presented data and numerous observations show a significant differentiation of the xylomycocomplex both in terms of the characteristic localization of different species within the substrate and the features of penetration into the wood. Species characteristic of whole-bark dead substrates, usually formed as a result of competitive drying of the oppressed part of the forest stand and penetrating through natural perforations of the bark, are designated as "transcortical". Species penetrating wood through mechanically damaged, barkless areas and characteristic of substrates with a root-leaf bond broken *in vivo* (broken for natural reasons or cut down), are designated as "non-cortical", or "wounded". The species of the intermediate group are designated as "semi-cortical". Typical stem parasites that cause chronic birch central rot (*Phellinus igniarius*, etc.) do not constitute a separate group in this regard but are in the range between transcortical and semicortical fungi.

Fungal species that carry out the first stages of wood decomposition (up to about 7 years) are designated as "primary"; species characteristic of heavily decomposed wood and lasting up to 15 years or more as "secondary". So-called "pioneering" species (in particular, *Chondrostereum purpureum*, *Cylindrobasidium laeve*), which appear soon after tree damage and rarely persist for more than two years, using mainly the protoplast of wood and bark cells for nutrition (Davydkina, 1974). There is no clear line between the distinguished groups of fungi. In incipient phases of wood degradation, there is no contact with the ground/soil; thus, decay develops very slowly. Therefore, in such states, the predominance of pathogenic saprotrophs with some other xerotolerant saprotrophs is seen.

On the destruction of roots and gross branches, the logs become slowly immersed into the ground, thus, contact establishes with capillary connection with water and or soil mycelium. Thus, such contact represents an important event in the decomposition of wood. In spruce forests, the pioneer groupings of wood decomposing fungi are substituted by decortical fungi as well as strong saprotrophic wood-colonizers (e.g., *Heterobasidion parviporum* is replaced by *Junghuhnia collabens*, *Perenniporia* spp. or *Dichostereum boreale*). A principal decayer like *Fomitopsis pinicola* is changed by *Amylocystis lapponica*, *Rhodofomes roseus* and *Pycnoporellus fugens* subsequently, the latter is changed with a rather large suite of tertiary successors fungi responsible for initiating the humification process (*Postia* spp., *Skeletocutis* spp., *Crustoderma* spp. or *Phlebia centrifuga*). Simultaneously, branch tipss and branchlets are decayed by some wood decaying fungi (*Peniophora pini*, *Punctularia strigosozonata, Vuilleminia alni*) and non-specific fungi (*Hyphodontia* spp., *Hyphoderma* spp., *Byssomerulius corium* and *Cylindrobasidium laeve*).

In the phase of humification, several ephemerous hygrophilic wood-inhabiting fungi (*Anomoporia* spp., *Ceriporiopsis* spp., *Ceriporia* spp., *Physisporinus* spp., *Trechispora* spp. and other corticioids) establish the rest of the cavernose wood mass. There are several threads and rhizomorph-forming wood fungi decompose fallen log remnants to transform amorphous and protruded by roots of young trees and shrubs are referred to as "soil bolsters". As a result of the humification of wood, the virgin boreal forests are categorized by fractured microrelief-formed by the "wood remnants-soil complexes" as well as a rather gross amorphous wood debris layer reaching 15 cm in depth (Smirnova, 2004). This layer is the main creator of humic and fulvic acids that migrate into the water bodies and accrue as colloids-linked Ca^{2+}, Mg^{2+} and K^{+} (Dighton et al., 2005). Thus, the boreal soils are acidic and dominated by oligotrophic moss communities.

Water does not reach the ground surface during the warm period in placore nemoral forests. In such forests, the rate of wood decomposition and humification is slow, although humus accumulates at a high rate, while humic acids production falls. *Ganoderma applanatum, Hypsizygus ulmarius, Pleurotus dryinus, Inonotus dryophilus, Aurantiporus fissilis, Phellinus* spp. and *Spongipellis* spp. initiate the decaying process of stands as well as stumps. Fallen logs as well as dry stumps are occupied by such key species *Bjerkandera adusta, Hapalopilus croceus* and or *Fomes fomentarius*. Transcortical fungi *Junghuhnia* spp., *Oxyporus* spp. and *Hypochnicium* spp. are leaders in the process of wood decortications. Many *Peniopora* sp. decomposes fallen terminal branches. After partial decortication and branch degradation, the rich wood-inhabiting fungal communities are revealed by marker fungi species (*Crepidotus* spp., *Pluteus* spp., *Phlebia* spp. and many other corticioids) (Malysheva and Malysheva, 2008).

In the south of the nemoral zone, the function of pathogenic saprotrophs (*Ganoderma* spp. and *Phellinus* spp.) in the wood decomposition process is increasing. The fallen wood decomposes very gradually and the role among non-pathogenic decayers is possible by *Trametes*, which represent in the southern areas more than 50 species (Zmitrovich et al., 2012).

13. Substrate Ecology and Some Problems in Biogeography

The fungi of the Polyporaceae family represent a central group of xylotrophic basidiomycetes capable of the most rapid and complete colonization of the woody substrate. The close relationship of Polyporaceae species in development with the substrate determines their coenogenetic conservatism, i.e., the predominance in their modern coenogenetic and geographical history the coenogenetic component over the phylocenogenetic one what is expressed in the existence of quite definite (statistically reliable) their coenospectral and substrate preferences (Zmitrovich, Spirin, 2005; Arefiev, 2010). Being embedded in the consortia of powerful stand-formers, these fungi consist the part of their cenogenetic formations.

The "fidelity" of each specific species to the coenogenetic suite is determined by its trophic amplitude. Highly specialized species are limited in distribution by the range of the stand-forming tree and usually radiate with it beyond the limits of the optimum range. Species with a wide trophic amplitude can be characterized by a certain coenocycle, either extremely wide (polychorous species) or limited to certain vegetation types. The coenospectrum complexes reflect the features of the distribution of species and genera within the limits of natural and climatic zones; their combinations are coenogeographic elements of mycobiota, a group of species that have a coenotic optimum in a certain vegetation zone, although they are able to irradiate as part of suites and intrazonal vegetation groupings (Zmitrovich et al., 2003).

On the one hand, unlimited possibilities for fungal expansion and their strict substrate and coenospectral confinement, are the main specific features of the distribution of these organisms (Zmitrovich et al., 2003). The "species information field" in the form of ubiquitous propagules spreads much faster and wider than the coenoses into which the corresponding species is included. During the period of extinction of dividogenetic processes and stabilization of the species, its exit beyond its coenogrouping most often turns out to be problematic due to the rigidity of the alien coenotic environment with its historically established links systems and ecological niche saturation.

In the case of fragmentation of fungal coenopopulations, the ubiquitous nature of fungal propagules determines the synchronism of the corresponding transformations of the mycobiota, the conjugation of processes occurring at the level of local and regional ecotypes. In the opposite case, during the spread, migration of a single grouping, fungi, as it were, precede its expansion, forming an outside fund of propagules that contributes to a fairly accurate reproduction of the structural features of the groupings in the event of the arrival, formation and development of the latter.

The aforementioned accurate reproduction is associated with a strong nature of the connection of xylotrophic fungi with the stands edificators (trophic connection) and a parallel "cementing" of these strong connections by a network of weaker links to coenotic environment, primarily insolation and humidity of the biotope (Zmitrovich, Spirin, 2005). The formation of mycobiota is connected with the fate of coenogenetic suites – their migration, overwhelming and historical transformations. Fungal groups turn out to be much more conservative than the phytocenotic formations of the anthropogen. Such concepts as "taiga species" or "meadow species" do not represent a historical and coenogenetic unity. The first group includes representatives of the relic mountain taiga, younger betuletal and pine forest complexes as well as a number of complexes included in the taiga flora from intrazonal coenogroupings. The second group includes representatives of the boreal and nemoral complexes of forest edge-glade species, intrazonal floodplain complexes and pine forest complexes, which are replaced in the forest steppe by complexes associated with aride mycobiota. The complexity of the vegetation cover also complicates the direct "adaptive" interpretation of ecomorphs, since coenogenetic formations overlap during large-time scaled coenogenetic processes (Zmitrovich et al., 2003). The reconstruction of the generic ranges of Polyporaceae is hampered by uneven knowledge of different continents. The polypores of the Holarctic are the most studied (Tables 10, 11).

Table 10. Distribution patterns of the genera of Polyporaceae.

Distribution pattern	Areas occupied	Genera
Cosmopolitan	All the continents and climatic zones	*Cerioporus, Epithele, Funalia, Ganoderma, Lentinus, Perenniporia, Pycnoporus, Trametes, Vanderbylia*
Panholarctic	Extratropical regions of the Holarctic	*Daedaleopsis, Dichomitus, Donkioporia, Fomes, Haploporus, Neofavolus* and *Pachykytospora*
American-East Asian	Extratropical regions of North America and East Asia	*Cryptoporus*
American irradiating	North, Central and South America	*Cubamyces, Fuscocerrena, Globifomes, Mollicarpus* and *Pogonomyces*
Bipolar	Cold and temperate zones of the northern and southern hemispheres	*Lenzites* and *Picipes*
Pantropical	Tropical (subtropical) belt	*Amauroderma, Cellulariella, Coriolopsis, Earliella, Grammothele, Hexagonia, Favolus, Truncospora, Porogramme, Pyrofomes* and *Sclerodepsis*
Paleotropical	Africa, South Asia, Northern Australia	*Microporus* and *Lignosus*
African	Africa	*Grammothelopsis* and *Xerotus*
Tropical Asian	Tropical regions of Asia	*Cerarioporia, Hymenogramme, Megasporoporiella* and *Sparsitubus*
Australian	Australia and New Zealand	*Australoporus*
Nothoboreal	Temperate regions of Australia and South America	*Phaeotrametes*
Neotropical	Tropical regions of South America	*Neofomitella* and *Perenniporiella*

In total, about 60 genera of Polyporaceae are known in the world, including 554 species (excluding taxa, the species status of which requires a critical study). These genera were divided into 6 tribes (*Polyporeae, Epitheleae, Lentineae, Ganodermateae, Pycnoporeae* and *Trameteae*) and 2 subfamilies (*Polyporoideae* and *Trametoideae*) (Zmitrovich, 2018). The most species-rich genera are *Perenniporia* (78), *Ganoderma* (67), *Amauroderma* (48), *Lentinus* (42), *Cerioporus* (26), *Truncospora* (23), *Epithele* (22), *Grammothele* (21). There are several monotypic genera (*Cladomeris, Earliella, Fuscocerrena, Mollicarpus, Mycobonia, Phaeotrametes, Pilatotrama, Pogonomyces* and *Xerotus, Yuchengia*), the identity of the genera *Fomes* and *Globifomes* remain questionable (Table 12).

The cosmopolitan and pantropical distribution ranges of modern genera (*Cerioporus, Epithele, Funalia, Ganoderma, Lentinus, Perenniporia, Pycnoporus, Trametes, Vanderbylia, Amauroderma, Cellulariella, Coriolopsis, Earliella, Grammothele, Hexagonia, Favolus, Truncospora, Porogramme, Pyrofomes* and

Table 11. Basic summaries on Polyporaceae from different regions of the world.

Major regions of the world	**Territories explored**	**Basic summaries**
Holarctic	European Russia, Caucasus	Bondartsev (1971)
	Caucasus	Ghobad-Nejhad (2011)
	Russia	Bondartseva (1998)
	Europe	Ryvarden and Gilbertson (1993, 1994); Ryvarden and Melo (2017)
	Israel	Ţura et al. (2011)
	China	Zhishu et al. (1993); Teng (1996)
	Japan	Imazeki et al. (1988)
	North America	Gilbertson and Ryvarden (1986, 1987)
Tropical regions	Central America	Fidalgo and Fidalgo (1966); Gilbertson and Ryvarden (1986, 1987)
	South America	Fidalgo and Fidalgo (1968); Corner (1981–1990); David, Rajchenberg (1985); Rajchenberg (1989); Ryvarden (2004, 2015, 2016); Nakasone (2003)
	South Asia	Corner (1981–1990); Parmasto (1986); Roy and De (1996); Núñez and Ryvarden (2001); Bhosle et al. (2010); Ranadive et al. (2011); Ranadive et al. (2018)
	Africa	Ryvarden and Johansen (1980); Härkonen et al. (2003); Härkonen et al. (2015)
	North Australia	Hood (2003)
Extratropical regions of the Southern Hemisphere	South America	Wright and Deschamps (1972, 1975); Rajchenberg (1989); Ryvarden (2004–2016)
	South Africa	Ryvarden and Johansen (1980)
	Australia	Hood (2003)
	New Zealand	Buchanan and Ryvarden (2000)

Sclerodepsis), having pantropical disjunction, can be interpreted as a Cretaceous relict. Also, highly characteristic is the genus *Pyrofomes* (*P. demidoffii*), associated with juniper, whose range covers the subtropics of North America, the Black Sea region and the Ethiopian highlands, areas that have been developing without floristic contacts for at least 60 million years.

Some relict genera have extended their range to the temperate and cold zones of the planet, where they are represented by species that hold the main volumes of the woody substrate and dominate ecosystems (*Ganoderma applanatum*, *Cerioporus squamosus*). The genera *Picipes* and *Lenzites* tend to have a bipolar distribution, although a number of species occur in the subtropics of the Northern (*Picipes subtropicus*) and Southern (*P. virgatus*) hemispheres. The bipolar nature of the ancestral area testifies to the initial cosmopolitanism of the genus and subsequent disjunction associated with cyclic fluctuations in the climate of the warm zone of the planet.

Table 12. Taxonomic proportions in the global biota of Polyporaceae.

Subfamilies	Tribes	Genera	Number of species
Polyporoideae	*Polyporeae*	*Cerarioporia*	1
		Cerioporus	26
		Cladomeris	1
		Echinochaete	5
		Favolus	13
		Megasporia	7
		Mycobonia	1
		Neofavolus	5
		Picipes	17
		Polyporus	1
	Epitheleae	*Epithele*	22
		Grammothele	21
		Hymenogramme	1
		Porogramme	7
	Lentineae	*Australoporus*	1
		Daedaleopsis	10
		Earliella	1
		Fomes	1
		Funalia	17
		Fuscocerrena	1
		Globifomes	1
		Grammothelopsis	7
		Hexagonia	8
		Lentinus	42
		Lignosus	8
		Microporus	4
		Mollicarpus	1
		Neofomitella	3
		Pogonomyces	1
		Xerotus	1
	Ganodermateae	*Amauroderma*	48
		Cryptoporus	2
		Dichomitus	22
		Donkioporia	2
		Ganoderma	67
		Haploporus	3
		Leifiporia	2

Table 12 contd. ...

...Table 12 contd.

Subfamilies	Tribes	Genera	Number of species
		Pachykytospora	8
		Perenniporia	78
		Perenniporiella	6
		Phaeotrametes	1
		Pyrofomes	7
		Sparsitubus	1
		Tomophagus	2
		Truncospora	23
		Vanderbylia	9
		Yuchengia	1
Trametoideae	*Pycnoporeae*	*Artolenzites*	1
		Cellulariella	2
		Cubamyces	4
		Pilatotrama	1
		Pycnoporus	3
	Trameteae	*Coriolopsis*	2
		Lenzites	5
		Sclerodepsis	3
		Trametes	18

Panholarctic genera (*Daedaleopsis*, *Fomes*, *Haploporus*, *Pachykytospora*) seem to be younger. The formation of such genera is associated with the events of the Oligocene, when the vegetation cover changed sharply in temperate and high latitudes, natural zonality shifted to the south and subtropical and warm temperate elements of ancient floras were preserved only in the East Asian, Central American and Near Asian refugia (Kamelin, 2007).

Known within Polyporaceae also more local generic range within the Holarctic, namely the American-East Asian (genus *Cryptoporus* with a single species of *Cryptoporus volvatus*), common in East Asia and the temperate zone of North America. This pattern of distribution may also indicate the influence of the Oligocene events on the warm-temperate vegetation of the Northern Hemisphere, which reduced its range and became separated much earlier than the separation of the Eurasian and American continents. The local generic ranges of fungi in the Southern Hemisphere are usually considered in connection with the events of the late Cretaceous and Paleocene (Rajchenberg, 1989; Ryvarden, 1991; Zmitrovich et al., 2015), when the African and South American continents as well as Hindustan and Madagascar, diverge. Paleotropical genera of Polyporaceae (*Microporus*, *Lignosus*) are common in Africa, South Asia and Northern Australia. They are clearly younger than cosmopolitan genera (their age does not exceed 100 million years) and could not expand their range a second time due to competitive limitations.

Neotropical (*Neofomitella*), African (*Grammothelopsis* and *Xerotus*), tropical Asian (*Cerarioporia, Hymenogramme, Megasporoporiella* and *Sparsitubus*), Australian (*Australoporus*) and Notoboreal (*Phaeotrametes*) genera of polypore fungi are even younger, i.e., separated during the Paleocene – Eocene. Thus, we can conclude that a significant part of the genera of Polyporaceae (having a cosmopolitan, pantropical, bioplar distribution) descends in their origin to Late Cretaceous. The formation of Neotropical, African, tropical Asian and Australian genera took place in the Paleocene-Eocene and, finally, the younger Holarctic and American-East Asian genera are the youngest, separated in the Oligocene.

Conclusion

The land woody vegetation as well as the ocean bioherms are confined to zones of an optimal balance of abiotic factors, which allows biomes to overproduce the biomass and to organize the space as cavernous "spongy surfaces". The land "spongy surface" formed by woody vegetation is the most dynamic formation, since it is updated in a regime of gap or fire dynamics and inside ecosystems, there is a downward flow of elements associated with the biodegradation of cellulose and lignin composites. In fact, we have a range of transitions from lignin into growing woody plants to soil humus, the main wealth of soils in arid areas. In the latter, the hydrophenyl lignin of herbaceous plants takes part in humus formation, which is utilized by fungal derivatives of xylotrophic fungi, so-called litter and humus saprotrophs. Thus, in nature, wood-inhabiting fungi are important agents involved in the decomposition of wood, soil humus formation and nutrient recycling in various ecosystems. In terms of substrate specialization, the various wood decomposing fungi belong to different groups. Some of them are restricted to colonizing one type of substrate followed by the distribution range of the substrate that they prefer (e.g., *Inonotus tamaricis*), whereas other fungal species evolved on other modes. For instance, fungal species are able to colonize wood at a fast rate, occurring on a wide range of substrata including living as well as dead hardwoods. At the early stages of evolution, the basidiomycetes developed the ability to lignin biodegradation, which predetermined their coevolutionary connection with the forest vegetation of the planet. We can only guess what developed earlier – ancestral "soft" rot or white rot with an abundance of produced peroxidases and laccases, leaning towards the second assumption. The substrate adaptation of wood-inhabiting basidiomycetes in the evolutionary timescale took place under conditions of climate fluctuations, vegetation transformation and ecosystem rearrangements that opened up new niches and allowed zonal-adapted species to expand their substrate range and geographic range.

In terms of a local recent slice, here we can distinguish at least three levels of substrate ecology of wood-inhabiting basidiomycetes. The first level is determined by the taxonomic specificity of woody substrates (coniferous, deciduous and others) and is associated with the genetically incorporated physicochemical properties of the wood (heartwood, sapwood and bark), which correspond, first of all, with the biochemical specificity of xylolysis in different basidiomycetes groupings. At this level, the arrangement of these fungi would be made from first-level euritrophs to stenotrophs. The second level is set by the vital state of the substrate, the rate and nature of the extinction of life processes and, accordingly, the processes of physiological and biochemical protection of the tree organism. At this level, various groups of fungi are distinguished that carry out pathogenic (usually, central rot) and saprotrophic pathways of wood decomposition and within the saprotrophic pathway – on the options for drying on the vine, or mechanical destruction of the trunk. Central rot involves the slowest chronic (during decades) extinction of the life processes of a tree, drying out on the vine – the average variant of extinction (about a year), mechanical destruction – in the limit, an instantaneous cessation of the life of a tree, but taking into account the inertia of resources, lasting from hours to months. Accordingly, the spectrum of wood-inhabiting fungi characteristic of the tree species varies from obligate pathogens to wound ruderal fungi. The third level is set by the ecological conditions of the decomposition of the wood substrate, determined by the interaction of environmental factors with such passive endogenous factors of the wood substrate as its size and the physical parameters of the bark (or its absence). Accordingly, the spectrum of species varies from violents (*Fomes fomentarius* and *Fomitopsis pinicola*), occupying large niches with an optimal hydrotheritic regime inside the trunk, to patients occupying various marginal niches with an unstable ecological regime. The mentioned levels of substrate ecology have a pronounced successional aspect, determined by changes in wood during its decomposition and the multiplication of the number of ecological niches of wood-destroying fungi at all levels. At the same time, if the first level of substrate ecology of wood-destroying species is traditionally widely discussed, then the differentiation of fungi at the third level, which largely explains their biological diversity, is usually not given befitting importance.

Acknowledgements

The work of I.V. Zmitrovich and M.A. Bondartseva was completed according to state task No. 122011900033-4. The work of S.P. Arefiev was completed according to the state task No. 121041600045-8, the project "Western Siberia in the context of Eurasian relations: human, nature, society". The work of A.G. Shiryaev was completed according to the state task No. 122021000092-9, the project "Biodiversity of flora and mycobiota and its dynamics under the influence of global, regional and local factors".

References

Andersson, L., Alexeeva, N. and Kuznetsova, E. (2009). Revealing and investigation of biologically important forests in North-West Russia. T. 2. Guide to species identification. Pobeda, St Petersburg, Russia, p. 217 (in Russian).

Arefiev, S.P. (1986). Complexes of species of xylotrophic fungi on birch wood. *In*: Botanical research in the Urals. Sverdlovsk (in Russian).

Arefiev, S.P. (1991). Xylotrophic fungi as causative agents of rot diseases of Siberian pine in the middle taiga Irtysh Region. Mikologiya i fitopatoologiya, 20: 419–425 (in Russian).

Arefiev, S.P. (1993). Consortive relationships of xylotrophic fungi with Siberian pine. Ekologiya, 2: 85–88 (in Russian).

Arefiev, S.P. (1997). The consortium structure of the community of xylotrophic fungi in the Tyumen city. Mikologiya i fitopatologiya, 5: 1–8 (in Russian).
Arefiev, S.P. (2002). An ecological co-ordination of wood-destroying fungi (on the example of a birch consortium). Mikologiya i fitopatologiya, 36: 1–14.
Arefiev, S.P. (2010). A system analysis of biota of xylotrophic fungi. Nauka, Novosibirsk, p. 260 (in Russian).
Arefiev, S.P. (2020). *Ganoderma lucidum*. In: Red data book of the Tyumen region: Animals, plants, fungi. Kemerovo, p. 395 (in Russian).
Aro, N. Pakula, T. and Penttilä, M. (2005). Transcriptional regulation of plant cell degradation by filamentous fungi. FEMS Microbiol. Rev., 29: 719–739.
Bandoni, R.J. (1961). The genus *Naematelia*. Am. Midland Nat., 66: 319–328.
Bandoni, R.J. (1987). Taxonomic overwiev of the Tremellales. Stud. Mycol., 30: 87–110.
Bhosle, S., Ranadive, K., Bapat, G., Garad, S., Deshpande, G. and Vaidya, J. (2010). Taxonomy and diversity of *Ganoderma* from the Western parts of Maharashtra (India). Mycosphere, 1(3): 249–262.
Binder, M. and Hibbett, D.S. (2006). Molecular systematics and biological diversification in Boletales. Mycologia, 98: 917–925.
Bon, M. (1991). Les tricholomes et ressemblants (Tricholomoideae et Leucopaxilloideae). Genres: *Tricholoma, Tricholomopsis, Callistosporium, Porpoloma, Floccularia, Leucopaxillus* et *Melanoleuca*. Fl. Mycol. Eur., 2: 1–163.
Bondartsev, A.S. (1956). A guide to the house fungi. Academy of Sciences, Lenindrad, p. 80 (in Russian).
Bondartsev, A.S. (1971). The Polyporaceae of the European USSR and Caucasia. Israel Program for scientific translations, Jerusalem, p. 896.
Bondartseva, M.A. (1965). The factors determined the distribution of aphyllophorous fungi by the forest types. *In*: The Problems of the Study of Fungi and Lichens. Tartu (in Russian).
Bondartseva, M.A. (1992). Species composition, distribution in forest biogeocoenoses and ecological function of wood-destroying polypores. *In*: Scientific Bases of Forest Resistance to Wood-destroying Fungi. Nauka, Moscow.
Bondartseva, M.A. (1998). The handbook on fungi of Russia. Order Aphyllophorales. Ser. 2. Nauka, St Petersburg, p. 391.
Bondartseva, M.A. and Parmasto, E.H. (1998). Clavs diagnosticum fungorum URSS. Ordo Aphyllophorales. Fasc.1.Familiae Hymenochaetacee, Lachnocladiaceae, Coniophoraceae, Schizophyllaceae. Nauka Publ., Leningrad, 1986, p. 192 (in Russian).
Buchanan, P.K. and Ryvarden, L. (2000). An annotated checklist of polypore and polypore-like fungi recorded from New Zealand. N.Z. J. Bot., 38: 265–323.
Burdsall, H.H. and Banik, M.T. (2001). The genus *Laetiporus* in North America. Harv. Pap. Bot., 6: 43–55.
Chadefaud, M. 1950. Les Psilotinées et l'évolution des Archégoniates. Bull. Soc. Bot. France., 97: 99–100.
Chamuris, G.P. (1988). The non-stipitate stereoid fungi in the Northern United States and adjacent Canada. Mycol. Mem., 14: 1–247.
Chen, D.M., Taylor, A.F.S., Burke, R.M. and Cairney W.G. (2001). Identification of genes for lignin peroxidases and manganese peroxidases in ectomycorhizal fungi. New Phytol., 152: 151–158.
Church, A.H. (1919). Thalassiophyta and the subaerial transmigration. Bot. Mem., 3: 1–95.
Cooke, R.C. and Whipps, J.M. (1980). The evolution of modes of nutrition in fungi parasitic on terrestrial plants. Biol. Rev., 55: 341–362.
Corner, E.J.H. (1981). The agaric genera *Lentinus*, *Panus* and *Pleurotus*, with particular reference to Malasyan species. Beih. Nova Hedw., 69: 1–169.
Corner, E.J.H. (1983). Ad Polyporaceas I. *Amauroderma* and *Ganoderma*. Beih. Nova Hedw., 75: 1–182.
Corner, E.J.H. (1984). Ad Polyporaceas II and III. *Polyporus*, *Mycobonia* and *Echinochaete*. *Piptoporus*, *Buglossoporus*, *Laetiporus*, *Meripilus* and *Bondarzewia*. Beih. Nova Hedw., 78: 1–222.
Corner, E.J.H. (1989). Ad Polyporaceas V. The genera *Albatrellus*, *Boletopsis*, *Coriolopsis* (dimitic), *Cristelloporia*, *Diacanthodes*, *Elmerina*, *Fomitopsis* (dimitic), *Gloeoporus*, *Grifola*, *Hapalopilus*, *Heterobasidion*, *Hydnopolyporus*, *Ischnoderma*, *Loweporus*, *Parmastomyces*, *Perenniporia*, *Pyrofomes*, *Steccherinum*, *Trechispora*, *Truncospora* and *Tyromyces*. Beih. Nova Hedw., 96: 1–218.
Corner, E.J.H. (1990). Ad Polyporaceas VI. The genus *Trametes*. Beih. Nova Hedw., 97: 1–206.

Cowling, E.G. (1961). Comparative biochemistry of the decay of sweetgum sapwood by white-rot and brown-rot fungi. US Dept. Agric. Tech. Bull., 258: 1–75.

D'Souza, T.M., Boominathan, K. and Reddy C.A. (1996). Isolation of laccase gene-specific sequences from white-rot and brown-rot fungi by PCR. Appl. Environ. Microbiol., 62: 3739–3744.

Davydkina, T.A. (1974). On the significance of biological characters for the taxonomy of the genus *Stereum* Pers. ex S.F. Gray s. lato. Mikologiya i fitopatologiya, 8: 78–81 (in Russian).

Davydkina, T.A. (1980). Stereaceous fungi of Soviet Union. Nauka, Leningrad., p.143 (in Russian).

Dighton, J., White, J.F. and Oudemans, P. (eds.). (2005). The Fungal Community. Its Organization and Role in the Ecosystem. Third edition. Taylor & Francis, L.; N.Y.; Singapore, p. 936.

Domański, S. (1988). Mała flora grzybów. Basidiomycetes (Podstawczaki). Aphyllophorales (Bezblaszkowce). 5. Corticiaceae: *Acanthobasidium–Irpicodon*. PWN, Warszawa–Krakow, p. 427 (in Polish).

Domański, S. (1991). Mała flora grzybów. I. Basidiomycetes (Podstawczaki). Aphyllophorales (Bezblaszkowce). Stephanosporales (Stefanosporowce). 6. Corticiaceae: *Kavinia-Rogersella*, Stephanosporaceae: *Lindtneria*. PWN, Warszawa–Krakow, p. 272 (in Polish).

Domański, S. (1992). Mała flora grzybów. I. Basidiomycetes (Podstawczaki). Aphyllophorales (Bezblaszkowce). 7. Corticiaceae: *Sarcodontia-Ypsilonidium*, *Christiansenia* and *Sygygospora*. W. Szafer Institute of Botany, Polish Academy of Sciences, Krakow, p. 258 (in Polish).

Elena, K. and Paplomatas, E.J. (2002). First report of *Fomitiporia punctata* infecting kiwifruit. Pl. Dis., 86: 1176.

Eriksson, J. and Ryvarden, L. (1973). The Corticiaceae of North Europe/With drawings by John Eriksson. Volume 2: *Aleurodiscus–Confertobasidium*. Fungiflora, Oslo, p. 261.

Eriksson, J. and Ryvarden, L. (1975). The Corticiaceae of North Europe/With drawings by John Eriksson. Volume 3: *Coronicium–Hyphoderma*. Fungiflora, Oslo, p. 546.

Eriksson, J. and Ryvarden, L. (1976). The Corticiaceae of North Europe/With drawings by John Eriksson. Volume 4: *Hyphodermella–Mycoacia*. Fungiflora, Oslo, p. 886.

Eriksson, J., Hjortstam, K. and Ryvarden L. (1978). The Corticiaceae of North Europe/With drawings by John Eriksson. Volume 5: *Mycoaciella–Phanerochaete*. Fungiflora, Oslo, p. 1047.

Eriksson, J., Hjortstam, K. and Ryvarden, L. (1981). The Corticiaceae of North Europ/With drawings by John Eriksson. Volume 6: *Phlebia–Sarcodontia*. Fungiflora, Oslo, p. 1276.

Eriksson, J., Hjortstam, K. and Ryvarden, L. (1984). The Corticiaceae of North Europe/With drawings by John Eriksson. Volume 7: *Schizopora–Suillosporium*. Fungiflora, Oslo, p. 1449.

Eriksson, K.E., Grunwald, A., Nilsson, T. and Vallander, L. (1980). A scanning electron microscopy study of the growth and attack on wood of three white-rot fungi and their cellulase-less mutants. Holzforschung, 34: 207–213.

Eriksson, K.E., Blanchette, R.A. and Ander, P. (1990). Microbial and Enzymatic Degradation of Wood Components. Springer, Berlin-Heidelberg p. 407.

Erland, S. and Taylor, A.F.S. (1999). Resupinate ectomycorrhizal fungal genera. *In*: Ectomycorrhizal Fungi: Key Genera in Profile. Springer Verl., Heidelberg, pp. 347–363.

Ershov, R.V. and Ezhov, O.N. (2009). Aphyllophoroid fungi of aspen on Nort-West of Russian Plain. Arkhangelsk., p. 123 (in Russian).

Ezhov, O.N., Ershov, R.V., Ruokolainen, A.V. and Zmitrovich, I.V. (2011). Aphyllophoraceous fungi of Pinega Reserve. Arkhangelsk., p. 147 (in Russian).

Fidalgo, O. and Fidalgo, M.E.P.K. (1966). Polyporaceae from Trinidad and Tobago. Mycologia, 58: 862–904.

Fidalgo, O. and Fidalgo, M.E.P.K. (1968). Polyporaceae from Venezuele. 1. Mem. N.Y. Bot. Gard., 17: 1–34.

Fischer, M. and Gonzalez Garcia, V. (2015). An annotated check-list of European basidiomycetes related to white rot of grapevine (*Vitis vinifera*). Phytopathologia Mediterranea, 54, 2: 281–298.

Gamauf, C., Metz, B. and Seiboth, B. (2007). Degradation of plant cell wall polymers by fungi. *In*: Esser K. (ed.). The Mycota. A Comprehensive Treatise of Fungi as Experimental Systems for Basic and Applied Research. In: Environmental and Microbial Relationships. 2nd Ed. Springer, Heidelberg.

Gensel, P.G., Glasspool, I., Castaldo, R.A., Libertin, M. and Kvacek, J. (2020). Back to the beginning: The Silurian – Devonian as a time of major innovation in plants and their communities. pp. 367–398. *In*: Martinetto, E. et al. (eds.). Nature through Time. Springer, Heidelberg.

Ghobad-Nejhad, M. and Kotiranta, H. (2008). The genus *Inonotus* sensu lato in Iran, with keys to *Inocutis* and *Mensularia* worldwide. Ann. Bot. Fennici, 45: 465–476.
Ghobad-Nejhad, M. (2011). Updated checklist of corticioid and poroid basidiomycetes of the Caucasus region. Mycotaxon, 117: 1–70.
Gilbert, G.S. and Sousa, P. (2002). Host specialization among wood-decay polypore fungi in a Caribbean mangrove forest. Biotropica, 34: 396–404.
Gilbert, G.S., Gorospe, J. and Ryvarden, L. (2008). Host and habitat preferences of polypore fungi in Micronesian tropical flooded forests. Mycological Research, 112: 674–680.
Gilbertson, R.L. (1980). Wood-rotting fungi of North America. Mycologia, 72: 1–49.
Gilbertson, R.L. and Ryvarden, L. (1986). North American polypores. Volume 1. Fungiflora, Oslo, p. 436.
Gilbertson, R.L. and Ryvarden L. (1987). North American polypores. Volume 2. Fungiflora, Oslo, p. 885.
González García, V., Portal Onco, M.A. and Rubio Susan, V. (2006). Review. Biology and systematics of the form genus *Rhizoctonia*. Spanish J. Agricultural Res., 4: 55–79.
Gordienko, P.V. (1986). Peculiarities of settlement of some xylotrophic species on the substrate with different parameters. Mikologiya i fitopatologiya, 20: 131–134 (in Russian).
Gottlieb, A.M., Wright, J.E. and Moncalvo, J.-M. (2002). *Inonotus* s. l. in Argentina – morphology, cultural characters and molecular analyses. Mycol. Prog., 1: 299–313.
Härkonen, M., Niemelä, T. and Mwasumbi, L. (2003). Tanzanian Mushrooms. Edible, Harmful and other Fungi. Helsinki, p. 200.
Härkonen, M., Niemelä, T., Mbindo, K., Kotiranta, H. and Piearce G.D. (2015). Zambian Mushrooms and Mycology. Helsinki, p. 207.
Heal, O.W. and Dighton, J. (1985). Resource quality and trophic structure of soil system. pp. 339–354. *In*: Fitter, A.H., Atkinson, D., Read, D.J. and Usher, M.B. (eds.). Ecological Interactions in Soil. Blackwell, Oxford.
Hibbett, D.S. and Donoghue, M.J. (1995). Progress toward a phylogenetic classification of the Polyporaceae through parsimony analysis of mitochondrial ribosomal DNA-sequences. Can. J. Bot., 73: S853–S861.
Hood, I. (2003). An introduction to fungi on wood in Queensland. University of New England School of Environmental Sciences and Natural Resources Management, Armidale., p. 388.
Imazeki, R., Otani Y. and Hongo, T. (1988). Nihon no Kinoko (Fungi of Japan). Tokyo, p. 623.
Isikov, V.P. and Konoplya, N.I. (2004). Dendromycology. Alma Mater, Lugansk, p. 347 (in Russian).
James, T.Y., Kauf, F., Schoch, C.L. et al. (2006). Reconstructing the early evolution of fungi using a six-gene phylogeny. Nature, 443: 818–822.
Jing, H.U.I., Wenjing, Z. and Zhiyan, Z. (2007). Changes in extracellular enzyme activities during submerged culture of *Tremella aurantialba*. Acta Edulis Fungi, 14: 33–36.
Jülich, W. (1984). Die Nichtblätterpilze, Gallertpilze und Bauchpilze. Aphyllophorales, Heterobasidiomycetes, Gastromycetes. Gustav Fischer, Jena, p. 626.
Kirk, P.M., Cannon, P.F., David, J.C. and Stalpers J.A. (2001). Ainsworth and Bisby's Dictionary of the Fungi. 9th edition. Oxford UniVolume Press, New York, p. 672.
Koenigs, J.W. (1972). Effects of Hydrogen peroxidase on cellulose and on its susceptibility to cellulose. Mater. Organismen, 7: 133–147.
Kotiranta, H. and Niemelä, T. (1993). Uhanalaiset käävät Suomessa (Threatened polypores in Finland), Painatuskeskus, Helsinki.Vesi-ja ympäristöhallinnon julkaisuja, sarja B, 17: 1–116.
Lange, M. (1986). Fungus succession on fallen logs of beech. Svampe, 13: 38–41.
Larsen, M. and Cobb-Poulle, L.A. (1990). *Phellinus* (Hymenochaetaceae). A survey of the world taxa. Synopsis Fungorum, 3: 1–206.
Larsson, K.-H., Parmasto, E., Fischer, M., Langer, E., Nakasone, K.K. and Redhead, S.A. (2006). Hymenochaetales: A molecular phylogeny for the hymenochaetoid clade. Mycologia, 98: 926–936.
Lindgren, M. (2001). Polypores (Basidiomycetes) species richness and community structure in natural boreal forest of NW Russian Karelia and adjacent areas in Finland. Acta Bot. Fennica, 170: 1–41.
Malysheva, V.F. and Malysheva, E.F. (2008). The higher basidiomycetes in forest and grassland communities of Zhiguli. St Petersburg, p. 242 (in Russian).
Manskaya, S.M. and Kodina, L.A. (1975). Geochemistry of lignin. Nauka, Moscow, p. 229 (in Russian).
Miettinen, O. and Larsson, K.-H. (2011). *Sidera*, a new genus in Hymenochaetales with poroid and hydnoid species. Mycol. Prog., 10: 131–141.

Miettinen, O., Niemelä, T. and Spirin, W. (2006). Northern *Antrodiella* species, the identity of A. semisupina and type studies of related taxa. Mycotaxon, 96: 211–239.

Mukhin, V.A. (1993). Biota of xylotrophic basidiomycetes of the West Siberian Plain. Nauka, Ekaterinburg, p. 232 (in Russian).

Nagy, L.G., Riley, R., Tritt, A., Adam, C., Daum, C., Floudas, D., Sun, H., Yadav, J.S., Pangilinan, J., Larsson, K.-H., Matsuura, K., Barry, K., Labutti, K., Kuo, R., Ohm, R.A., Bhattacharya, S.S., Shirouzu, T., Yoshinaga, Y., Martin, F.M., Grigoriev, I.M. and Hibbet D.S. (2015). Comparative genomics of early-diverging mushroom-forming fungi provides insights into the origins of lignocellulose decay capabilities. Mol. Biol. Evol., 33(4): 959–970. 10.1093/molbev/msv337.

Nakasone K.K. and Gilbertson R.L. (1978). Cultural and other studies of fungi that decay Ocotillo in Arizona. Mycologia, 70(2): 266–299.

Nobles, M.K. (1958). Cultural characters as a guide to the taxonomy and phylogeny of the Polyporaceae. Can. J. Bot., 36: 883–926.

Núñez, M. and Ryvarden, L. (2001). East Asian Polypores. Volume 2. Fungiflora, Oslo, pp. 170–522.

Olive, L.S. (1946). New or rare Heterobasidiomycetes from Norh Carolina 2. J. Elisha Mitchell Sci. Soc., 62: 65–71.

Parmasto, E. (1986). Preliminary list of vietnamense Aphyllophorales and Polyporaceae s. str. Scripta Mycol., 14: 1–88.

Parmasto, E. (1998). *Athelia arachnoidea*, a lichenicolous basidiomycete in Estonia. Folia Cryptogamica Estonica, 32: 63–66.

Rabinovich, M.L., Bolobova, A.V. and Kondrashchenko, V.I. (2001). Theoretical bases for biotechnology of wood composites. Book 1: Wood and wood-destroying fungi. Nauka, Moscow, p. 264 (in Russian).

Ragan, M.A. and Chapman, D.J. (1978). Biochemical phylogeny of the protists, New York, p. 127.

Rajchenberg, M. (1989). Polyporaceae (Aphyllophorales, Basidiomycetes) from Southern South America: A mycogeographical view. Sydowia, 41: 277–291.

Ranadive, K. (2013). An overview of Aphyllophorales (wood rotting fungi) from India. Int. J. Curr. Microbiol. App. Sci., 2(12): 112–139.

Ranadive, K.R., Vaidya, J.G., Jite, P.K., Ranade, V.D., Bhosale, S.R. et al. (2011). Checklist of Aphyllophorales from the Western Ghats of Maharashtra State, India. Mycosphere, 2(2): 91–114.

Ranadive, K.R., Jite, P.K. and Ranade, V.D. (2015). Wood-rotting fungal taxonomy and Indian Aphyllo-Fungal Database, a study from the Western Ghats of Pune District. Academic Publishers.

Ranadive, K., Jagtap, N. and Khare, H. (2018). Fungifromindia: The first online initiative to document fungi from India. IMA Fungus, 8(2): 67–69. 10.1007/BF03449465.

Renvall, P. (1995). Community structure and dynamics of wood-rotting Basidiomycetes on decomposing conifer trunks in northern Finland. Karstenia, 35: 1–51.

Renwall, P. and Niemelä, T. (1992). Fungi on fungi. Wood-rotting basidiomycetes and their threatened successors. *In*: Fungi of Europe. Invesigation, Recording, Conservation: Abstr. XI Congr. Eur. Mycol., Kew.

Rinaldi, A.C., Comandini, O. and Kuyper, T.W. (2008). Ectomycorrhizal fungal diversity: Separating the wheat from the chaff. Fungal Diversity, 33: 1–45.

Ripachek, R. (1967). Biology of wood-destroying fungi. Lesnaya Promyshlennost, Moscow, p. 276 (in Russian).

Roy, A. and De, A.B. (1996). Polyporaceae of India. R.P. Singh Gahlot, Dahra Dun, p. 287.

Runge, A. (1969). Pilzsukzession auf Eichenstümpfen. Abh. Landesmus. Naturk. Münster Westfalen, 32: 3–10.

Runge, A. (1975). Pilzsukzession auf Laubholzstümpfen. Z. Pilkz., 47: 1–38.

Ryvarden, L. (1991). Genera of polypores. Nomenclature and taxonomy. Synopsis Fung., 5: 1–363.

Ryvarden, L. (2004). Neotropical polypores. Pt 1. Introduction, *Ganodermataceae* and *Hymenochaetaceae*. Synopsis Fung., 19: 1–229.

Ryvarden, L. (2005). The genus *Inonotus*, a synopsis. Synopsis Fung., 21: 1–149.

Ryvarden, L. (2015). Neotropical polypores. Pt 2. Polyporaceae. *Abundisporus–Nigroporus*. Synopsis Fung., 34: 233–443.

Ryvarden, L. (2016). Neotropical polypores. Pt 3. Polyporaceae. *Obba–Wrightoporia*. Synopsis Fung., 36: 447–613.

Ryvarden, L. and Johansen, I. (1980). A preliminary polypore flora of East Africa. Fungiflora, Oslo, p. 225.

Ryvarden, L. and Gilbertson, R.L. (1993). European polypores. Part 1. *Abortiporus–Lindtneria*. Fungiflora, Oslo, p. 387.

Ryvarden, L. and Gilbertson, R.L. (1994). European polypores. Part 2. *Meripilus–Tyromyces*. Fungiflora: Oslo, p. 743.

Ryvarden, L. and Melo, I. (2017). Poroid fungi of Europe / with photos by T. Niemelä and drawings by I. Melo and T. Niemelä. Fungiflora, Oslo, p. 455.

Schmidt, O. (2006). Wood and tree fungi. Biology, damage, protection and use. Springer, Berlin - Heidelberg, p. 334.

Shiryaev, A.G., Arefiev, S.P. and Kotiranta, H. (2013). Aphyllophoroid, heterobasidioid and exobasidioid fungi of the Russian part of Dahuria. Mikologiya i fitopatologiya, 47: 36–45 (in Russian).

Shiryaev, A.G., Zmitrovich, I.V. and Shiryaeva, O.S. (2021). Species richness of Agaricomycetes of hedge vines in Ekaterinburg city (Russia). Mikologiya i fitopatologiya, 55(5): 340–352.

Shiryaev, A.G., Zmitrovich, I.V., Bulgakov, T.S., Shiryaeva, O.S. and Dorofeyeva, L.M. (2022). Warming favors the development of rich and heterogeneous mycobiota on alien vines in a boreal city under continental climate. Forests, 13(2): 323. 10.3390/f13020323.

Sinadsky, Yu.V. (1983). Pine: Their Diseases and Destroyers. Moscow, p. 335 (in Russian).

Smirnova, O.V. (ed.). (2004). East European Forests: Holocene History and Modern State. Nauka, Moscow, p. 479 (in Russian).

Spirin, W.A. (2002). Aphyllophoraceous fungi of Nizhegorod Region: species composition and ecological peculiarities. Thesis. St Petersburg, p. 242 (in Russian).

Spirin, W.A., Zmitrovich, I.V. and Malysheva, V.F. (2006). On the systematics of *Inonotus* s.l. and *Phellinus* s.l. (Mucronoporaceae, Hymenochaetales). NoVolume Syst. Pl. non Vasc., 40: 153–188 (in Russian).

Stavishenko, I.V. (2000). Successions of xylotrophic fungi in the forest formations of the Visim Reserve. In: Ecology of processes of biological decomposition of the wood. Ekaterinburg, pp. 16–30 (in Russian).

Strid, A. (1975). Wood-inhabiting fungi of alder forests in North-Central Scandinavia 1. Aphyllophorales (Basidiomycetes). Taxonomy, ecology and distribution. University of Umea, Umea, p. 237.

Stubblefield, S.P., Taylor, T.N. and Beck, C.B. (1985). Studies of Paleozoic fungi. V. Wood-decaying fungi in *Callixylon newberryi* from the Upper Devonian. Am. J. Bot., 72: 1765–1174.

Takhtajan, A.L. (1950). Phylogenetic bases of vascular plants system. Botanical Journal, 13: 135–139.

Teng, S.C. (1996). Fungi of China. Ithaka: Mycotaxon Ltd., p. 586.

Torkelsen, A.-E. (1997). Tremellaceae Fr. *In*: Hansen, L. and Knudsen, H. (Eds.) Nordic Macromycetes 3. Heterobasidioid, Aphyllophoroid and Gastromycetoid Basidiomycetes. Nordsvamp, Copenhagen.

Ţura, D., Zmitrovich, I.V., Wasser, S.P., Spirin, W.A. and Nevo E. (2011). Biodiversity of Heterobasidiomycetes and non-gilled Hymenomycetes (former Aphyllophorales) of Israel. A.R.A. Gantner Verlag K.-G., Ruggell, p. 566.

Vakhmistrova, T.V. (1987). Specialization of *Fomes fomentarius* and *Fomitopsis pinicola* to the diameter of the woody substrate. Mikologiya i fitopatologiya, 21: 505–508 (in Russian).

Wright, J.E. (1985). Los hongos xilofagos: una revista. Anal. Acad. Nac. Cs. Ex. Fis. Nat., Buenos Aires, 37: 121–135.

Wright, J.E. and Deschamps, J.R. (1972). Basidiomycetes xilofilos de los Bosques Andinopatagonicos. ReVolume Invest. Agrop. INTA, 9: 111–195.

Wright, J.E. and Deschamps, J.R. (1975). Basidiomycetes xilofilos de la region mesopotamica. II. Los generous *Daedalea, Fomitopsis, Heteroporus, Laetiporus, Nigroporus, Rigidoporus, Perenniporia* and *Vanderbylia*. ReVolume Invest. Agrop. INTA, 12: 127–204.

Yurchenko, E.O. (2001). Corticioid fungi on mosses in Belarus. Mycena, 1: 71–91.

Yurchenko, E.O. (2006). Natural substrata for corticioid fungi. Acta Mycol., 41: 113–124.

Yurchenko, E.O. and Zmitrovich, I.V. (2001). Variability of *Hyphoderma setigerum* (Corticiaceae s. l., Basidiomycetes) in Belarus and northwest Russia. Mycotaxon, 78: 423–434.

Yurchenko, E.O. and Golubkov, V.V. (2003). The morphology, biology and geography of a necrotrophic basidiomycete *Athelia arachnoidea* in Belarus. Mycol. Prog., 2: 275–284.

Zhishu, B., Guoyang, Z. and Taihui L. (1993). The Macrofungus Flora of China's Guangdong Province. Chinese University Press, Hong Kong, p.734.

Zmitrovich, I.V. (2008). Definitorium fungorum Rossicum. Familia Atheliaceae et Amylocorticiaceae. KMK, Petropolis, p. 278 (in Russian).

Zmitrovich, I.V. (2010). Epimorphology and tectomorphology of higher fungi. Folia Cryptogamica Petropolitana, 5: 1–279.

Zmitrovich, I.V. (2012). Features of structure and dynamics of floodland gray alder forests in North-Wester European Russia. *In*: Human and North: Anthropology, Archaeology and Ecology. Tjumen' (in Russian).

Zmitrovich, I.V. (2018). Conspectus systematis Polyporacearum Volume 1.0. Folia Cryptogamica Petropolitana, 6: 3–145.

Zmitrovich, I.V., Malysheva, V.F. and Malysheva, E.F. (2003). Some concepts and terms of mycogeography: A critical review. Bulletin of Ecology, Forest Management and Landscape Management, 4: 173–188 (in Russian).

Zmitrovich, I.V. and Spirin, V.A. (2005). Ecological aspects of speciation in higher fungi. Bulletin of Ecology, Forest Science and Landscape Science, 6: 46–68 (in Russian).

Zmitrovich, I.V., Wasser, S.P. and Ezhov, O.N. (2012). A survey of species of genus *Trametes* Fr. (higher basidiomycetes) with estimation of their medicinal source potential. International Journal of Medicinal Mushrooms, 14(3): 307–319. 10.1615/intjmedmushr.v14.i3.70.

Zmitrovich, I.V., Wasser, S.P. and Ţura, D. (2015). Wood-inhabiting fungi. *In*: Misra, J.K., Tewari, J.P., Deshmukh, S.K. and Vágvölgyi, C. (eds.). Fungi from different substrates. CRC Press, Taylor and Francis Group, New York.

Zugmaier, W., Bauer, R. and Oberwinkler, F. (1994). Mycoparasitism of some *Tremella* species. Mycologia, 86: 49–56.

9

Wood-Rot Polypores of Kerala, India

*TK Arun Kumar** and *N Vinjusha*

1. Introduction

Kerala is a state occupying an area of about 39,000 km, located within the megadiverse Western Ghats hill ranges in the southwestern part of India (coordinates 8°18'–12°48'N, 74°52'–77°22'E). With a maritime tropical climate and two principal rainy seasons (southwest in June-August and northeast in October-December), the state experiences alternating wet and dry conditions yearly (Kumar et al., 2019). Because of the unique climate and geography, the state harbours rich biodiversity. The flora is paleotropic. According to the Kerala forests and wildlife department (www.old.forest.kerala.gov.in) there are about 1272 endemics among the 3800 angiosperm species in Kerala. Per published compilations (Farook et al., 2013; Kumar et al., 2019), Kerala has an equally rich mycobiota, with more than 900 macromycete species documented thus far.

Polypores are a polyphyletic group of basidiomycotan fungi belonging to the class Agaricomycetes. They are generally characterized by basidiomata with a poroid hymenium, though members having gills or tubes are also present. Polypores function as major wood decomposers and contribute to large-scale nutrient recycling in forests. Many species are economically important as plant pathogens. With their ability to degrade plant materials like cellulose and lignin, and to produce various secondary metabolites, many polypores are industrially valuable. As an ecologically and economically important group of fungi, detailed studies on the diversity and ecology of polypores are vital. Polyporoid fungi have been reported worldwide.

Department of Botany, The Zamorin's Guruvayurappan College (University of Calicut), Kerala 673014, India.

* Corresponding author: tkakumar@gmail.com

The earliest available records of polyporoid fungi from India are those of Kotzch (1832, 1833). Berkeley (1850, 1851, 1851a, 1851b, 1851c, 1854, 1854a, 1854b, 1854c, 1867) reported many taxa based on his studies on the collections made by Hooker from Sikkim Himalayan regions of India. Bose in his series of publications (1918, 1919, 1919a, 1920, 1920a, 1921, 1921a, 1921b, 1922, 1922a, 1922b, 1923, 1923a, 1924, 1925, 1928, 1928a, 1934, 1937, 1938, 1944, 1946) provided a comprehensive record of 143 species from Bengal. Eleven polyporoid species were reported from Madras by Sundararaman and Marudarajan (1925). In the revised edition of Butler and Bisby's (1931) "Fungi of India" by Vasudeva (1960), 400 polyporoid species belonging to 21 genera were reported. As evident from a few of the recent publications (Hembrom et al. (2015), Prasher (2015), Senthilarasu (2015), Sharma and Atri (2015), Ranadive and Jagtap (2016), Hembrom et al. (2017), Pongen et al. (2018), and Singh et al. (2019), polyporoid fungi continued to be studied and reported from different regions of India.

Two major documentary works published so far on polyporoid fungi of Kerala are those of Leelavathy and Ganesh (2000), and Mohanan (2011). Leelavathy and Ganesh (2000) described 79 species belonging to 32 genera, placed in 3 families; Polyporaceae, Ganodermataceae and Hymenochaetaceae. Mohanan (2011) gave an account of 91 species of 46 genera, belonging to six families (Polyporaceae, Hymenochaetaceae, Meruliaceae, Meripilaceae, Fomitopsidaceae and Ganodermataceae). Adarsh et al. (2018) published a checklist, compiling the 145 species of polypores published from Kerala. A systematic revision of the polyporoid fungi of Kerala (Vinjusha, 2020, Vinjusha and Kumar, 2022b), reveals that the present diversity is represented by131 species belonging to 63 genera of 16 families. This includes members of the Polyporales and the Hymenochaetales.

Based on extensive collecting of basidiomata, especially during the monsoon seasons, from sites including forests, sacred groves, and human-inhabited areas, we studied the wood-decaying polyporoid fungi of Kerala during a five-year period. Morphological and molecular methods were employed for taxonomic identification. An account of the ecology of the wood rotting polypores of Kerala based on our observations during this study period is presented here.

2. Distribution

The distribution of polyporoid fungi is generally associated with latitude, altitude, rainfall, and diversity of habitats (Lodge and Cantrell, 1995). There are cosmopolitan as well as climate-dependent genera among polyporoid fungi. *Abortiporus*, *Antrodiella*, *Bjerkandera*, *Ceriporia*, *Daedalea*, *Datronia*, *Dichomitus*, *Ganoderma*, *Gloeophyllum*, *Phellinus*, *Polyporus*, *Phylloporia*, *Trametes*, and *Tyromyces* are some of the polypore genera, which are cosmopolitan in distribution. Climate dependent genera among polyporoid fungi include both tropical (e.g., *Amauroderma*, *Cyclomyces*, *Earliella*, *Hexagona*, *Leucophellinus*, *Microporellus*, *Microporus*, *Grammothelopsis*, *Theleporus*, *Paratrichaptum* and *Echinopora*) and boreal or temperate genera (*Albatrellus*, *Anomoporia*, *Cerrena*, *Fistulina*, *Fomes*, *Meripilus*, *Perenniporia*, *Phaeolus*, *Podofomes*, *Pycnoporellus*, and *Spongipellis*) (Ryvarden, 1991).

Sacred groves of Kerala were found to be good habitats for polyporoid fungi, especially Vallikkattu Kavu, Poyil Kavu and Thurayil Kavu of Kozhikode district, Neeliyar Kottam, Poongottu Kavu of Kannur district, and Iringole Kavu of Ernakulam district. Among the sacred groves visited for the study, species diversity of polypores was higher in Thurayil Kavu of Kozhikode district and Iringole Kavu of Ernakulam district. Species of *Phellinus* were common on trees in most of the sacred groves. In addition to this, some species such as *Favolus grammocephalus*, *Rigidoporus lineatus* and *Microporus xanthopus* were also frequently seen. Interestingly, some polypores like *Daedalea radiata*, *Neofomitella guangxiensis*, and *Podoscypha involuta* were commonly observed on trees in different sacred groves throughout Kerala, but rarely seen in other deep forest areas. *H. ochromarginata* is a common species in Mannampurath Kavu (Kasaragod district), Poyil Kavu, Bayankavu (Kozhikode district) and Iringole Kavu (Ernakulam district), but other than these sacred groves, *H. ochromarginata* was only obtained from the campus of Kerala Forest Research Institute campus, Malappuram district. In all sacred groves for study, polyporoid fruit bodies growing on soil are comparatively less than those occurring on wood.

Cellulariella acuta, *Earliella scabrosa*, *Favolus grammocephalus*, *Inonotus pachyphloeus*, *Microporus affinis*, *M. xanthopus*, *Phellinus gilvus*, *P. rimosus*, *Rigidoporus lineatus*, and *Trametes villosa* are some of the widely occurring taxa among polypores in most of the forest areas of Kerala. Among the frequently visited collection localities, Janaki forest, Kakkayam forest and Peruvannamuzhy forest of Kozhikode district and Muthanga forest of Wayanad district were found to be rich in species diversity of polypores throughout the seasons. Altitude seemed to have an influence in the diversity and fruiting pattern of polypores. We noticed that with increasing altitude, occurrence and fruiting decreased. This is an observation that corresponds with many earlier studies (Schmit et al., 2005; Zhang et al., 2010; Andrew et al., 2013) that compared the correlation between altitude and fruit body formation in macrofungi. A study carried out by Adarsh et al. (2018a) also reported a decreasing pattern in the diversity of polypores along with increasing altitudinal gradients in Silent Valley National park of Kerala. The decrease in diversity of tree species in higher altitudes might be a major reason for the corresponding decrease in diversity of wood-rotting polypores. Wet evergreen forests have a different tree species composition than the shola forests characteristic of the higher altitude areas of Idukki and Wayanad districts. At higher altitudes, we encountered species like *Postia tephroleuca*, *Spongiporus floriformis*, *Fomitopsis palustris*, *Bjerkandera adusta*, *Terana coerulea*, and *Trichaptum biforme*, which were not found at lower elevations. *Ganoderma*, *Lentinus*, *Microporus*, *Panus*, *Phellinus* and *Trametes*, were the dominant genera of wet evergreen forests and regions at lower altitudes. Relatively higher temperature and humidity at regions around the sea level seem to be more congenial for the growth and fruiting of polypores than the lower temperature and increasing dryness at regions nearing higher altitudes. Since most polyporoid fungi occur in less disturbed regions, they were considered useful indicators of disturbance in a forest ecosystem (Sverdrup-Thygeson et al., 2003; Gibertoni et al., 2007). Species of *Microporus* and *Phellinus* that produce hardy annual to perennial basidiomata, except *Microporus xanthopus* and *Phellinus gilvus*, could be suggested as good disturbance indicator species in forests of Kerala.

Apart from the deep forest areas, many local collections were also obtained from live trees as well as dead woods along the roadside, parks and campuses. Basidiomata of *Earliella scabrosa*, *Ganoderma* species, *Lentinus dicholamellatus*, *L. squarrosulus*, *Pseudofavolus tenuis*, and *Trametes meyenii* frequented these sites with increased human interference. These taxa seemed to be more tolerant to environmental stresses like pollution.

3. Environmental Conditions

Many polyporoid species have a cosmopolitan distribution whereas others are climate dependent (tropical, temperate, or boreal) (Ryvarden, 1991). Diversity is higher in tropical forests through the abundance of fruit bodies is less. This is true with the Kerala collections as well with a diversity that ranged to about 131 species belonging to 63 genera. The tropical humid climate of Kerala has an influence on the growth and production of basidiomata of polypores. Although polypore basidiomata are encountered throughout the year in Kerala, there is a season dependent fluctuation in the number of species that produce basidiomata. Many of the species infrequently produced basidiomata, and were low in numbers. These species are season dependent and develop basidiomata only during the two monsoon seasons in June-August and October–December. The production of basidiomata is dependent on rainfall and humidity. The humidity level of Kerala is approximately 100% throughout the year and is a favourable factor for the growth of fungi. Intense rainfall after drier periods was always found to stimulate basidiomatal initiation in polypores. A direct proportionality was noted between basidiomatal productivity and the amount of rainfall received prior to basidiomatal initiation. The monsoon-dependent species mostly possess annual basidiomata. Annual basidiomata are typical of species of *Lentinus* as their fruit bodies mature and decay within a specified time. However, species that produced perennial basidiomata, like some *Phellinus* species, are less season-dependent. Four types of basidiomatal phenology, according to the criteria followed by Schigel et al. (2006), can be recognized in polypores of Kerala (Fig. 1).

(1) Short-lived but fast-growing small basidiomata (ephemeral) selectively produced in monsoon seasons. These species appear immediately with the receipt of adequate rainfall. Members of *Postia*, *Spongiporus*, *Flavadon*, *Irpex*, *Butyrea*, *Bjerkandera*, *Oxychaete*, *Pappia*, *Phlebia*, *Sebipora*, *Terana* and *Phlebiopsis* belong to this group.

(2) Larger and robust basidiomata that are fertile for a considerably long period (annual sturdy) and require an extended rainfall and high humidity. Species of *Innonotus*, *Coltricia*, *Hymenochaete*, *Ganoderma*, *Lentinus*, *Panus*, *Pycnoporus* and *Favolus* are examples.

(3) Conspicuous basidiomata that are relatively long-lasting, become dormant and re-grow with every favourable season (annual hibernating). *Trichaptum biforme* and *T. byssogenum* are species that cease and resume growth depending on seasons. The receipt of adequate humidity during the monsoon seasons favour resumption of growth.

Fig. 1. Examples of types of basidiomatal phenology recognized: (A) Ephemeral–*Bjerkandera adusta*; (B) Annual sturdy–*Panus neostrigosus*; (C) Annual hibernating–*Trichaptum biforme*; (D) Perennial–*Phylloporia gabonensis*.

(4) Perennial species produce hard and sturdy basidiomata that have distinctly stratified pore tube layers. They are season-independent and basidiomata are fertile throughout the year. Members of *Daedalea*, *Fuscoporia*, *Phellinus*, *Perenniporia*, *Phylloporia* and *Tropicoporus* produce perennial basidiomata.

Some polypores are capable of tiding over unfavourable environmental conditions by producing characteristic hard and resistant resting structures. Two such species that possessed resting structures to grow in Kerala. *Lignosus rhinoceros* has a large underground sclerotium at the base of the stipe that helps the fungus regenerate every season. Similarly, the pseudosclerotium at the base of *Panus similis* is inserted deep into the substrate from which growth is revived and fruit bodies are produced with the arrival of favourable conditions.

4. Substrates

Tropical macromycetes are known to be less substrate-specific (Shmit, 2005). Many tropical wood-rotting polypores derive nutrition from a wide range of host species and are mostly host generalists (Carranza-Morse, 1991, 1992, 1993; Ryvarden, 1992). Polypores use plant parts for their nutritional requirements, and these plant parts may be herbaceous to woody in nature. Plant parts are rich sources of various nutrients, including carbon. The help of various enzymes that may be having an exo- and or endo-digesting activity brings about the decomposition of these plant-based substrates. Based on the enzyme apparatus available for decay even within the different generalist species, there is a gradient in the efficacy of substrate utilization. Cellulose, hemicellulose, lignin, and pectin are the major polymers subjected to breakdown by an assortment of enzymes. The quality and quantity of plant material available for breakdown partially decide the diversity of the wood decayers. Quality of substrate available for polypore species to feed on depends on the vegetational diversity of an area and the stage of decay of the plant part. Based also on the stage of decay of a substrate different species can even be found to produce basidiomata somewhat successively. Since plant diversity is more in the tropics, the availability of a wide range of substrates must have paved the way for tropical polypores to a more generalized substrate utilization mode. Although visible evidence of wood decay on live or dead plant material is evident from the characteristic decay type, the presence and identity of a particular species or a species community is evident only when basidiomata appear with the arrival of favourable environmental conditions. This implies that the mere presence or absence of basidiomata on dead/live plants cannot be taken as a measure of polypore species diversity. Although tropical polypores are generally less selective regarding substrate utilization, taxa that are specialized can also be encountered in tropical forests. *Wolfiporia extensa* is an example of a species that specialize in the wood of palms. *Amauroderma* species selectively decompose dead and buried tree roots and is an example of a species specializing in the microecology of the habitat (Kost, 2004). Polypore fungi are also known to form different insect associations (Thunes, 1994; Komonen, 2001; Schigel et al., 2006; Schigel, 2011).

Most polypore species are saprotrophic on dead wood, whereas many are serious parasites on different tree species. Species of the group degrade both cellulose and lignin of wood and causes white rot, whereas, others incompletely degrade wood leaving lignin untouched, causing brown rot (Ryvarden and Melo, 2014). According to Ryvarden and Johansen (1980), 75 percent of the fungal species involved in wood decay belong to the family Polyporaceae. Basidiomatal collections of polypores from Kerala were from either live trees or dead and decayed logs or branches, whereas some fruit bodies were obtained from soil, often attached to small roots and leaf litters of nearby trees or plants. The majority of species collected from Kerala belonged to Polyporaceae. Most polypore species were found on dead wood of diverse angiosperms. *Ganoderma keralense* and *G. pseudoapplanatum* were found growing on both live and dead coconut palms as well as on dead hardwood of unidentified plants. Host specificity was particularly observed in *Panus bambusinus*, which grew only on roots of bamboo. Adarsh et al. (2015) report that *Fomitopsis palustris*, *Hexagonia sulcata*, *Rigidoporus lineatus* showed a narrow host range. *Phellinus* species were collected from standing live trees. Many species of *Lentinus, Microporus, Phlebia* and *Irpex* were collected from the forest floor on fallen twigs of large trees. *Sanguinoderma rugosum* and *Amauroderma fuscoporia* basidiomata were consistently collected from soil, and their mycelia were observed growing on dead roots of nearby plants. A study on the polypores of the Peechi-Vazhani wildlife sanctuary of Kerala (Florence et al., 2000) shows that polypores of the area were mostly infecting wood of *Terminalia paniculata*, *Tectona grandis*, and *Xylia xylocarpa.* Polyporoid fruit bodies growing on soil are comparatively less frequent than those occurring on wood. *Podoscypha thozetii* was collected from soil whereas many other species of *Podoscypha* were collected from dead logs.

5. Decomposers of Wood and Other Materials

Polyporoid fungi are considered one of the major wood decomposers and they play an important role in nutrient cycling. This group includes white rot as well as brown rot fungi. The white-rot fungi degrade cellulose and lignin of the wood leaving a white powdery mass. Enzymes like quinone oxidoreductase, lignin peroxidase, and laccases perform wood decay. Examples of white-rot fungi are species of *Ganoderma*, *Trametes* and *Microporus*. Brown rot fungi are capable of degrading only cellulose and hemicellulose, thus breaking the wood into cubical fragments. As a result of the decay process, the wood shows a brown discoloration. Examples of brown rot fungi are species of *Antrodia*, *Fomitopsis* and *Postia* (Ryvarden and Melo, 2014). Mycelium of the rot-causing species initially gets ramified in the woody tissues and later develops characteristic fruiting structures, which indicates the infection (Leelavathy and Ganesh, 2000).

The presence of cellulolytic and ligninolytic enzymes in polyporoid fungi makes them useful in bioremediation processes like clearing various industrial effluents, oil-containing crop wastes, degradation of polycyclic aromatic hydrocarbons and decolourization of synthetic dyes (Peláez et al., 1995; Balan and Monteiro, 2001; Hmd, 2011; Choi et al., 2013; Krastanov et al., 2013; Singh et al., 2019). The white-rot fungi *Abortiporus biennis* is used in pilot-scale bioreactors for large-scale production of laccase enzyme (Erden et al., 2009). Willow sawdust after pretreatment with *A. biennis* is reported to have increased biogas production (Alexandropoulou et al., 2015). *Bjerkandera adusta* is reported to degrade kraft pulp lignin, polycyclic aromatic hydrocarbons, and effluents containing heavy metals (Heinfling et al., 1998; Haritash and Kaushik, 2009; Heinfling et al., 2011). Another white-rot species, *Flavodon flavus* (Klotzsch) Ryvarden from marine habitat is found effective in decolorizing brown-colored pigments from the effluents of molasses-based alcohol distilleries and in removal of their toxicity (Raghukumar and Rivonkar, 2001; Raghukumar et al., 2004). This species is used in the dye industry for large-scale extraction of yellow pigments (http://indiasendangered.com/researchers-find-thousands-of-fungi-species-insinghadforests/).

Pycnoporus sanguineus is capable of degrading plastic (Cesarino et al., 2019). *Pycnoporus* species also have many industrial and biotechnological applications, which include large-scale degradation of biomass by the activity of laccase enzyme, degradation and decolourization of various dyes, clarification of wastewater from oil mills and production of coloured pigments (Alexopoulos et al., 1996; Falconnier et al., 1994; Oddou et al., 1999; Schliephake et al., 1999, 2000; Eugenio et al., 2009; Lomascolo et al., 2011; Göçenoğlu and Pazarlioglu, 2014; Zimbardi et al., 2016; Wang et al., 2019). *Nigroporus* species are also capable of degrading organic wastes (Kondo, 2005) and polyaromatic hydrocarbons (PAHs) and polychlorinated biphenyls (PCBs) in culture media (Siripong et al., 2009).

Except for members of the families, Dacryobolaceae and Fomitopsidaceae, all polypore specimens collected during the present study from Kerala belonged to the white-rot wood decay group. Species of Dacryobolaceae and Fomitopsidaceae, which formed approximately 8% of the total polypore species in Kerala, caused brown rot. The white rot polypore *Abortiporus biennis* known for its laccase production and activity was collected from Kerala. Other economically important polypores like *Bjerkandera adusta*, *Flavadon flavus*, *Pycnoporus sanguineus*, and *Nigroporus* species known for decaying and utilizing wood as well as various organic wastes and pollutants were collected from various forests of Kerala. This indicates the richness of polypore resources for bioprospecting.

6. Pathogens

Many polypore species infect and become serious pathogens on living host plants. *Ganoderma* species are considered one of the serious plant and tree pathogens as they affect plantation crops like oil palm, coconut, rubber, betelnut, tea and other forest trees and result in considerable yield loss (Naidu et al., 1966). According to Naidu et al. (1966), 48 plant species belonging to 36 genera and 19 families are known to be infected by species of *Ganoderma*. *Ganoderma* rot or basal stem rot is a serious disease affecting coconut palms, and different species of *Ganoderma* such as *G. applanatum*, *G. boninense*, *G. lucidum* and *G. zonatum* had been treated as the pathogens of the disease (Vinjusha and Kumar, 2022a). The polypore *Amauroderma parasiticum* and *A. rude* are known to cause root rot disease in plantations of *Acacia mangium* (Glen et al., 2009). Genus *Bjerkandera* is known to cause timber damage and negatively affects the cultivation of edible mushrooms (Jung et al., 2014). *Trametes hirsuta* and *T. versicolor* cause white spongy rot on different landscape trees (Hickman et al., 2011). *Rigidoporus* is considered as a serious pathogen of *Hevea brasiliensis* (Peries, 1969; Jayasinghe et al., 1995; Jayasuriya, 1996; Jayasuriya and Deacon, 1996). *Perenniporia fraxinea* and *P. robiniophila* are economically important species, which cause white-rot on different hardwood trees, either as a parasite or as a saprotroph (Szczepkowski, 2004; Kuo, 2016).

Hymenochaetaceae members are also well-known forest pathogens (Dai et al., 2007; Dai et al., 2010). Species of *Phellinus* are serious pathogens causing heart rot disease and cankers on living trees (Sunhede and Vasiliauskas, 2002; Wagner and Fischer, 2002; Miklaševičs, 2019; Ranadive et al., 2012). According to Larsen and Cobb-Poulle (1990), *Phellinus* species causes more timber loss than other groups of wood-destroying fungi, thereby causing huge economic loss to the wood industry. Species of *Phellinus* cause rot disease of *Artocarpus heterophyllus*. Species of *Inonotus* are also considered as important primary decayers of forest ecosystem, as well as serious pathogens of forest trees, and urban landscapes (Lindner et al., 2006; Robles et al., 2011). *Fuscoporia torulosa* is considered a serious pathogen that causes white pocket rot on broad-leaved hardwood trees and conifers (Motta et al., 1996; Wagner and Fischer, 2001; Tomšovský and Jankovský, 2007; Campanile et al., 2008; Tzean et al., 2016). Similarly, *Phylloporia* species are also reported as serious forest pathogens possessing host specificity (Esquivel and Carranza-Morse, 1996; Wagner and Ryvarden, 2002; Dai, 2010; Rajchenberg and Robledo, 2013; Yombiyeni et al., 2015; Zhou, 2016; Chen et al., 2017).

Ganoderma species were frequently encountered in many collection localities of Kerala; especially on living or dead coconut palms affected with stem rot disease. During the study, two types of *Ganoderma* fruit bodies that could be recognized macroscopically were consistently found associated with diseased palms of coconut in a field. The symptoms of disease seen in the infected coconut plantations in Kerala were comparable to those of earlier *Ganoderma* rot reports from India. The two *Ganoderma* species accompanying the stem-rot disease of coconut in Kerala were identified as *G. keralense* and *G. pseudoapplanatum* (Vinjusha and Kumar, 2022a). *G. enigmaticum* is known to cause root rot of *Jacaranda mimosifolia. Ganoderma orbiforme* is known to be pathogenic on palms. Hymenochaetaceae member *Inonotus pachyphloeus* is pathogenic *on Peltophorum ferrugineum* and *Terminalia* species.

7. Medicinal Species

Polyporoid taxa belonging to 22 genera with proven medicinal activities were collected and recorded from different regions of Kerala. Some important medicinal species from Kerala are, *Abortiporus biennis*, *Bjerkandera adusta*, *Earliella scabrosa*, *Flavodon flavus*, *Irpex lacteus*, *Lignosus rhinoceros*, *Podoscypha petalodes*, and *Sanguinoderma rugosum*. Species belonging to the following potentially medicinal genera were recorded: *Amauroderma*, *Cellulariella*, *Coriolopsis*, *Daedalea*, *Ganoderma*, *Fomitopsis*, *Panus*, *Perenniporia*, *Phellinus*, *Polyporus*, *Postia Pycnoporus*, *Rigidoporus* and *Trametes.*

Among members of Polyporales, the species of *Ganoderma*, especially *G. lucidum* is one of the major medicinal mushrooms. As per Upton (2006), they are the most significant of all Japanese medicinal polypores. In China, it is variously known as "mushroom of immortality", "ten thousand year mushroom", "mushroom of spiritual potency" and "spirit plant" (Huang, 1993; Liu and Bau, 1994). Apart from *G. lucidum*, *G. applanatum*, *G. japonicum*, and *G. tsugae* are used as medicine and are traded as *reishi* mushrooms in the USA (Upton, 2006). According to Chen et al. (2006), the presence of various bioactive compounds in basidiomata of *Ganoderma* provides them antitumour, antiinflammatory, antichronic bronchitis, immunoenhancing, cardiovascular regulating and hepatoprotective properties. *Ganoderma* species are used in treatment of different types of cancers (Zhang et al., 2010; Hapuarachchi et al., 2016), type 2 diabetes mellitus (Seto et al., 2009), gastric ulcers (Rony et al., 2011), hepatitis (Li and Wang 2006), hyperlipidemia (Chen et al., 2005) and hypertension (Morigiwa et al., 1986). The anticancer activity of *G. lucidum* is due to the presence of methanol soluble triterpenoid extracts, known as ganoderic acids (GAs) in their fruit body (Radwan et al., 2011). This fungus is also known to possess anti-HIV activity (El-Mekkawy et al., 1998). Because of the immense medicinal properties, a large number of products of *G. lucidum* are commercially accessible as nutraceuticals in the form of tonics, pills, or powders (Bishop et al., 2015). A similar genus *Amauroderma* also has potential medicinal value and has been ethnically used by the Chinese to cure inflammation, indigestion and cancer (Dai and Yang, 2008). This fungus possesses good antioxidant and anti-inflammatory properties, and has been used by some native people of Malaysia for preventing epilepsy in children (Chang and Lee, 2004; Azliza et al., 2012; Chan et al., 2013). *Sanguinoderma rude* is well known for its anticancer (Jiao et al., 2013; Pan et al., 2015; Hapuarachchi et al., 2018) and antioxidant activities (Wang and Qi, 2016). This species is reported to be more efficient than the medicinal fungus, *Ganoderma lucidum* in curing various types of cancers (Jiao et al., 2013). Mycelium of *Sanguinoderma rugosum* is nutritious (Chan et al., 2013), and also shows antimicrobial activity against *Clostridium difficile, Escherichia coli, Pseudomonas aeruginosa, Staphylococcus aureus* and *S. pyogenes* (Liew et al., 2015).

Laccase from *Abortiporus biennis* is also known to possess anticancer and anti-tumour activities (Zhang et al., 2011; Ivanova et al., 2014; Jayalakshmi et al., 2015). The polypore *Bjerkandera adusta* contains a good amount of unsaturated fatty acids, ergosterols of medicinal value, and phenolic organic compounds and is highly recommended as a supplementary food or nutraceutical (Küçükaydi and Duru, 2017). The *B. adusta* also has antibacterial, antiradical and pro-oxidant properties (Shintani et al., 2002; Korneichik and Kapich, 2011; Sugawara et al., 2019). Fruit bodies of another polypore, *Fomitopsis pinicola* are dried and applied to wounds to stop bleeding (Rogers, 2012). The species *Trametes versicolor* is reported to have good immunomodulating activities (Hobbs, 2004). According to Kobayashi et al. (1994), the presence of *T. versicolor* polysaccharopeptides increases *in vitro* anticancer activity of the chemotherapy drug, cisplatin. Some *Polyporus* species have anti-inflammatory and immune-enhancing properties and are used in curing various ailments (Ying et al., 1987; Yuan et al., 2004; Liu and Liu, 2009; Kawashima et al., 2012). The sclerotia of *P. umbellatus* possess rich medicinal properties (Ying et al., 1987; Xing et al., 2013). The polypore *Earliella scabrosa* is reported to have *in vitro* antifungal activity (Peng and Don, 2013). *Irpex lacteus* has been used as a medicine for many years in China against inflammation, bacterial and fungal infections and urinary problems (Dong et al., 2017; Chen et al., 2018).

Various kinds of terpenoids and biologically active metabolites have been reported from *I. lacteus* (Silberborth et al., 2000; Tang et al., 2018). Sclerotia of *Lignosus rhinocerus* (Cooke) possess anti oxidative, anti tumour and immunomodulatory effects (Lai et al., 2008; Wong et al., 2011; Vinjusha and Kumar, 2021). This species is consumed as a traditional medicine for treating various illnesses like asthma, breast cancer, cough, fever and food poisoning (Lee et al., 2012). Some of the other species of order Polyporales having medicinal properties are *Cellulariella warnieri* (Savino et al., 2014, 2016), *Coriolopsis gallica* (Fakoya and Oloketuyi, 2012; Doskocil et al., 2016), *Daedalea quercina* (Bal et al., 2017), *Flavodon flavus* (*Fernando* et al., 2016), *Panus* species (Smith et al., 2002; Boa, 2006), *Perenniporia fraxinea* (Kim et al., 2008), *Podoscypha petalodes* (Fernando et al., 2015), *Postia ptychogaster* (Doskocil et al., 2016), *Pycnoporus* species (Correa et al., 2006; Viecelli et al., 2009; Jena and Thatoi, 2019) and *Rigidoporus ulmarius* (Chao et al., 2011). *Phellinus* species are used in the treatment of wounds, and throat ailments. It is known to possess antitumor, antimicrobial and anti-inflammatory properties (Vaidya and Lamrood, 2000).

8. Edible Species

Mycophagy has been a part of many civilizations, and different cultural practices have evolved in human societies around the consumption of fungi and fungi-derived products. A number of macromycetes have been known to be edible, and humans of all ages have depended on macroscopic fruiting bodies of fungi (mushrooms) for their nutritional needs. Mushrooms have been considered an alternate source of nutrition (Karun and Sridhar, 2017). Although gilled and fleshy agarics have been the most preferred, polypore species have also been used for consumption. The less preferable nature of polypores as food is because of palatability issues with basdiomata being of a generally hard woody or corky nature. Even then, young basidiomatal stages of several polypore species have been hunted in the wild or cultivated and consumed by human populations worldwide.

A few polypore species that were collected from Kerala were found to be edible. The polypore *Bjerkandera adusta* contains a good amount of unsaturated fatty acids, ergosterols of medicinal value, and phenolic organic compounds and is highly recommended as a supplementary food or nutraceutical (Küçükaydi and Duru, 2017). Many members in the genus *Lentinus* are edible (Burkill, 1966; Chin, 1981; Karunaratha et al., 2011; Seelan et al., 2015). The species *Lentinus squarrosulus* Mont. is widely consumed in Central Africa as a nutrient source (Watling, 1993). In some parts of Kerala, basidiomata of *L. dicholamellatus*, a close relative of the commercially cultivated *L. sajor-caju*, are consumed. Basidiomata of the species are edible only during earlier stages of development when it is soft and fleshy. Species of *Polyporus* such as *P. tenuiculus* (P. Beauv.) Fr., *P. squamosus* (Huds.) Fr. and *P. umbellatus* (Pers.) Fr. are edible (Gomes-Silva et al., 2012; Ergönül et al., 2013; Bandara et al., 2015). Because of the presence of various macro and micronutrients, the species *Microporus xanthopus* is recommended as a protein supplementary diet (Meghalatha et al., 2014). The mycelium of *Sanguinoderma rugosum* is a source of proteins, dietary fibre, carbohydrates, sodium potassium and phosphorus (Chan et al., 2013).

9. Mutualistic Associations

Polyporoid fungi are well known for various insect associations. Basidiocarps of many polypore species like that of *Fomes fomentarius* and *Fomitopsis pinicola* host various insect communities (Thunes, 1994; Jonsell and Nordlander, 1995; Nilsson, 1997; Fossli and Andersen, 1998; Jonsell, 1998; Rukke and Midtgaard, 1998; Hågvar, 1999). Thirty-three species of beetles have been recorded from species like *Fomes fomentarius*, *Fomitopsis pinicola* and *Trametes pubescens* (Selonen et al., 2005). In a study by Schigel (2011), 176 spp. of *Coleoptera* were documented from 116 species of polypores belonging to both Polyporales (*Ganoderma applanatum*), and Hymenochaetales (*Phellinus igniarius* and *Phellinus populicola*). A symbiotic association between ambrosia beetles and basidiomycotan fungi was reported first in *Flavadon ambrosias* (Li et al., 2015; Simmons et al., 2016). According to Li et al. (2017), *F. ambrosias* is associated with two genera of ambrosia beetles. Beetles

Fig. 2. (A) Basidiomata of *Trametes meyenii*; (B) Colonies of unicellular algae growing on the tomentose hairs of the pileus of *T. meyenii*.

often act as agents of fungal spore dispersal as they move about basidiomata attracted by the volatile compounds emitted. Basidiomata of polypores that are relatively long-lasting than those of other mushrooms are often used by the insects as sources of nutrition and as safe hatcheries of eggs and shelter for developing young ones. Most of the basidiomata of polypores collected from Kerala harboured diverse communities of insects at different developmental stages. The diversity of these insect communities has not been assessed. Ectomycorrhizal associations are reported in some species of polyporoid fungi (*Coltricia* and *Coltriciella* Murrill) (Tedersoo et al., 2007; Hibbett et al., 2014). Colonies of unicellular algae were always found inhabiting the tomentose hairs that covered the pileus of basidiomata of *Trametes meyenii* collected from Kerala (Fig. 2). According to Ryvarden and Johansen (1980), this is a consistent feature and the presence of algal colonies is responsible for the zonate greenish colouration at the pileus of *T. meyenii*.

10. Conclusion and Outlook

Polyporoid fungi are an ecologically and economically important group of macromycetes. Their diversity and distribution has been relatively well documented, and species with wide and restricted distribution ranges are known. Climatic conditions influence growth and distribution patterns of polypores. Environmental conditions, including substrate preferences play a role in deciding the type and nature of polypore basidiomata. Many species are adapted to a sabrobic mode of nutrition with primary dependence on dead plant parts, though live plants are also attacked, and a few form associations with other organisms. Categories of wood decomposers are recognized based on the enzymatic apparatus possessed and wood degradation mode adopted. The cellulolytic and lignolytic enzymes of polyporoid fungi for wood degradation also make them suitable for degrading a variety of other natural and synthetic substrates. Unique metabolic pathways and their products in polypores have been recognized as medicinally valuable. There are edible species of polypores that are encountered in the wild, and also commercially cultivated. The chapter discusses the ecological characteristics and economic importance of polyporoid fungi of Kerala. A better understanding of the process of wood decay in forest plants and crops, and data on the presence and mode of action of the pathogenic polypores will be of immense help in agriculture and forestry. Exploration of the metabolic products and utilization of complex polymers and synthetic substrates will be of interest in bioremediation. Information on the biology, chemistry, diversity, and ecology of the polypores of an area should be considered while formulating governmental policies regarding bioprospecting, conservation of biowealth and maintenance of environmental health.

References

Andrew, E.E., Kinge, T.R., Tabi1, E.M., Thiobal N. and Mih A.M. (2013). Diversity and distribution of macrofungi (mushrooms) in the Mount Cameroon Region. J. Ecol. Nat. Environ., 5: 318–334.

Adarsh, C.K., Kumar, V., Vidyasagaran, K. and Ganesh, P.N. (2015). Decomposition of wood by polypore fungi in tropics – biological, ecological and environmental factors—A case study. Res. J. Agric. For. Sci., 3: 15–37.

Adarsh, C.K., Vidyasagaran, K. and Ganesh, P.N. (2018). A checklist of polypores of Kerala state, India. Stud. Fungi, 3: 202–226.

Adarsh, C.K., Vidyasagaran, K. and Ganesh, P.N. (2018). Distribution of polypores along the altitudinal gradients in Silent Valley National Park, Southern Western Ghats, Kerala, India. Curr. Res. Environ. Appl. Mycol., 8: 380–403.

Alexandropoulou, M., Antonopoulou, G., Ntaikou, I. and Lyberatos, G. (2015). Using *Leiotrametes menziesii* and *Abortiporus biennis* for biological pretreatment of willow sawdust for enhancing biogas production, 3rd International Conference on Sustainable Solid Waste Management, 2–4 July 2015, Tinos island, Greece.

Alexopoulos, C.J., Mims, C.W. and Blackwell, M. (1996). Phylum Basidiomycota order Aphyllophorales, polypores, Chantharelles, tooth fungi, coral fungi and corticioids. *In*: Harris, D. (ed.). Introductory Mycology, 4th edn. New York, USA, Wiley and sons Inc.

Azliza, M.A., Ong, H.C., Vikineswary, S., Noorlidah, A. and Haron, N.W. (2012). Ethno-medicinal resources used by the Temuan in Ulu Kuang Village. Stud. Ethno-Med., 6: 17–22.

Bal, C., Akgül, S., Sevindik, M., Akata, I. and Yumrutas, O. (2017). Determination of the anti-oxidative activities of six mushrooms. Fres. Environ. Bull., 26: 6246–6252.

Balan, D.S.L. and Monteiro, R.T.R. (2001). Decolorization of textile indigo dye by ligninolytic fungi. J. Biotechnol., 89: 141–145.

Bandara, A., Rapior, S., Bhat, J., Kakumyan, P., Chamyuang, S., Xu, J. and Hyde, K.D. (2015). *Polyporus umbellatus*, an edible-medicinal cultivated mushroom with multiple developed health-care products as food, medicine and cosmetics: A review. Crypt. Mycol., 36: 3–42.

Berkeley, M.J. (1850). Decades of fungi. Decades XXV to XXX. Sikkim Himalaya fungi, collected by Dr. JD Hooker. Hooker's Journal of Botany and Kew Garden Miscellany, 2: 76–88.

Berkeley, M.J. (1851). Decades XXXII–XXXIII. Sikkim-Himalayan Fungi, collected by Dr. JD Hooker. Hooker's Journal of Botany and Kew Garden Miscellany, 3: 39–49.

Berkeley, M.J. (1851a). Decade XXXIV. Sikkim-Himalayan Fungi, collected by Dr. Hooker. Hooker's Journal of botany and Kew Garden Miscellany, 3: 77–84.

Berkeley, M.J. (1851b). Decade XXXV. Sikkim-Himalayan Fungi, collected by Dr. Hooker. Hooker's Journal of Botany and Kew Garden Miscellany, 3: 167– 172.

Berkeley, M.J. (1851c). Decade XXXVI. Sikkim-Himalayan Fungi, collected by Dr. Hooker. Hooker's Journal of Botany and Kew Garden Miscellany, 3: 200– 206.

Berkeley, M.J. (1854). Decades XLI, XLIII. Indian fungi. Hooker's Journal of Botany and Kew Garden Miscellany, 6: 129–143.

Berkeley, M.J. (1854a). Decades XLIV, XLVI. Indian fungi. Hooker's Journal of Botany and Kew Garden Miscellany, 6: 161–174.

Berkeley MJ. (1854b). Decades XLVII, XLVIII. Indian fungi. Hooker's Journal of Botany and Kew Garden Miscellany, 6: 204–212.

Berkeley, M.J. (1854c). Decades XLIX, L. Indian fungi. Hooker's Journal of Botany and Kew Garden Miscellany, 6: 225–235.

Berkeley, M.J. (1867). Fungi of the plains of India. Intellectual Observer, 12: 18–21.

Bishop, K.S., Kao, C.H., Xu, Y., Glucina, M.P., Paterson, R.R.M. and Ferguson, L.R. (2015). From 2000 years of *Ganoderma lucidum* to recent developments in nutraceuticals. Phytochemistry, 114: 56–65.

Boa, E. (2006). Champignons comestibles sauvages. Vue d'ensemble sur leurs utilisations et leur impor tance pour les populations. Produits Forestiers Non Ligneux 17.

Bose, S.R. (1918). Descriptions of the fungi in Bengal: I. *In*: Proceedings of the Indian Association for the Cultivation of Science, Jadavpur, Kolkata, p. 137.

Bose, S.R. (1919). Descriptions of the fungi in Bengal: I. *In*: Proceedings of the Indian Association for the Cultivation of Science, Jadavpur, Kolkata, pp. 109–114.

Bose, S.R. (1919a). Descriptions of the fungi in Bengal: II. *In*: Proceedings of the Indian Association for the Cultivation of Science, Jadavpur, Kolkata, pp. 136–143.

Bose, S.R. (1920). Fungi of Bengal. Polyporaceae of Bengal-III. Bull. Carmichael Med. Coll., 1: 1–5.

Bose, S.R. (1920a). Records of Agaricaceae from Bengal. Journal of Asiatic Society Bengal, 16: 347–354.

Bose, S.R. (1921). Polyporaceae of Bengal-IV. Bull. Carmichael Med. Coll., 2: 1–5.

Bose, S.R. (1921a). One new species of Polyporaceae and some Polypores new to Bengal. Ann. Mycol., 19: 129–131.

Bose, S.R. (1921b). Two new species of Polyporaceae. J. Ind. Bot., 2: 300–301.

Bose, S.R. (1922). Polyporaceae of Bengal-V. Bull. Carmichael Med. Coll., 3: 20–25.

Bose, S.R. (1922a). *Une* Polyporaceae *nouvelle du Bengala*. Bull. Soc. Mycol. Fr., 38: p. 173.

Bose, S.R. (1922b). Polyporaceae of Bengal: Polyporaceae of Bengal VI. pp. 55–62. *In*: Proceedings of the Indian Association for the Cultivation of Science, Jadavpur, Kolkata.

Bose, S.R. (1923). Polyporaceae of Bengal-VII. pp. 27–36. *In*: Proceedings of the Indian Association for the Cultivation of Science for the year 1920–21, Jadavpur, Kolkata.

Bose, S.R. (1923a). *Une Polyporaceae nouvelle d'Inde*. Bulletin de la Société mycologique de France 39: 226.

Bose, S.R. (1924). *Les Polyporacees du Bengale*. Rev. Path. Veg. Ent. Agric. Fr., 11: 134–149.

Bose, S.R. (1925). A new species of Polyporaceae from Bengal. Ann. Mycol., 23: 179–181.

Bose, S.R. (1928). Polyporaceae of Bengal-VIII. J. Dep. Sci. Calcutta Univ., 9: 27–34.

Bose, S.R. (1928a). Polyporaceae of Bengal-IX. J. Dep. Sci. Calcutta Univ., 9: 35–44.

Bose, S.R. (1934). Polyporaceae of Bengal-X. J. Dep. Sci. Calcutta Univ., 11: 1–18.

Bose, S.R. (1937). Polyporaceae of Lokra Hills (Assam). Ann. Mycol., 35: 119–137.

Bose, S.R. (1938). Polyporaceae of Bengal X. Bull. Carmichael Med. Coll., 11: 1–18.

Bose, S.R. (1944). Importance of anatomy in the systematics of Polyporaceae. J. Ind. Bot. Soc., 23: 153–157.

Bose, S.R. (1946). Polyporaceae of Bengal-XI. J. Dep. Sci. Calcutta Univ., 2: 53–87.

Burkill, I.H. (1966). A Dictionary of the Economic Products of the Malay Peninsula. Volumes I and II. Ministry of Agriculture, Kuala Lumpur.

Butler, E.J. and Bisby, G.R. (1931). The Fungi of India, Scientific monograph # I. Indian Council of Agricultural Research, Delhi.

Campanile, G., Schena, L. and Luisi, N. (2008). Blackwell Publishing Ltd Real-time PCR identification and detection of *Fuscoporia torulosa* in *Quercus ilex*. Pl. Patho., 57: 76–83.

Carranza-Morse, J. (1991). Pore fungi of Costa Rica I. Mycotaxon, 41: 345–370.

Carranza-Morse, J. (1992). Pore fungi of Costa Rica II. Mycotaxon, 43: 351–369.

Carranza-Morse, J. (1993). Pore fungi of Costa Rica III. Mycotaxon, 48: 45–57.

Chan, P.M, Kanagasabapathy, G., Tan, Y.S. and Sabaratnam, V. (2013). *Amauroderma rugosum* (Blume & T. Nees) Torrend: nutritional composition and antioxidant and potential anti inflammatory properties. Evid. Based Compl. Alt. Med., 304713. 10.1155/2013/304713.

Chang, Y.S. and Lee, S.S. (2004). Utilisation of macrofungi species in Malaysia. Fungal Diversity, 15: 15–22.

Chao, C.H., Wu, H.J. and Lu, M.K. (2011). Promotion of fungal growth and underlying physiochemical changes of polysaccharides in *Rigidoporus ulmarius*, an edible Basidiomycete mushroom. Carbohy. Polym., 85: 609–614.

Chen, W., Luo, S., LI, H. and Yang, H. (2005). Effects of *Ganoderma lucidum* polysaccharides on serum lipids and lipoperoxidation in experimental hyperlipidemic rats. J. Chinese Mat. Med., 30: 1358–60.

Chen, X., Hu, Z.P, Yang, X.X., Huang, M., Gao, Y., Tang, W., Chan, S.Y., Dai, X., Ye, J., Ho, P.C.L., Duan, W., Yang, H.Y., Zhu, Y.Z. and Zhou, S.H. (2006). Monitoring of immune responses to a herbal immuno-modulator in patients with advanced colorectal cancer. Int. Immunopharmacol., 6: 499–508.

Chen, Y.Y., Zhu, L., Xing, J. and Cui, B.K. (2017). Three new species of *Phylloporia* (Hymenochaetales) with dimitic hyphal systems from tropical China, Mycologia, 109: 951–964.

Chen, H.P., Zhao, Z.Z., Li, Z.H., Feng, T. and Liu, J.K. (2018). Seco-tremulane sesquiterpenoids from the cultures of the medicinal fungus *Irpex lacteus* HFG1102. Nat. Prod. Bioprospect., 8: 113–119.

Chin, F.H. (1981). Edible and poisonous fungi from the forest of Sarawak. Part 1. Sarawak Museum Journal, 50: 211–225.

Choi, Y.S., Long, Y., Kim, M.J., Kim, J.J. and Kim, G.H. (2013). Decolorization and degradation of synthetic dyes by *Irpex lacteus* KUC8958. J. Environ. Sci. Health Part A, 48: 501–508.

Correa, E., Cardona, D., Quinones, W., Torres, F., Franco, A.E., Vélez, I.D., Robledo, S., Echeverri, F. (2006). Leishmanicidal activity of *Pycnoporus sanguineus*. Phytother. Res., 20: 497–499.

Dai, Y.C. (2010). Hymenochaetaceae (Basidiomycota) in China. Fungal Diversity, 45: 131–343. 10.1007/s13225-010-0066-9.

Dai, Y.C. and Yang, Z.L. (2008). A revised checklist of medicinal fungi in China. Mycosystema, 27: 801–824.

Dai, Y.C., Cui, B.K., Yuan, H.S. and Li, B.D. (2007). Pathogenic wood-decaying fungi in China. For. Pathol., 37: 105–120.

Dai, Y.C., Zhou, L.W., Cui, B.K., Chen, Y.Q. and Decock, C. (2010). Current advances in *Phellinus sensu lato*: medicinal species, functions, metabolites and mechanisms. Appl. Microbiol. Biotechnol., 87: 1587–1593.

Doskocil, I., Havlik, J., Verlotta, R., Tauchen, J., Vesela, L., Macakova, K., Opletal, L., Kokoska, L. and Rada, V. (2016). *In vitro* immunomodulatory activity, cytotoxicity and chemistry of some central European polypores. Pharmaceut. Biol., 54: 2369–2376.

Dong, X.M., Song, Y.H. and Dong, C.S. (2017). Nutritional requirements of the mycelial growth of *Irpex lacteus* (Basidiomycetes) in submerged culture. Int. J. Med. Mushrooms, 19: 829−838.

El-Mekkawy, S., Meselhy, M.R., Nakamura, N., Tezuka, Y., Hattori, M. and Kakiuchi, N. (1998). Anti-HIV-1 and anti-HIV-1-protease substances from *Ganoderma lucidum*. Phytochem., 49: 1651–1657.

Erden, E., Ucar, M.C., Gezer, T. and Pazarlioglu, N.K. (2009). Screening for ligninolytic enzymes from autochthonous fungi and applications for decolorization of remazole marine blue. Braz. J. Microbiol., 40: 346–353.

Ergönül, P.G., Ilgaz A., Kalyoncu, F. and Ergönül, B. (2013). Fatty acid compositions of six wild edible mushroom species. The Sci. World J., 163964. 10.1155/2013/163964.

Esquivel, R.E. and Carranza-Morse, J. (1996). Pathogenicity of *Phylloporia chrysita* (Aphyllophorales: Hymenochaetaceae) on *Erythrociton gymnanthus* (Rutaceae). Rev. Biol. Trop., 44: 137–145.

Eugenio, M.E., Carbajo, J.M., Martin, A., González, A.E. and Villar, J.C. (2009). Laccase production by *Pycnoporus sanguineus* under different culture conditions. J. Basic Microbiol., 49: 433–40.

Falconnier, B., Lapierre, C., Lesage-Meessen, L., Yonnet, G., Brunerie, P., Colonna, C.B., Corrieu, G. and Asther, M. (1994). Vanillin as a product of ferulic acid biotransformation by the white-rot fungus *Pycnoporus cinnabarinus* I-937: Identification of metabolic pathways. J. Biotechnol., 37: 123–132.

Fakoya, S. and Oloketuyi, S.F. (2012). Antimicrobial Efficacy and phytochemical screening of Mushrooms, *Lenzites Betulinus* and *Coriolopsis Gallica* extracts. TAF Prev. Med. Bull., 11: 695–698.

Farook, V.A., Khan, S.S. and Manimohan, P. (2013). A checklist of agarics (gilled mushrooms) of Kerala State, India. Mycosphere, 4: 97–131.

Fernando, M.D.M., Wijesundera, R.L.C., Soysa, P., Silva, E.D. and Nanayakkara, C.M. (2015). Strong radical scavenging macrofungi from the dry zone forest reserves in Sri Lanka. Front. Environ. Microbiol., 1: 32–38.

Fernando, M.D.M., Wijesundera, R.L.C., Soysa, P., Silva, E.D. and Nanayakkara, C.M. (2016). Antioxidant potential and content of the polyphenolic secondary metabolites of white rot macrofungi; *Flavodon flavus* (Klotzsch.) and *Xylaria feejeensis* (Berk.). J. Pl. Sci., 1: 10–15.

Florence, E.J.M. and Yesodharan, K. (2000). Macrofungal flora of PeechiVazhani Wildlife Sanctuary, Research Report 191, Kerala Forest Research Institute, Peechi, Kerala, India, 43.

Florence, E.J.M. and Balasundaran, M. (2000). Mushroom cultivation using forest litter and waste wood. Kerala Forest Research Institute Research Report # 195.

Fossli, T.E. and Andersen, J. (1998). Host preference of *Cisidae* (Coleoptera) on tree- inhabiting fungi in nothern Norway. Entomol. Fen., 9: 65–78.

Gibertoni, T.B., Santos, P.J.P. and Cavalcanti, M.A.Q. (2007). Ecological aspects of Aphyllophorales in the Atlantic rain forest in Northeast Brazil. Fungal Diversity, 25: 49–67.

Glen, M., Bougher, N.L., Francis, A.A., Nigg, S.Q., Lee, S.S., Irianto, R., Barry, K.M., Beadle, C.L. and Mohammed, C.L. (2009). *Ganoderma* and *Amauroderma* species associated with root–rot disease of *Acacia mangium* plantation trees in Indonesia and Malaysia. Aust. Pl. Pathol., 38: 345–356.

Göçenoğlu, A. and Pazarlioglu, N. (2014). Cinnabarinic acid: Enhanced production from *Pycnoporus cinnabarinus*, characterization, structural and functional properties. J. Biol. Chem., 42: 281–290.

Gomes-Silva, A.C. and Gibertoni, T.B. (2012). Neotypification of *Amauroderma picipes* Torrend, 1920 (Ganodermataceae, Agaricomycetes). Mycosphere, 3: 23–27.

Hågvar, S. (1999). Saproxylic beetles visiting living sporocarps of *Fomitopsis pinicola* and *Fomes fomentarius*. Nor. J, Entomol., 46: 25–32.

Hapuarachchi, K.K., Wen, T.C., Jeewon, R., Wu, X.L., Kang, J.C. and Hyde, K.D. (2016). Mycosphere Essays 7: *Ganoderma lucidum*-are the beneficial anti-cancer properties substantiated? Mycosphere, 7: 305–332.

Hapuarachchi, K.K., Karunarathna, S.C., Phengsintham, P., Kakumyan, P., Hyde, K.D. and Wen, T.C. (2018). *Amauroderma* (Ganodermataceae, Polyporales), bioactive compounds, beneficial properties and two new records from Laos. Asian J. Mycol., 1: 121–136.

Haritash, A.K and Kaushik, C.P. (2009). Biodegradation aspects of polycyclic aromatic hydrocarbons (PAHs): A review. J. Haz. Mat., 169: 1–15.

Heinfling, A., Martınez, M.J., Martınez, A.T., Bergbauer, M. and Szewzyk, U. (1998). Purification and characterization of peroxidases from the dye-decolorizing fungus *Bjerkandera adusta*. FEMS Microbiol. Lett., 165: 43–50.

Heinfling, A., Martínez, M.J., Martínez, A.T., Bergbauer, M. and Szewzyk, U. (2011). Transformation of industrial dyes by manganese peroxidases from *Bjerkandera adusta* and *Pleurotus eryngii* in a manganese independent reaction. Appl. Environ. Microbiol., 64: 2788–93.

Hembrom, M.E., Parihar, A. and Das, K. (2015). Three interesting wood rotting macrofungi from Jharkhand (India). J. Threat. Taxa, 8: 8518–8525.

Hembrom, M.E., Das, K., Parihar, A. and Sengupta, C. (2017). *Hymenochaete conchata* (Hymenochaetaceae), a new record for Indian mycobiota. Ind. J. Pl. Sci., 6: 1–9.

Hibbett, D.S., Bauer, R., Binder, M., Giachini, A.J., Hosaka, K. et al. (2014). Agaricomycetes. Systematics and Evolution, 2nd Edition, The Mycota Part VII. McLaughlin, A.D.J. and Spatafora, J.W. (eds.). Springer-Verlag, Berlin and Heidelberg.

Hickman, G.W., Perry, E.J. and Davis, R.M. (2011). Wood decay fungi in landscape Trees. UC ANR Publication 74109, Oakland, CA.

Hmd, R.F.K. (2011). Degradation of some textile dyes using biological and physical Treatments. MSC Dissertation, Microbiology Department, Faculty of Science, Ain-Shams University, Cairo, Egypt, p. 188.

Hobbs, C. (2004). Medicinal value of Turkey Tail Fungus *Trametes versicolor* (L.:Fr.) Pilat (Aphyllophoromycetideae). A literature review. Int. J. Med. Mushrooms, 6: 195–218.

Huang, K.C. (1993). The Pharmacology of Chinese Herbs. CRC Press, Boca Raton, USA.

Ivanova, T.S., Krupodorova, T.A., Barshteyn, V.Y., Artamonova, A.B. and Shlyakhovenko, V.A. (2014). Anticancer substances of mushroom origin. Exp. Oncol., 36: 58–66.

Jayalakshmi, R., Mallika, D.S., Amos, S.J. and Kasturi, K. (2015). A noval approach for treating cancer by using laccases from marine fungi. Int. J. Pharmaceut. Sci. Rev. Res., 34: 124–129.

Jayasinghe, C.K., Jayasuriya, K.E. and Fernando, T.H.P.S. (1995). Pentachlorophenol-effective and economical fungicide for the management of white root disease caused by *Rigidoporus lignosus* in Sri Lanka. J. Rubber Res. Inst. Sri Lanka, 75: 61–70.

Jayasuriya, K.E. (1996). Biological control of plant pathogens and possible biocontrol approaches against *Rigidoporus lignosus*, the cause of white root disease of rubber. Bull. Rubber Res. Inst. Sri Lanka, 33: 1–12.

Jayasuriya, K.E. and Deacon, J.W. (1996). A possible role for 2-Furaldehyde in the biological control of *Rigidoporus lignosus*, the cause of white root disease of rubber. J. Rubber Res. Inst. Sri Lanka, 77: 15–27.

Jena, R. and Thatoi, H.N. (2019). Screening and evaluation of phytochemicals and anticancer activities of *Pycnoporus coccineus*—A medicinal mushroom from similipal Biosphere Reserve, Odisha. J. Basic Appl. Res. Int., 25: 271–277.

Jiao, C., Xie, Y.Z., Yang, X. and Li, H. (2013). Anticancer activity of *Amauroderma rude*. PLoS ONE, 8: e66504.

Jonsell, M. (1998). A new anobiid-beetle for Sweden: Dorca toma minor Zahradnik (Coleoptera: Anobiidae) and its host preference. Entomologisk Tidskrift, 119: 105–109.

Jonsell, M. and Nordlander, G. (1995). Field attraction of Coleoptera to odours of the wood-decaying polypores *Fomitopsis pinicola* and *Fomes fomentarius*. Ann. Zool. Fennici, 32: 391–402.

Jung, P., Fong, J.J., Park, M.S., Oh, S.Y., Kim, C. and Lim, Y.W. (2014). Sequence Validation for the identification of the white-rot fungi *Bjerkandera* in public sequence databases. J. Microbiol. Biotechnol., 24: 1301–1307.

Karun, N.C. and Sridhar, K.R. (2017). Edible mushrooms of the Western Ghats: Data on the ethnic knowledge. Data in Brief, 14: 320–32.

Karunarathna, S.C., Yang, Z.L., Zhao, R.L., Vellinga, E.C., Bahkali, A.H., Chukeatirote, E. and Hyde, K.D. (2011). Three new species of *Lentinus* from northern Thailand. Mycol. Prog., 10: 389–398.

Kawashima, N., Deveaux, T.E., Yoshida, N., Matsumoto, K. and Kato, K. (2012). Choreito, a formula from Japanese traditional medicine (Kampo medicine), for massive hemorrhagic cystitis and clot retention in a pediatric patient with refractory acute lymphoblastic leukemia. Phytomedicine, 19: 1143–1146.

Kim, J.S., Kim, J.E., Choi, B.S., Park, S.E., Sapkota, K., Kim, S., Lee, H.H., Kim, C.S., Park, Y., Kim, M.K., Kim, Y.S. and Kim, S.J. (2008). Purification and characterization of fibrinolytic metalloprotease from *Perenniporia fraxinea* mycelia. Mycol. Res., 112: 990–998.

Klotzsch, J.F. (1832). Mycologische Berichtigungen. Linnaea, 7: 193–204.

Klotzsch, J.F. (1833). Fungi exoticie collectionibus britannorum. Linnaea, 7: 478–490.

Kobayashi, Y., Kariya, K., Saigenji, K. and Nakamura, K. (1994). Enhancement of anti cancer activity of cisdiaminedichloroplatinum by the protein-bound polysaccharide of *Coriolus versicolor* QUEL (PS-K) *in vitro*. Canc. Biother., 9: 351–8.

Komonen, A. (2001). Structure of insect communities inhabiting old-growth forest specialist bracket fungi. Ecol. Entomol., 26: 63–75.

Kondo, R. (2005). Evaluation of white rot fungi for treatment of organic wastes without enviromental impact. The 2005 World Sustainable Building Conference, Tokyo, 27–29 September 2005 (SB05Tokyo).

Korneichik, T.V. and Kapich, A. (2011). Prooxidant activity of *Bjerkandera adusta* BIMF-260 and *Pleurotus ostreatus* BIMF-247 fungi in a solid-state culture. Moscow Univ. Biol. Sci. Bull., 66: 60–62.

Kost, G. (2004). Ecology and morphology of some tropical fungi. *In*: Agerer, R., Piepenbring, M. and Blanz, P. (eds.). Frontiers in Basidiomycote Mycology. IHW Press, Verlag.

Krastanov, A., Koleva, R., Alexieva, Z. and Stoilova, I. (2013). Decolorization of industrial dyes by immobilized mycelia of *Trametes versicolor*. Biotechnol. Biotechnol. Equip., 27: 4263–4268.

Küçükaydin, S. and Duru, M.E. (2017). Lipid compositions of *Bjerkandera adusta* (Willd) P Karst. Sigma J. Eng. Nat. Sci., 35: 405–410.

Kumar, T.K.A., Thomas, A., Kuniyil, K., Nanu, S. and Nellipunath, V. (2019). A checklist of the non-gilled felshy fungi of Kerala. Mycotaxon, 134: 221.

Kuo, M. (2016). *Perenniporia robiniophila*. Retrieved from the MushroomExpert.Com. http://www.mushroomexpert.com/perenniporia_ robiniophila.html.

Lai, K.M., Wong, K.H. and Cheung, P.C.K. (2008). Antiproliferative effects of sclerotial polysaccharides from *Polyporus rhinocerus* Cooke (Aphyllophoromycetideae) on different kinds of leukemic Cells. Int. J. Med. Mushrooms, 10: 255–264.

Larsen, M.J. and Cobb-Poulle, L.A. (1990). *Phellinus* (Hymenochaetaceae). A survey of the world taxa. Synopsis Fungorum 3, Fungiflora, Oslo.

Lee, M.L., Tan, N.H., Fung, S.Y., Tan, C.S. and Ng, S.T. (2012). The antiproliferative activity of sclerotia of *Lignosus rhinocerus* (tiger milk mushroom). Evid. Based Compl. Alt. Med., 697603. 10.1155/2012/697603.

Leelavathy, K.M. and Ganesh, P.N. (2000). Polypores of Kerala. Daya Publishing House, New Delhi, p. 175.

Li, Y., Simmons, D.R., Bateman, C.C., Short, D.P.G., Kasson, M.T., Rabaglia, R.J. and Hulcer, J. (2015). New fungus-insect symbiosis: culturing, molecular and histological methods determine saprophytic Polyporales mutualists of *Ambrosiodmus Ambrosia* beetles. PLoS ONE, 10: e0137689.

Li, Y., Bateman, C.C., Skelton, J., Jusino, M.A., Nolen, Z.J., Simmons, D.R. and Hulcr, J. (2017). Wood decay fungus *Flavodon ambrosius* (Basidiomycota: Polyporales) is widely farmed by two genera of ambrosia beetles. Fungal Biol., 21: 984–989.

Li, Y.Q. and Wang, S.F. (2006). Anti-hepatitis B activities of ganoderic acid from *Ganoderma lucidum*. Biotechnol. Lett., 28: 837–41.

Liew, G.M., Khong, H.Y., Kutoi, C.J. and Sayok, A.K. (2015). Phytochemical screening, antimicrobial and antioxidant activities of selected fungi from Mount Singai, Sarawak, Malaysia. Int. J. Res. Stud. Biosci., 3: 191–197.

Lindner, D.L., Burdsall, H.H. and Stanosz, G.R. (2006). Species diversity of polyporoid and corticioid fungi in northern hardwood forests with differing management histories. Mycologia, 98: 195–217.

Liu, B. and Bau, Y.S. (1994). Fungi Pharmacopoeia (Sinica). Kinoko Co., Oakland, California.

Liu, Z. and Liu, L. (2009). Essentials of Chinese Medicine, Springer Verlag, London.

Lodge, D.J. and Cantrell, S. (1995). Fungal communities in wet tropical forests: Variation in time and space. Can. J. Bot., 73: S1391–S1398.

Lomascolo, A., Uzan-Boukhris, E., Herpoël-Gimbert, I., Sigoillot, J.C. and Lesage-Meessen, L. (2011). Peculiarities of *Pycnoporus* species for applications in biotechnology. Appl. Microbiol. Biotechnol., 92: 1129–1149.

Meghalatha, R., Ashok, C., Nataraja, S. and Krishnappa, M. (2014). Studies on chemical composition and proximate analysis of wild mushrooms. World J. Pharmaceut. Sci., 2: 357–363.

Mohanan, C. (2011). Macrofungi of Kerala. Kerala Forest Research Institute, Kerala, p. 597.

Morigiwa, A., Kitabatake, K., Fujimoto, Y. and Ikekawa, N. (1986). Angiotensin converting enzyme-inhibitory triterpenes from *Ganoderma lucidim*. Chem. Pharmaceut. Bull., 34: 3025–3028.

Motta, E., Annesi, T. and Biocca, M. (1996). *Processi di carie a carico di Eucalyptus* spp. *in un parco cittadino*. Montie Boschi, 1: 55–7.

Miklašēvičs, Z. (2019). Evaluation of heartrot caused *Phellinus pini* and related yield loss in *Pinus sylvestris* stands. Environment. Technology. Resources. Rezekne, Latvia, Proceedings of the 12th International Scientific and Practical Conference. Volume III, 166–171.

Naidu, G.V.B., Kumar, S.N.S. and Sannamarappa, M. (1966). Anabe roga, *Ganoderma lucidum* (Leys.) Karst. on arecanut palm: A review and further observations. J. Mysore Hort. Soc., 11: 14–20.

Nilsson, T. (1997). Spatial population dynamics of the black tinder fungus beetle *Bolitophagus reticulatus* (Coleoptera: Tenebrionidae). Acta Universitatis Upsaliensis. Comprehensive Summaries of Uppsala Dissertations from the Faculty of Science and Technology, 311: 1–44.

Oddou, J., Stentelaire, C., Lesage-Meessen, L., Asther, M. and Colonna-Ceccaldi, B. (1999). Improvement of ferulic acid bioconversion into vanillin by use of high-density cultures of *Pycnoporus cinnabarinus*. Appl. Microbiol. Biotechnol., 53: 1–6.

Pan, H., Han, Y., Huang, J. and Yu, X. (2015). Purification and identification of a polysaccharide from medicinal mushroom *Amauroderma rude* with immunomodulatory activity and inhibitory effect on tumor growth. Oncotarget, 10: 17777–17791.

Peng, T.V. and Don, M.M. (2013). Antifungal activity of *in vitro* grown *Earliella scabrosa*, a Malaysian fungus on selected wood-degrading fungi of rubberwood. J. Phy. Sci., 24: 21–33.

Peláez, F., Martinez, M.J. and Martinez, A.T. (1995). Screening of 68 species of basidiomycetes for enzymes involved in lignin degradation. Mycol. Res., 99: 37–42.

Peries, O.S. (1969). Economics of control of the white root disease (*Fomes lignosus*) of *Hevea brasiliensis* in Ceylon. *In*: Toussoun, T.A., Bega, R.V. and Nelson, P.A. (eds.). Root diseases and soil-borne pathogens. Second International Symposium on Factors Determining the Behaviour of Plant Pathogens in Soil, July 14–28, 1968. Imperial College, London, University of California Press, p. 252.

Pongen, A.S., Chuzho, K., Harsh, N.S.K., Dkhar, M.S. and Kumar, M. (2018). *Coltriciella dependens* (Berk. & M.A. Curtis) Murrill, a new addition to wood-rotting fungi of India. J. Threat. Taxa, 10: 12140–12143.

Prasher, I.B. (2015). Wood-rotting non-gilled Agaricomycetes of Himalayas. Springer, Netherlands, p. 653.

Radwan, F.F., Perez, J.M. and Haque, A. (2011). Apoptotic and immune restoration effects of ganoderic acids define a new prospective for complementary treatment of cancer. J. Clin. Cell. Immunol., S3: 4.

Raghukumar, C. and Rivonkar, G. (2001). Decolorization of molasses spent wash by the white-rot fungus *Flavodon flavus*, isolated from a marine habitat. Appl. Microbiol. Biotechnol., 55: 510–4.

Raghukumar, C., Mohandas, C., Kamat, S. and Shailaja, M.S. (2004). Simultaneous detoxification and decolorization of molasses spent wash by the immobilized white-rot fungus *Flavodon flavus* isolated from a marine habitat. Enz. Microb. Technol., 35: 197–202.

Rajchenberg, M. and Robledo, G. (2013). Pathogenic polypores in Argentina. For. Pathol., 43: 171–184.

Ranadive, K.R. and Jagtap, N.V. (2016). Preliminary checklist of fungal flora of Kas lateritic plateau and surroundings from the North Western Ghats of Maharashtra State. Appl. Bot., 60: 16637–16640.

Ranadive, K.R., Jagtap, N. and Vaidya, J. (2012). Host diversity of genus *Phellinus* from world. Elixir Appl. Bot., 52: 11402–11408.

Robles, C.A., Carmaran, C. and Lopez, S.E. (2011). Screening of xylophagous fungi associated with *Platanus acerifolia* in urban landscapes: Biodiversity and potential biodeterioration. Landsc. Urb. Plan., 100: 129–135.

Rogers, R. (2012). Three under-utilized medicinal polypores. J. Am. Herbalists Guild, 12: 15–21.

Rony, K.A., Mathew, J., Neenu, P.P. and Janardhanan, K. (2011). *Ganoderma lucidum* (Fr.) P. Karst occurring in South India attenuates gastric ulceration in rats. Ind. J. Nat. Prod. Resour., 2: 19–27.

Rukke, B.A. and Midtgaard, F. (1998). The importance of scale and spatial variables for the fungivorous beetle *Bolitophagus reticulatus* (Coleoptera: Tenebrionidae) in a fragmented forest landscape. Ecography, 21: 561–572.

Ryvarden, L. (1991). Genera of polypores. Nomencl. Taxon, Syn. Fungorum, 5: 1–363.

Ryvarden, L. (1992). Tropical polypores. Abstracts to Tropical Mycology Symposium April 1992.

Ryvarden, L. and Johansen, I. (1980). A preliminary polypore flora of East Africa. Fungiflora, Oslo.

Ryvarden, L. and Johansen, I. (1980). A preliminary polypore flora of East Africa. Oslo, Norway, p. 636.

Ryvarden, L. and Melo, I. (2014). Poroid fungi of Europe. Synopsis Fungorum, 31: 1–455.

Savino, E., Girometta, C., Chinaglia, S., Guglielminetti, M.L., Rodolfi, M. et al. (2014). Medicinal mushrooms in Italy and their *ex situ* conservation through culture collection. Proceedings of the 8th International Conference on Mushroom Biology and Mushroom Products (ICMBMP8) 2014.

Savino, E., Girometta, C., Staleva, J., Kostadinova, A. and Krumova, E. (2016). Wood decay macrofungi: Strain collection and studies about antioxidant properties. Comptes rendus de l'Académie bulgare des Sciences, 69: p. 747.

Schigel, D.M. (2011). Polypore–beetle associations in Finland. Ann. Zool. Fennici, 48: 319–348.

Schigel, D.S., Niemela, T. and Kinnunen, J. (2006). Polypores of western Finnish Lapland and seasonal dynamics of polypore beetles. Karstenia, 46: 37–64.

Schliephake, K., Lonergan, G.T., Jones, C.L. and Mainwaring, D.E. (1999). Decolourisation of a pigment plant effluent by *Pycnoporus cinnabarinus* in a packed-bed bioreactor. Biotechnol. Lett., 15: 1185–1188.

Schliephake, K., Mainwaring, D.E., Lonergan, G.T., JFones, I.K. and Baker, W.L. (2000). Transformation and degradation of the disazo dye Chicago Sky Blue by a purified laccase from *Pycnoporus cinnabarinus*. Enz. Microbial. Biotechnol., 27: 100–107.

Schmit, J.P. (2005). Species richness of tropical wood-inhabiting macrofungi provides support for species-energy theory. Mycologia, 97: 751–761.

Schmit, J.P., Mueller, G.M., Leacock, P.R. and Mata, J.L. (2005). Assessment of tree species richness as a surrogate for macrofungal species richness. Biol. Conser., 121: 99–110.

Seelan, J.S.S., Justo, A., Nagy, L.G., Grand, E.A., Readhead, H.A. and Hibbett, D. (2015). Phylogenetic relationships and morphological evolution in *Lentinus*, *Polyporellus* and *Neofavolus*, emphasizing southeastern Asian taxa. Mycologia, 107: 460–474.

Senthilarasu, G. (2015). The lentinoid fungi (*Lentinus* and *Panus*) from Western Ghats, India. IMA Fungus, 6: 119–128.

Selonen, V.A.O., Ahlroth, P. and Kotiaho, J.S. (2005). Anthropogenic disturbance and diversity of species: Polypores and polypore-associated beetles in forest, forest edge and clear-cut. Scan. J. For. Res., 20: 49–58.

Seto, S.W., Lam, T.Y., Tam, H.L., Au, A.L., Chan, S.W., Wu, J.H., Yu, P.H., Leung, G.P., Ngai, S.M., Yeung, J.H., Leung, P.S., Lee, S.M. and Kwan, Y.W. (2009). Novel hypoglycemic effects of *Ganoderma lucidum* water-extract in obese/diabetic (+db/+db) mice. Phytomedicine: Int. J. Phytother. Phytopharmacol., 16: 426–436.

Sharma, S.K. and Atri, N.S. (2015). The genus *Lentinus* (Basidiomycetes) from India—An annotated checklist. J. Threat. Taxa, 7: 7843–7848.

Shintani, N., Sugano, Y. and Shoda, M. (2002). Decolorization of kraft pulp lignin bleaching effluent by a newly isolated fungus, *Geotrichum candidum*. J. Wood Sci., 48: 402–408.

Silberborth, S., Erkel, G., Anke, T. and Sterner, O. (2000). The irpexans, a new group of biologically active metabolites produced by the basidiomycete *Irpex* sp. 93028. J. Antibiot. (Tokyo), 53: 1137–44.

Simmons, D.R., Li, Y., Bateman, C.C. and Hulcr, J. (2016). *Flavodon ambrosius sp. nov.*, a basidiomycetous mycosymbiont of *Ambrosiodmus ambrosia* beetles. Mycotaxon, 131: 277–285.

Singh, R.P., Kashyap, A.S., Pal, A., Singh, P. and Tripathi, N.N. (2019). Macrofungal diversity of North-Eastern part of Uttar Pradesh (India). Int. J. Curr. Microbiol. Appl. Sci., 8: 823–838.

Siripong, P., Oraphin, B. and Duanporn, P. (2009). The ability of five fungal isolates from nature to degrade of polyaromatic hydrocarbons (PAHs) and polychlorinated biphenyls (PCBs) in Culture Media. Aust. J. Basic Appl. Sci., 3: 1076–1082.

Smith, J.E., Rowan, N.J. and Sullivan, R. (2002). Medicinal Mushrooms: Their Therapeutic Properties and Current Medical Usage with Special Emphasis on Cancer Treatments. Cancer Research UK, University of Strathclyde, UK, p. 253.

Sugawara, K., Igeta, E., Amano, Y., Hyuga M. and Sugano, Y. (2019). Degradation of antifungal anthraquinone compounds is a probable physiological role of DyP secreted by *Bjerkandera adusta*. AMB Express, 9: 56. 10.1186/s13568-019-0779-4.

Sundararaman, S. and Marudarajan, D. (1925). Some polyporaceae of the Madras Presidency. Madras Agriculture Deppartment Year Book.

Sunhede, S. and Vasiliauskas, R. (2002). Ecology and decay pattern of *Phellinus robustus* in old-growth *Quercus robur*. Karstenia, 42: 1–11.

Sverdrup-Thygeson, A. and Lindenmayer, D.B. (2003). Ecological continuity and assumed indicator fungi in boreal forest: The importance of the landscape matrix. For. Ecol. Manage., 174: 353.

Szczepkowski, A. (2004). *Perenniporia fraxinea* (fungi, Polyporales), a new species for Poland. Polish Bot. J., 49: 73–77.

Tang, Y., Zhao, Z.Z., Li, Z.H., Feng, T., Chen, H.P. and Liu, JK. (2018). Irpexoates A–D, four triterpenoids with malonyl modifications from the fruiting bodies of the medicinal fungus *Irpex lacteus*. Nat. Prod. Bioprospect., 8: 171–176.

Tedersoo, L., Suvi, T., Beaver, K. and Saar, I. (2007). Ectomycorrhizas of *Coltricia* and *Coltriciella* (Hymenochaetales, Basidiomycota) on Caesalpiniaceae, Dipterocarpaceae and Myrtaceae in Seychelles. Mycol. Prog., 6: 101–107.

Thunes, K.H. (1994). The coleopteran fauna of *Piptoporus betulinus* and *Fomes fomentarius* (Aphyllophorales: Polyporaceae) in northern Norway. Entomol. Fenn., 5: 157–168.

Tomšovský, M. and Jankovský, L. (2007). DNA sequence analysis of extraordinary fruiting specimens of *Fuscoporia torulosa* (*Phellinus torulosus*) on *Pyrus* spp. Czech Mycol., 59: 91–99.

Tzean, Y., Shu, P.Y., Liou, R.F. and Tzean, S.S. (2016). Development of oligonucleotide microarrays for simultaneous multi-species identification of *Phellinus* tree-pathogenic fungi. Microb. Biotechnol., 9: 235–244.

Upton, R. (2006). Reishi Mushroom, *Ganoderma lucidum*, Standards of analysis, quality control and therapeutics. California, United States of America: American Herbal Pharmacopoeia and Therapeutic Compendium.

Vaidya, J.G. and Lamrood, P.Y. (2000). Traditional medicinal mushrooms and Fungi of India. Int. J. Med. Mushrooms, 2: 209–214.

Vasudeva, R.S. (1960). The fungi of India (revised edition of Butler and Bisby's work). ICAR, New Delhi, 552.

Viecelli, C.A., Stangarlin, J.R., Kuhn, O.J. and Schwan-Estrada, K.R.F. (2009). Induction of resistance in beans against *Pseudocercospora griseola* by culture filtrates of *Pycnoporus sanguineus*. Trop. Pl. Pathol., 34: 87–96.

Vinjusha, N. (2020). Systematic studies on the polyporoid fungi (Agaricomycetes, Basidiomycota) of Kerala. PhD Thesis. University of Calicut, Kerala, India.

Vinjusha, N. and Kumar, T.K.A. (2021). A rare medicinal fungus, *Lignosus rhinocerus* (Polyporales, Agaricomycetes), new to India. Stud. Fungi, 6: 151–158.

Vinjusha, N. and Kumar, T.K.A. (2022a). Revision of *Ganoderma* species associated with stem rot of coconut palm. Mycologia, 114: 157–174.

Vinjusha, N. and Kumar, T.K.A. (2022b). The Polyporales of Kerala. SporePrint Books, Calicut, India, p. 229.

Wang, Q. and Qi, Y. (2016). Antioxidant activity of *Amauroderma rudis*-roots of *Lentinus* solid fermentation compounds. J. Food Saf. Qual., 7: 682–685.

Wang, M., Yin, H., Peng, H., Feng, M., Lu, G. and Dang, Z. (2019). Degradation of 2,2′,4,4′-tetrabromodiphenyl ether by *Pycnoporus sanguineus* in the presence of copper ions. J. Environ. Sci., 83: 133–143.

Wagner, T. and Fischer, M. (2001). Natural groups and a revised system for the European poroid Hymenochaetales (Basidiomycota) supported by nLSU rDNA sequence data. Mycol. Res., 105: 773–782.

Wagner, T. and Fischer, M. (2002). Proceedings towards a natural classification of the worldwide taxa *Phellinus s.l.* and *Inonotus s.l. and* phylogenetic relationships of allied genera. Mycologia, 94: 998–1016.

Wagner, T. and Ryvarden, L. (2002). Phylogeny and taxonomy of the genus *Phylloporia* (Hymenochaetales). Mycol. Prog., 1: 105–116.

Watling, R. (1993). Comparison of the macromycete biotas in selected tropical areas of Africa and Australia. *In*: Isaac, S., Frankland, J., Watling, R. and Whalley, A.J.S. (eds.). Aspects of Tropical Mycology. Cambridge Univ. Press, Cambridge.

Wong, K.H., Lai, C.K.M. and Cheung, P.C.K. (2011). Immunomodulatory activities of mushroom sclerotial polysaccharides. Food Hydrocolloids, 25: 150–158.

Xing, Y.M., Zhang, L.C., Liang, H.Q., Lv, J., Song, C., Guo, S.X., Wang, C.L., Lee, T.S. and Lee, M.W. (2013). Sclerotial formation of *Polyporus umbellatus* by low temperature treatment under artificial conditions. PLoS ONE, 8: e56190.

Ying, J., Mao, X., Ma, Q., Zong, Y. and Wen, H. (1987). Icones of Medicinal Fungi from China, Science Press, Beijing, China, p. 575.

Yombiyeni, P., Balezi, A., Amalfi, M. and Decock, C. (2015). Hymenochaetaceae from the Guineo-Congolian rainforest: three new species of *Phylloporia* based on morphological, DNA sequences and ecological data. Mycologia, 107: 996–1011.

Yuan, D., Mori, J., Komatsu, K.I., Makino, T. and Kano, Y. (2004). An anti-aldosteronic diuretic component (drain dampness) in *Polyporus sclerotium*. Biol. Pharmaceut. Bull., 27: 867–870.

Zhang, Y., Hyde, K.D., Zhou, D.Q., Zhao, Q. and Zhou, T.X. (2010). Diversity and ecological distribution of macrofungi in the Laojun Mountain region, southwestern China. Biodivers. Conser., 19: 3545–3563.

Zhang, G., Tianc, T., Liua, Y.P., Wangb, H. and Chena, Q. (2011). A laccase with anti-proliferative activity against tumor cells from a white root fungus *Abortiporus biennis*. Proc. Biochem., 46: 2336–2340.

Zhou, L.W. (2016). *Phylloporia minutipora* and *P. radiata spp. nov.* (Hymenochaetales, Basidiomycota) from China and a key to worldwide species of *Phylloporia*. Mycol. Prog., 15: 57.

Zimbardi, A.L., Camargo, P.F., Carli, S., Neto, S.A., Meleiro, L.P., Rosa, J.C., Andrade, A.R.D., Jorge, J.A. and Furriel, R.P.M. (2016). A high redox potential laccase from *Pycnoporus sanguineus* RP15: Potential application for dye decolorization. Int. J. Mol. Sci., 17: 672.

10

Host Preferences of *Pinus*-dwelling Hymenochaetaceae

Balázs Palla,[1,*] *Yuan Yuan*,[2] *Yu-Cheng Dai*[2] and *Viktor Papp*[1,*]

1. Introduction

Hymenochaetaceae is a family of Agaricomycetes, in the order Hymenochaetales; its type genus, *Hymenochaete*, was described by Léveillé (1846) based on *H. rubiginosa* as type. According to recent classifications, the family is comprised of 47 genera and ca. 1000 species of Hymenochaetaceae worldwide (He et al., 2019; Wijayawardene et al., 2020; Wu et al., 2022). Though the genus Tubulicirinis was also interpreted as a member of Hymenochaetaceae in many of these works (Korotkin et al., 2018; He et al., 2019; Wijayawardene, 2020), most recent studies exclude the genus from the family (Wang et al., 2021). As the taxonomic position of this group has not been clearly revised until present and there are a vast number of species from this genus that inhabit *Pinus*, the present chapter includes Tubulicrinis in the characterization of *Pinus*-dwelling Hymenochaetaceae. Most species of Hymenochaetaceae are saprotrophic, producing white-rot (Dai, 2010), but phytopathogenic (e.g., *Coniferiporia* spp., *Onnia* spp., *Fomitiporia* spp., *Phellinus resupinatus*) and pharmacologically relevant (e.g., *Inonotus obliquus*, *Sanghuangporus* spp.) species are also present in the family, having significant economic effects (Dai et al., 2007; Cloete et al., 2016; Zhou, Vlasák and Dai 2016; Lin et al., 2017).

[1] Department of Botany, Hungarian University of Agriculture and Life Sciences, Budapest, Hungary.
[2] Institute of Microbiology, School of Ecology and Nature Conservation, Beijing Forestry University, Beijing 100083, China.
* Corresponding authors: palla.balazs@phd.uni-mate.hu/papp.viktor@uni-mate.hu

Species in this family are characterized by annual to perennial basidiocarps with stipitate, pileate to resupinate attachment, coriaceous to woody texture and mostly brownish color, with a xanthocroic, darkening reaction to KOH (Dai, 2010). Hymenophore of the basidiocarps could be poroid, hydnoid or smooth. Hyphal system is monomitic to dimitic, without clamps. A particular characteristic of the family is the setae (sterile elements in the hymenium or in the trama), which could occur in many of the species. Basidiospores are dextrinoid in some cases, becoming reddish brown with Melzer's reagent, but usually are non-dextrinoid and inamyloid (Dai, 2010; Ryvarden and Melo, 2014; Yuan et al., 2020).

Due to the widespread distribution of coniferous forests on earth, hymenochaetoid species dwelling on coniferous hosts have a particular ecological role. Of the approximately 1000 species of gymnosperms on Earth, two-thirds are conifers, that encompass more than 39% of the world's forests (Armenise et al., 2012; Wang and Ran, 2014). Out of these, around 230 species belong to the Pinaceae family, which consists of 12 genera and are naturally distributed in North America, Europe, Asia and North Africa. The *Pinus* genus is represented the most in Pinaceae (more than 110 species) which is also naturally present in the Northern Hemisphere (Farjón, 2010; Wang and Ran, 2014; Simpson, 2019). Only exception from this is *P. merkusii*, which has natural occurrences as far as central Sumatra, south from the Equator (Santisuk, 1997). At the same time, more than 30 pine taxa have naturalized distributions as well, being introduced as non-indigenous species of interest to forestry, from which 28 can be found in the Southern Hemisphere (Procheş et al., 2012). The family is economically important as lumber/timber trees and cultivated ornamentals and act as a source of wood pulp, turpentine, gums, resin, oils, food and other products (Simpson, 2019).

Due to the specific substrate-fungus relationships of hymenochaetoid species dwelling on coniferous hosts, biotic relations in this regard can have a peculiar set of aspects in terms of host preference, type and means of decay and the geographic distribution of both the hosts and the fungi. The latter could be further influenced by the indigeneity of the hosts in the geographical region of the occurrence. As the geographic and ecological distribution of fungal species could be broadly affected by these factors, we aim to clarify and pronounce these features, in order to get a better hold of some of the main driving forces behind the current ecological state of Hymenochaetaceae.

2. Ecological Background of Hymenochaetoid Fungal Host Preference on Gymnosperms and on *Pinus* spp.

Hymenochaetoid fungi are found on a wide range of substrata, habitats and ecological conditions. From terrestrial and wood-inhabiting species of Neotropical regions (Baltazar et al., 2010; Drechsler-Santos et al., 2010) through species on temperate European broadleaved hosts (Ryvarden and Melo, 2014) to spruce-inhabiting taxa in boreal forests (Halme et al., 2008), hymenochaetoid fungi could be found in various kinds of ecological settings. Aside the climatic qualities of the habitat, characteristics of the colonized substrata are main factors that drive the abundance of lignicolous fungi (e.g., availability of dead wood, size of substrata and others). Among these, host species are clearly one of the most important factors (Küffer et al., 2008; Krah, Seibold et al., 2018; Elo et al., 2019). The chemical composition of the host substratum and the enzymatic-chemical components of the attached fungi collectively determine whether the two are suitable for each other. Thus, differences in lignin, cellulose and hemicellulose content of the wood and the convergent evolution of fungal enzymes lead to the diversification of host preferences (Green and Highley, 1997; Presley and Schilling, 2017; Krah, Bässler et al., 2018). As a result, a commonly observed trend in natural settings is that a specific group of wood-dwellers – brown-rotting fungi – are more likely to be found on softwood, gymnosperm hosts (Käärik, 1983; Green and Highley, 1997; Krah, Bässler et al., 2018). In this regard, a clear distinction between coniferous and broadleaf species specialization can be drawn (Küffer et al., 2008).

2.1 Evolution of Wood Decomposition and Gymnosperm-Specialization

Specialization to gymnosperm hosts was seen as interconnected to the emergence of brown-rotting, a means of decay when cell walls of dead wood are decomposed with the exception of lignin, which is not appreciably degraded, promoting a brownish-red color and cubical fragments in the residual wood (Gilbertson, 1980; Hibbett and Donoghue, 2001; Daniel, 2016). After the establishment of the brown-rot fungal mycelia on the substratum, the hyphae penetrate the cell lumina in close contact with the S3 cell wall layer and secrete enzymes (e.g., cellulases, mannanase, xylanase) and non-enzimatic agents (e.g., hydrogen peroxide – Fenton reagent, oxalic acid and hydroquinones) which start to degrade the S2 layer (Käärik, 1983; Green and Highley, 1997; Hatakka and Hammel, 2011; Presley and Schilling, 2017). The non-enzimatic hydroxyl radicals act in spatial and/or temporal segregation from the hydrolytic enzymes, due to the deactivating effect of the former group on the latter. The activity and amount of certain fungal hemicellulase enzymes could also depend on the biochemical composition of the host wood (Presley and Schilling, 2017). Hemicellulose utilization is key in the initiation of brown-rotting, as they act as an envelope around the cellulose microfibrils. As the fungi rapidly depolymerize the cellulose and hemicellulose content of the cell wall of woody tissues, the degradation products are accumulated much faster, then they are utilized. Though essentially undigested, lignin is also modified by demethylation and oxidation, but its presence in even advanced stages of decay leave wood cell forms intact and more pliable (Cowling, 1961; Kirk and Adler, 1970; Daniel, 2016). Lignin type in the host substrata has minor influence on the rate of brown-rot degradation, suggesting a well-developed mechanism in brown rotting fungi to evade lignin barriers. Thus, hardwoods (containing syringyl and guiaicyl lignin) and softwoods (containing only guiaicyl lignin) can be degraded at the same rate (Nilsson et al., 1988; Daniel, 2016). The depolymerization of carbohydrates result in elevated amounts of alkali solubility products and decreased strength properties, with the eventual browning, shrinking and cubical fragmentation of the wood (Green and Highley, 1997).

The other most significant way of wood decomposition by basidiomycetous polypores is the white rot decay. Contrary to brown-rotting, white rot has the ability to delignify the substratum, resulting in a bleached appearance of the residual material. Hyphae of white-rotting fungi grow through ray parenchyma radially into the xylem and establish itself in the tracheids and tracheas of the wood through natural pathways (e.g., pits) or with specialized bore hyphae. As in the case with brown-rots, white rot fungus usually cause decay from the cell lumina to an outward direction, though mycelia can also be present in the middle lamella between cells and decompose the cells from there (Käärik, 1983; Daniel, 2016). According to the speed and order of degradation and the micromorphological character of wood fibers during decay, two decay patterns are distinguished. During "simultaneous" white rot, cellulose, hemicelluloses (xylan and mannan) and lignin are degraded in the same time in a conspicuous but thin decay zone, from the cell lumen to an outward direction (Eriksson et al., 1990; Daniel, 1994). In case of "selective" or "preferential" white rot, hemicelluloses and lignin are alternatively degraded, leaving cellulose fibers only slightly modified in advanced stages of the decay. The decay zone in the selective form is more pronounced, due to the different decomposition state of hemicellulose and lignin between the outer and inner perimeter of the decay zone. In both simultaneous and preferential cases, the corner regions of the cells are the most resistant to degradation, which could remain in the advanced stages of decay. The selective decay type is often present on coniferous wood (Käärik, 1983; Eriksson et al., 1990; Daniel, 1994, 2016). On the biochemical level, white-rot fungi also include enzymatic and non-enzymatic means of biodegradation. In the presence of the lignocellulosic carbon of wood, a wide range of hemicellulose-, pectin- and carbohydrate-active enzyme coding genes (CAZ-ymes) become up-regulated in the fungi (Miyauchi et al., 2020). In order to delignify cells, extracellular ligninolytic peroxidases (e.g., lignin-, manganese- and versatile peroxidases) and laccases are secreted. Non-enzymatic agents include redox-active glycopeptides and other agents dependent on enzymatic systems such as oxalate, veratryl alcohol or reactive oxygen forms (Akamatsu et al., 1990; Aust, 1995; Hatakka and Hammel, 2011; Huang et al., 2021). As the white rot proceeds, a wide range of decay patterns and discolorations could be produced depending on fungal species, type of wood, rate of decay and physiological state. Eventually, the wood becomes bleached as the consequence of lignin degradation (Daniel, 2016).

In nature, white rotting fungi usually colonize hardwoods (Daniel, 2016). As ancestors they preceded the emergence of brown-rotters (Hibbett and Donoghue, 2001), suggesting an evolutionary reduction in the extracellular ligninolytic enzyme systems with the parallel development of the ability to selectively and rapidly digest the carbohydrates of the substrata. The ability to rapidly utilize cellulose and hemicelluloses favored the establishment of brown-rot fungi in boreal and mountainous coniferous forests with short growing seasons (Gilbertson, 1980). This concept suggests that the evolution of brown rot was parallelly correlated with the evolution of specificity to gymnosperm hosts. Recent findings challenge this view, stating that most brown rotting lineages switched to angiosperm-gymnosperm generalism during their evolution, while white rot lineages switched mostly to angiosperm specialism. After the angiosperm diversification in the Cretaceous, white rotting fungi were able to evolve and adapt more rapidly to the newly available niche of angiosperm host species, while brown-rotters were limited in exploiting angiosperm resources. The reasons for such trends are not yet understood, but the loss of decay-related genes could be one explanation (Kohler et al., 2015; Krah, Bässler et al., 2018).

3. Hymenochaetoid Fungi on Gymnosperms and on *Pinus* spp.

Species in Hymenochaetaceae are generally a white-rotting group of taxa (Dai, 2010), as only around 20% of species in Hymenochaetales are gymnosperm specialists and the rest are generalists (occurring on both gymnosperm and angiosperm hosts) or angiosperm specialists (Krah, Bässler et al., 2018). Until now, monographs, regional species checklists, phylogenetic studies and other related literature have outlined 121 species in 27 genera (including *Tubulicrinis*) in the family Hymenochaetaceae that have gymnosperm hosts (Table 1). Most of these inhabit predominantly the Pinaceae family (in case of 103 species), but Cupressaceae (48 spp.), Taxaceae (10 spp.), Ephedraceae (1 sp.) and Podocarpaceae (1 sp.) are also represented as host families (Fig. 1a). Some species (e.g. in *Fomitiporia*, *Fuscoporia*, *Hymenochaete* and *Tubulicrinis*) could be found on the representatives of two or three host families as well, representing mainly generalist species (those occurring also on angiosperms), but some of these are restricted to gymnosperms (which are considered as gymnosperm specialists) (e.g., *Coniferiporia sulphurascens*, *Fomitiporia repanda*, *Inonotopsis subiculosa* and others). From the 121 species on gymnosperms, 37 are solely restricted to one genus; thus, around one-third of gymnosperm-dwelling Hymenochaetaceae species are genus specialists (Fig. 1b). Together with the remaining gymnosperm-specialists, 66 – more than a half – of the species are solely specialized to gymnosperms. Sometimes angiosperm specialist hymenochaetoid fungi were reported on only one certain gymnosperm genera, which forms a particular group that also were recorded on gymnosperms (20). There are 35 generalist species that occur both on a wide range of gymnosperm and angiosperm taxa (Fig. 1b). Thus, among all 888 species in Hymenochaetaceae (Wijayawardene, 2020), only around ~ 7% of the species are specialized to inhabit only gymnosperm taxa and around ~ 14% of them are having gymnosperm hosts. This trend in Hymenochaetaceae falls short of the approximately 20% gymnosperm specialism in all Hymenochaetales. The strict sense of 'gymnosperm specialism' – in the context of an organism solely being found on gymnosperms – will consequently mean lower species numbers in this group, if compared to the 90% threshold approach of gymnosperm association in the related literature observed on Hymenochaetales. At the same time, this observation in its magnitude corresponds well to the fact, that the white rotting species in Hymenochaetales are generally angiosperm specialists (Krah, Bässler et al., 2018).

The most frequent gymnosperm host in Hymenochaetaceae is *Pinus*; 66 species in Hymenochaetaceae utilize it as substratum (Fig. 2a). *Picea*, *Abies*, *Larix* and *Tsuga* in Pinaceae are also inhabited by a larger number of hymenochaetoid species, while in Cupressaceae, *Juniperus* is the most relevant host. In case of Pinaceae-related hymenochaetoid species, generally around 50–60% of the species are gymnosperm specialists, from which 5–20% are solely restricted to only one conifer genus (Fig. 2b). Such is the case in *Pinus*-dwelling Hymenochaetaceae, where altogether 33 species are gymnosperm-specialists and 11 are only restricted to *Pinus*. Probably

Table 1. Checklist of genera in Hymenochaetaceae with respective global and gymnosperm-dwelling species numbers and references within the source literature.

Genera	Number of all species	Number of species found on gymnosperms	Source literature
Asterodon	1	1	Volk et al., 1994; Renvall, 1995; Høiland and Bendiksen, 1996; Yurchenko, 2003; Dai, 2010; Ghobad-Nejhad, 2011; Karadelev et al., 2014; Ezhov and Zmitrovich, 2015; Viner, 2015; Zhou, Vlasák, and Dai, 2016; Viner et al., 2016
Coltricia	40	5	Gilbertson and Ryvarden, 1986; Cardoso et al., 1992; Zervakis et al., 1998, 2002; Buchanan and Ryvarden, 2000; Wagner and Fischer, 2001; May et al., 2003; Ryvarden, 2004; Grand and Vernia, 2005; Wright and Wright, 2005; Borgen et al., 2006; Bernicchia et al., 2007; Tedersoo et al., 2007; Gibertoni et al., 2007; Baltazar and Gibertoni, 2009; Dai, 2010, 2012; Ghobad-Nejhad, 2011; Lee et al., 2012; Prasher and Ashok, 2013; Prasher and Lalita, 2013; Ranadive, 2013; Vasco Palacios and Franco Molano, 2013; Ryvarden and Melo, 2014; Kotiranta and Shiryaev, 2015; Ezhov and Zmitrovich, 2015; Kotiranta et al., 2016; Zhou, Nakasone, et al., 2016; Zhou, Vlasák, and Dai, 2016; Hubregtse, 2019; Kinge et al., 2020; Park et al., 2020; Piepenbring et al., 2020; Wu et al., 2022
Coltriciella	13	3	Gilbertson and Ryvarden, 1986; Buchanan and Ryvarden, 2000; May et al., 2003; Ryvarden, 2004; Grand and Vernia, 2005; Tedersoo et al., 2007; Baltazar and Gibertoni, 2009; Dai, 2010, 2012; Dai et al., 2011; Lee et al., 2012; Zhou, Nakasone, et al., 2016; Zhou, Vlasák, and Dai, 2016; Hubregtse, 2019; Wu et al., 2022
Coniferiporia	4	4	Gilbertson and Ryvarden, 1987; Wagner and Fischer, 2002; Dai et al., 2007; Dai, 2010, 2012; Doğan et al., 2011; Ghobad-Nejhad, 2011; Filippova and Zmitrovich, 2013; Ryvarden and Melo, 2014; Ezhov and Zmitrovich, 2015; Zhou, Nakasone, et al., 2016; Zhou, Vlasák, and Dai, 2016; Piepenbring et al., 2020; Wu et al., 2022
Cyanotrama	1	1	Gilbertson and Ryvarden, 1986; Ghobad-Nejhad and Dai, 2010; Ghobad-Nejhad, 2011; Ghobad-Nejhad and Hallenberg, 2012; Prasher and Lalita, 2013; Ranadive, 2013; Prasher, 2015; Zhou, Nakasone, et al., 2016
Cyclomyces	5	1	Dai et al., 2004; Dai, 2010, 2012
Fomitiporella	13	1	Gilbertson and Ryvarden, 1987; Buchanan and Ryvarden, 2000; May et al., 2003; Dai, 2010, 2012; Dai et al., 2011; Ghobad-Nejhad and Hallenberg, 2012; Ranadive, 2013; Zhou, Nakasone, et al., 2016; Wu et al., 2022

Table 1 contd. ...

...Table 1 contd.

Genera	Number of all species	Number of species found on gymnosperms	Source literature
Fomitiporia	46	11	(Lombard et al., 1972; Gilbertson et al., 1974; Gilbertson and Ryvarden, 1987; Tortić, 1987; Cardoso et al., 1992; Zervakis et al., 1998, 2002; Buchanan and Ryvarden, 2000; Ortega and Lorite, 2000; Wagner and Fischer, 2001, 2002; May et al., 2003; Fischer and Binder, 2004; Lizoň and Kautmanová, 2004; Ryvarden, 2004; Dai et al., 2004, 2007; Wright and Wright, 2005; Robledo and Rajchenberg, 2007; Decock et al., 2007; Ţura et al., 2008, 2010; Baltazar and Gibertoni, 2009; Dai, 2010, 2012; Ghobad-Nejhad, 2011; Doğan et al., 2012; Ghobad-Nejhad and Hallenberg, 2012; Brazee, 2013; Ranadive, 2013; Vasco Palacios and Franco Molano, 2013; Ryvarden and Melo, 2014; Ezhov and Zmitrovich, 2015; Kotiranta and Shiryaev, 2015; Viner, 2015; Doğan and Kurt, 2016; Kotiranta et al., 2016; Viner et al., 2016; Zhou, Nakasone, et al., 2016; Zhou, Vlasák, and Dai, 2016; Kinge et al., 2020; Piepenbring et al., 2020; de la Fuente et al., 2020; Wu et al., 2022)
Fulvifomes	33	2	Gilbertson and Ryvarden, 1987; Hattori, 1999; Ryvarden, 2004; Robledo and Rajchenberg, 2007; Baltazar and Gibertoni, 2009; Dai, 2010, 2012; Dai et al., 2011; Prasher and Lalita, 2013; Ranadive, 2013; Prasher, 2015; Zhou, Nakasone et al., 2016; Wu et al., 2022
Fuscoporia	62	9	Gilbertson et al., 1974; Piirto et al., 1984; Gilbertson and Ryvarden, 1987; Cardoso et al., 1992; Volk et al., 1994; Renvall, 1995; Høiland and Bendiksen, 1996; Zervakis et al., 1998; Hattori, 1999; Buchanan and Ryvarden, 2000; Ortega and Lorite, 2000; Wagner and Fischer, 2001, 2002; May et al., 2003; Rizzo et al., 2003; Ryvarden, 2004; Dai et al., 2004, 2011; Küffer and Senn-Irlet, 2005; Wright and Wright, 2005; Bernicchia et al., 2007; Decock et al., 2007; Gibertoni et al., 2007; Küffer et al., 2008; Ţura et al., 2008, 2010; Baltazar and Gibertoni, 2009; Dai, 2010, 2012; Doğan et al., 2011; Ghobad-Nejhad, 2011; Ghobad-Nejhad and Hallenberg, 2012; Lee et al., 2012; Filippova and Zmitrovich, 2013; Prasher and Ashok, 2013; Prasher and Lalita, 2013; Ranadive, 2013; Raymundo et al., 2013; Vasco Palacios and Franco Molano, 2013; Ryvarden and Melo, 2014; Urbizu et al., 2014; Ezhov and Zmitrovich, 2015; Prasher, 2015; Viner, 2015; Doğan and Kurt, 2016; Bolshakov et al., 2016; Kotiranta et al., 2016; Viner et al., 2016; Zhou, Nakasone, et al., 2016; Zhou, Vlasák, and Dai, 2016; Kunttu et al., 2019; de la Fuente et al., 2020; Piepenbring et al., 2020; Chen et al., 2020; He et al., 2021; Wu et al., 2022

Hymenochaete	149	11	Gilbertson et al., 1974; Piirto et al., 1984; Cardoso et al., 1992; Renvall, 1995; Høiland and Bendiksen, 1996; Ortega and Lorite, 2000; Greslebin and Rajchenberg, 2003; May et al., 2003; Wojewoda, 2003; Yurchenko, 2003; Küffer and Senn-Irlet, 2005; Wright and Wright, 2005; Borgen et al., 2006; Gibertoni et al., 2007; Ţura et al., 2008, 2010; Baltazar and Gibertoni, 2009; Gorjón et al., 2009; Dai, 2010; Ghobad-Nejhad, 2011; Doğan et al., 2011; Ghobad-Nejhad and Hallenberg, 2012; Prasher and Ashok, 2013; Ranadive, 2013; Prasher, 2015; Viner, 2015; Ezhov and Zmitrovich, 2015; Bolshakov et al., 2016; Doğan and Kurt, 2016; Corfixen and Parmasto, 2017; Kunttu et al., 2019; Miettinen et al., 2019; Kinge et al., 2020; Park et al., 2020; Běťák et al., 2021
Hydnoporia	13	4	Gilbertson et al., 1974; Piirto et al., 1984; Cardoso et al., 1992; Volk et al., 1994; Ortega and Lorite, 2000; May et al., 2003; Yurchenko, 2003; Lizoň and Kautmanová, 2004; Wright and Wright, 2005; Borgen et al., 2006; Baltazar and Gibertoni, 2009; Dai, 2010; Ghobad-Nejhad, 2011; Lee et al., 2012; Ghobad-Nejhad and Hallenberg, 2012; Ranadive, 2013; Filippova and Zmitrovich, 2013; Kotiranta and Shiryaev, 2015; Prasher, 2015; Kotiranta et al., 2016; Viner et al., 2016; Zhou, Vlasák, and Dai, 2016; Corfixen and Parmasto, 2017; Miettinen et al., 2019; Kinge et al., 2020; He et al., 2021
Inocutis	9	2	Gilbertson and Ryvarden, 1986; Buchanan and Ryvarden, 2000; Wagner and Fischer, 2001, 2002; May et al., 2003; Ţura et al., 2008, 2010; Dai, 2010, 2012; Ghobad-Nejhad, 2011; Ghobad-Nejhad and Hallenberg, 2012; Lee et al., 2012; Ranadive, 2013; Ryvarden and Melo, 2014; Ezhov and Zmitrovich, 2015; Viner et al., 2016; Zhou, Nakasone et al., 2016; Zhou, Vlasák, and Dai, 2016; Kotiranta et al., 2016; Park et al., 2020; Wu et al., 2022
Inonotopsis	1	1	Wagner and Fischer, 2001, 2002; Dai, 2010, 2012; Ryvarden and Melo, 2014; Zhou, Nakasone, et al., 2016; Zhou, Vlasák and Dai, 2016; Wu et al., 2022
Inonotus	120	1	Gilbertson and Ryvarden, 1986; Zhou, Nakasone et al., 2016; Wu et al., 2022
Mensularia	6	1	Gilbertson and Ryvarden, 1986; Ortega and Lorite, 2000; Wagner and Fischer, 2001, 2002; May et al., 2003; Grand and Vernia, 2005; Borgen et al., 2006; Baltazar and Gibertoni, 2009; Dai, 2010, 2012; Ghobad-Nejhad, 2011; Ryvarden and Melo, 2014; Viner et al., 2016; Zhou, Nakasone, et al., 2016; Zhou, Vlasák and Dai, 2016; Kotiranta et al., 2016; Wu et al., 2022
Neomensularia	4	1	Gilbertson and Ryvarden, 1986; Wagner and Fischer, 2002; Ryvarden, 2004; Zhou, Nakasone et al., 2016; Wu et al., 2022

Table 1 contd. ...

...Table 1 contd.

Genera	Number of all species	Number of species found on gymnosperms	Source literature
Onnia	8	7	Gilbertson et al., 1974; Gilbertson and Ryvarden, 1986; Tortić, 1987; Volk et al., 1994; Renvall, 1995; Zervakis et al., 1998, 2002; Wagner and Fischer, 2001, 2002; May et al., 2003; Grand and Vernia, 2005; Dai et al., 2007; Decock et al., 2007; Dai, 2010, 2012; Tomšovský et al., 2010; Ghobad-Nejhad, 2011; Brazee, 2013; Prasher and Ashok, 2013; Prasher and Lalita, 2013; Ranadive, 2013; Filippova and Zmitrovich, 2013; Karadelev et al., 2014; Ryvarden and Melo, 2014; Kotiranta and Shiryaev, 2015; Prasher, 2015; Viner, 2015; Ezhov and Zmitrovich, 2015; Zhou, Nakasone, et al., 2016; Zhou, Vlasák, and Dai, 2016; Ji et al., 2017; Zhou and Wu, 2018; Wu et al., 2019, 2022; Park et al., 2020
Phellinidium	5	2	Gilbertson et al., 1974; Gilbertson and Ryvarden, 1987; Renvall, 1995; Høiland and Bendiksen, 1996; Wagner and Fischer, 2001, 2002; Küffer and Senn-Irlet, 2005; Dai, 2010, 2012; Ghobad-Nejhad, 2011; Filippova and Zmitrovich, 2013; Ryvarden and Melo, 2014; Viner, 2015; Ezhov and Zmitrovich, 2015; Kotiranta and Shiryaev, 2015; Viner et al., 2016; Zhou, Nakasone, et al., 2016; Zhou, Vlasák, and Dai, 2016; Kotiranta et al., 2016; Park et al., 2020; Wu et al., 2022
Phellinopsis	10	1	Wu et al., 2022
Phellinus	202	3	Wagner and Fischer, 2002; May et al., 2003; Baltazar and Gibertoni, 2009; Vlasák et al., 2011; Lee et al., 2012; Prasher and Ashok, 2013; Ranadive, 2013; Prasher, 2015; Zhou, Nakasone, et al., 2016; Zhou, Vlasák, and Dai, 2016; Zhou, Vlasák, Qin, et al., 2016; de la Fuente et al., 2020; Wu et al., 2022
Phellopilus	1	1	Gilbertson et al., 1974; Gilbertson and Ryvarden, 1987; Volk et al., 1994; Renvall, 1995; Høiland and Bendiksen, 1996; Zervakis et al., 1998; Wagner and Fischer, 2002; Dai, 2010, 2012; Ghobad-Nejhad, 2011; Filippova and Zmitrovich, 2013; Ryvarden and Melo, 2014; Viner, 2015; Ezhov and Zmitrovich, 2015; Viner et al., 2016; Zhou, Nakasone et al., 2016; Zhou, Vlasák and Dai, 2016; Kotiranta et al., 2016; Park et al., 2020; Běťák et al., 2021; Wu et al., 2022

Phylloporia	38	1	Gilbertson and Ryvarden, 1987; Cardoso et al., 1992; Ortega and Lorite, 2000; Wagner and Fischer, 2001, 2002; May et al., 2003; Dai et al., 2004, 2007, 2011; Ţura et al., 2008, 2010; Dai, 2010, 2012; Sell and Kotiranta, 2011; Ghobad-Nejhad, 2011; Ghobad-Nejhad and Hallenberg, 2012; Prasher and Ashok, 2013; Prasher and Lalita, 2013; Ranadive, 2013; Ryvarden and Melo, 2014; Prasher, 2015; Zhou, Nakasone et al., 2016; Zhou, Vlasák and Dai, 2016; Piepenbring et al., 2020; Wu et al., 2022
Porodaedalea	21	21	Gilbertson et al., 1974; Larsen et al., 1979; Gilbertson and Ryvarden, 1987; Tortić, 1987; Cardoso et al., 1992; Volk et al., 1994; Renvall, 1995; Høiland and Bendiksen, 1996; Zervakis et al., 1998; Larsen, 2000; Ortega and Lorite, 2000; Wagner and Fischer, 2001, 2002; May et al., 2003; Lizoň and Kautmanová, 2004; Dai et al., 2004, 2007; Bernicchia et al., 2007; Decock et al., 2007; Dai, 2010; Tomšovský et al., 2010; Dai, 2012; Ghobad-Nejhad, 2011; Doğan et al., 2012; Filippova and Zmitrovich, 2013; Prasher and Lalita, 2013; Ranadive, 2013; Ryvarden and Melo, 2014; Kotiranta and Shiryaev, 2015; Prasher, 2015; Viner, 2015; Ezhov and Zmitrovich, 2015; Kotiranta et al., 2016; Viner et al., 2016; Zhou, Nakasone, et al., 2016; Zhou, Vlasák, and Dai, 2016; Doğan and Kurt, 2016; Wu et al., 2019, 2022; Park et al., 2020
Pseudoinonotus	8	3	Gilbertson and Ryvarden, 1986; Cardoso et al., 1992; Ortega and Lorite, 2000; Wagner and Fischer, 2001, 2002; May et al., 2003; Grand and Vernia, 2005; Dai et al., 2007; Decock et al., 2007; Ţura et al., 2008, 2010; Dai, 2010, 2012; Ghobad-Nejhad, 2011; Ghobad-Nejhad and Hallenberg, 2012; Ranadive, 2013; Ryvarden and Melo, 2014; Prasher, 2015; Zhou, Nakasone, et al., 2016; Zhou, Vlasák, and Dai, 2016; Wu et al., 2022
Tropicoporus	12	1	Wu et al., 2022
Tubulicrinis	34	22	Thind and Rattan, 1970; Gilbertson et al., 1974; Piirto et al., 1984; Tortić, 1987; Cardoso et al., 1992; Volk et al., 1994; Renvall, 1995; Høiland and Bendiksen, 1996; Zervakis et al., 1998, 2002; Ortega and Lorite, 2000; Greslebin and Rajchenberg, 2003; Wojewoda, 2003; Yurchenko, 2003; Lizoň and Kautmanová, 2004; Küffer and Senn-Irlet, 2005; Bernicchia et al., 2007; Gibertoni et al., 2007; Ryvarden, 2007; Ghobad-Nejhad et al., 2008; Küffer et al., 2008; Ţura et al., 2008, 2010; Baltazar and Gibertoni, 2009; Gorjón et al., 2009, 2013; Ghobad-Nejhad, 2011; Sell and Kotiranta, 2011; Doğan et al., 2011, 2012; Ghobad-Nejhad and Hallenberg, 2012; Prasher and Ashok, 2013; Ranadive, 2013; Vasco Palacios and Franco Molano, 2013; Filippova and Zmitrovich, 2013; Urbizu et al., 2014; Kotiranta and Shiryaev, 2015; Prasher, 2015; Viner, 2015; Ezhov and Zmitrovich, 2015; Kotiranta et al., 2016; Kunttu, 2016; Bolshakov et al., 2016; Saitta and Losi, 2016; Viner et al., 2016; Doğan and Kurt, 2016; Shevchenko, 2018; Kunttu et al., 2019, 2021; He et al., 2020, 2021; Park et al., 2020; Běťák et al., 2021; Maekawa, 2021

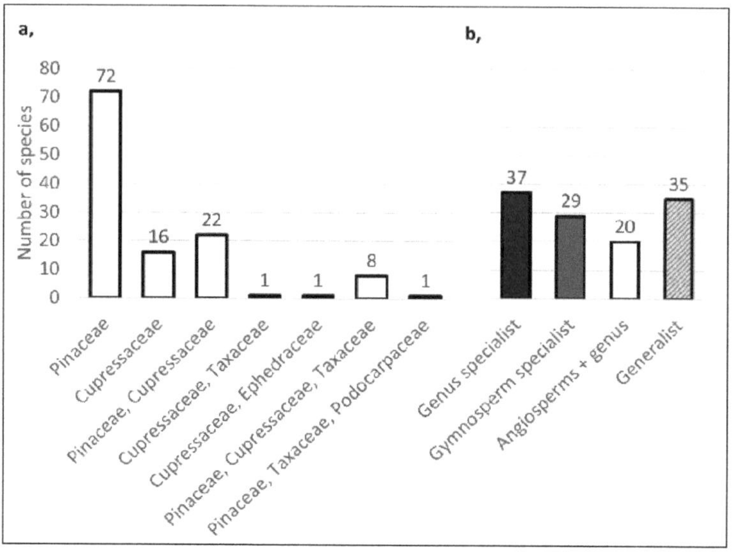

Fig. 1. Frequency of species in Hymenochaetaceae (a) collected from different gymnosperm host families and (b) in different groups of host specificity.

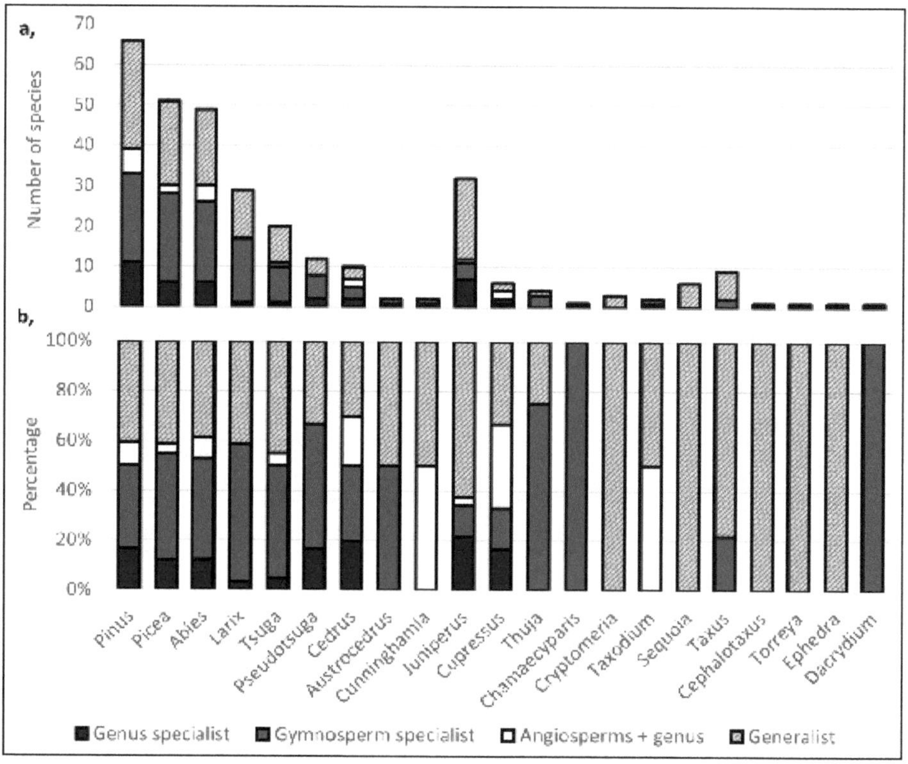

Fig. 2. Frequency (**a**) and host specificity relations (b) of Hymenochaetaceae species found on different gymnosperm hosts.

due to the less widespread distribution and low frequencies of their host species on Earth and the Northern Hemisphere, hymenochaetoid fungi inhabiting Cupressaceae and Taxaceae are more commonly generalists or angiosperm specialists (the latter only occurring only on one gymnosperm genus). The only exception is *Juniperus*, which attracts a higher number of genus-specialist hymenochaetoid species (Fig. 2a) but in terms of ratios of host specialism, the relative amount of specificity differs between these genera (Fig. 2b). As *Juniperus* is the fourth largest gymnosperm genus of generally temperate conifers on Earth (Debreczy and Rácz, 2000), it serves as a larger potential niche of substrata for hymenochaetoid fungi.

Among the 66 species inhabiting *Pinus* as a host, *Tubulicrinis* (18 spp.), *Porodaedalea* (9 spp.), *Fuscoporia* (8 spp.), *Onnia* (7 spp.) and *Hymenochaete* (4 spp.) are the most represented hymenochaetoid genera (Fig. 3a). Most of *Tubulicrinis*, *Fuscoporia* and *Hymenochaete* are generalists or angiosperm specialists, while *Porodaedalea* and *Onnia* are restricted to gymnosperms (Fig. 3b), with *Porodaedalea* having three (*P. chinensis*, *P. kesiyae*, *P. yunnanensis*)

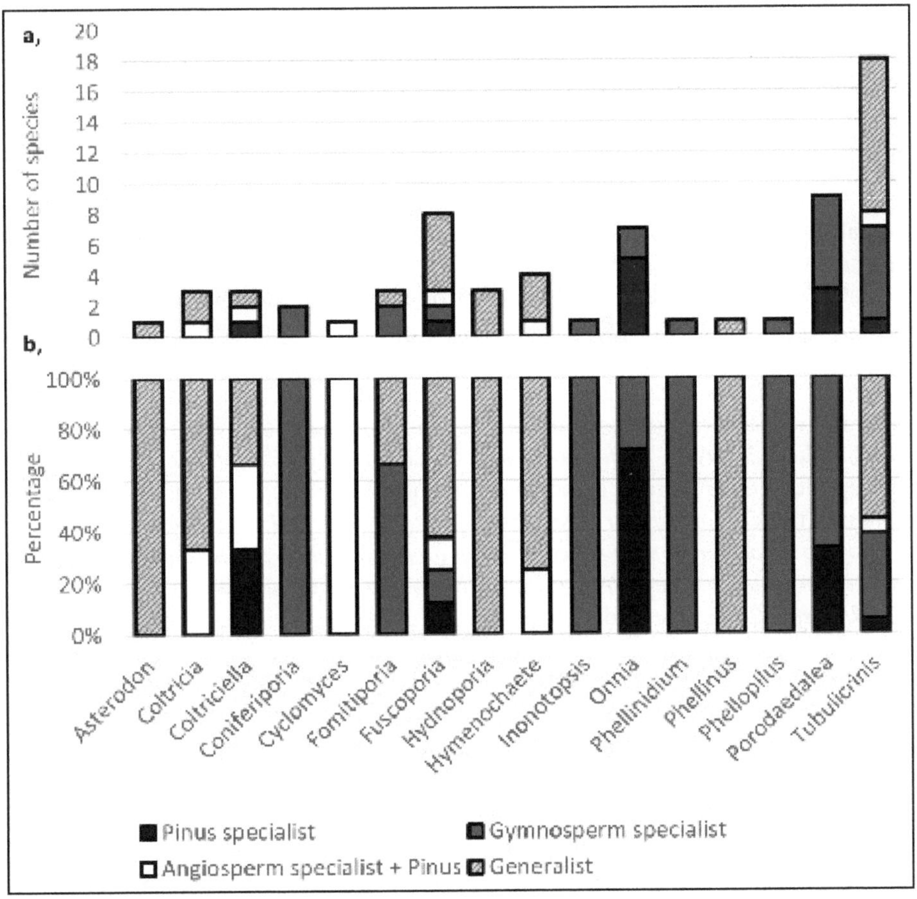

Fig. 3. Frequency (**a**) and host specificity relations (**b**) of Hymenochaetaceae species collected from *Pinus*.

and *Onnia* having five species (*O. kesiya*, *O. microspora*, *O. subtriquetra*, *O. tibetica*, *O. triquetra*), that are exclusively inhabit *Pinus* (*O. triquetra* was reported on two occasions on other genera, but these species is basically also *Pinus*-specialist). Besides these, *Coltriciella tasmanica*, *Fuscoporia contiguiformis* and *Tubulicrinis xantha* were also only reported from *Pinus*. *Coniferiporia sulphurascens*, *C. weirii*, *Fomitiporia hartigii* (mentioned only once on angiosperm host), *F. repanda*, *Fuscoporia coronadensis*, *Inonotopsis subiculosa* (mentioned only once on *Betula*), *Phellinidium ferruguineofuscum* (mentioned only once on angiosperm host) and *Phellopilus nigrolimitatus* are found also on other gymnosperms then *Pinus*, but are restricted to conifers, while *Cyclomyces xeranthicus*, *Fuscoporia chinensis*, *Hymenochaete canescens* and *Tubulicrinis inornatus* are angiosperm-specialists, which have been only reported on *Pinus* from gymnosperms. *Coltricia perennis* and *Coltriciella oblectabilis* are basically terrestrial species, but they also could grow on stumps; *C. perennis* usually occurs in pine forests, but it has also been reported in deciduous ones, while *C. oblectabilis* could be present in both deciduous and pine-oak forests (Fig. 3).

3.1 *Ecological and Physical Relationships of Gymnosperm- and Pinus-Dwelling Hymenochaetaceae*

In terms of the ecological relation to the host and the means of nutrition, mycorhizzal, saprotrophic and pathogenic groups could be distinguished in Hymenochaetaceae (Fig. 4). Some species in *Coltricia* and *Coltriciella* has been reported to form mutualistic ectomycorhizza with native hosts (Tedersoo et al., 2007), but only *Coltricia perennis* and *Coltriciella dependens*, that are also connected to *Pinus*, were indicated on species level as being able to form mycorrhizzal connections with their hosts (Fig. 4c). Most of the hymenochaetoid species are having saprotrophic nutrition, utilizing solely the organic material of the dead, decayed hosts (Fig. 4a, b). Almost all genera in Hymenochaetaceae have taxa that are solely saprotrophic and some genera have been reported mostly or only on dead substrata (e.g., *Hymenochaete* and *Tubulicrinis*). All *Coniferiporia* species and most of *Porodaedalea* and *Onnia* are reported to be purely pathogenic, in most cases occuring on living hosts. Amongst these, *Coniferiporia sulphurascens*, *C. weirii*, *Porodaedalea chinensis*, *P. kesiyae*, *P. laricis*, *P. mongolica*, *P. orientoamericana*, *P. pini*, *P. yunnanensis* and *Onnia leporina*, *O. tomentosa*, *O.triquetra* utilize *Pinus* also as host, from which *P. chinensis*, *P. kesiyae* and *O.triquetra* are basically *Pinus* specialist pathogens (Fig. 4c). Some species (e.g., the remainder of *Porodaedalea* and *Onnia*, except *O. kesiya*; some *Fomitiporia*, *Fuscoporia*, *Hydnoporia*, *Hymenochaete*, *Phellinus* and *Phellopilus* species) have been reported on both living and dead substrata, thus sharing both saprotrophic and pathogenic means of decay on the host wood (Fig. 4a, c). From this ecological group *Onnia microspora*, *O. subtriquetra* and *O. tibetica* are exclusively *Pinus* specialists.

Patterns of the basidiocarp position on the host and the size of the colonized substrata could be also characteristic of the inhabiting genera. Based on the available literature and the corresponding field notes for *Pinus*-related genera, some specific preferences could be highlighted. One is that corticioid *Hydnoporia* and

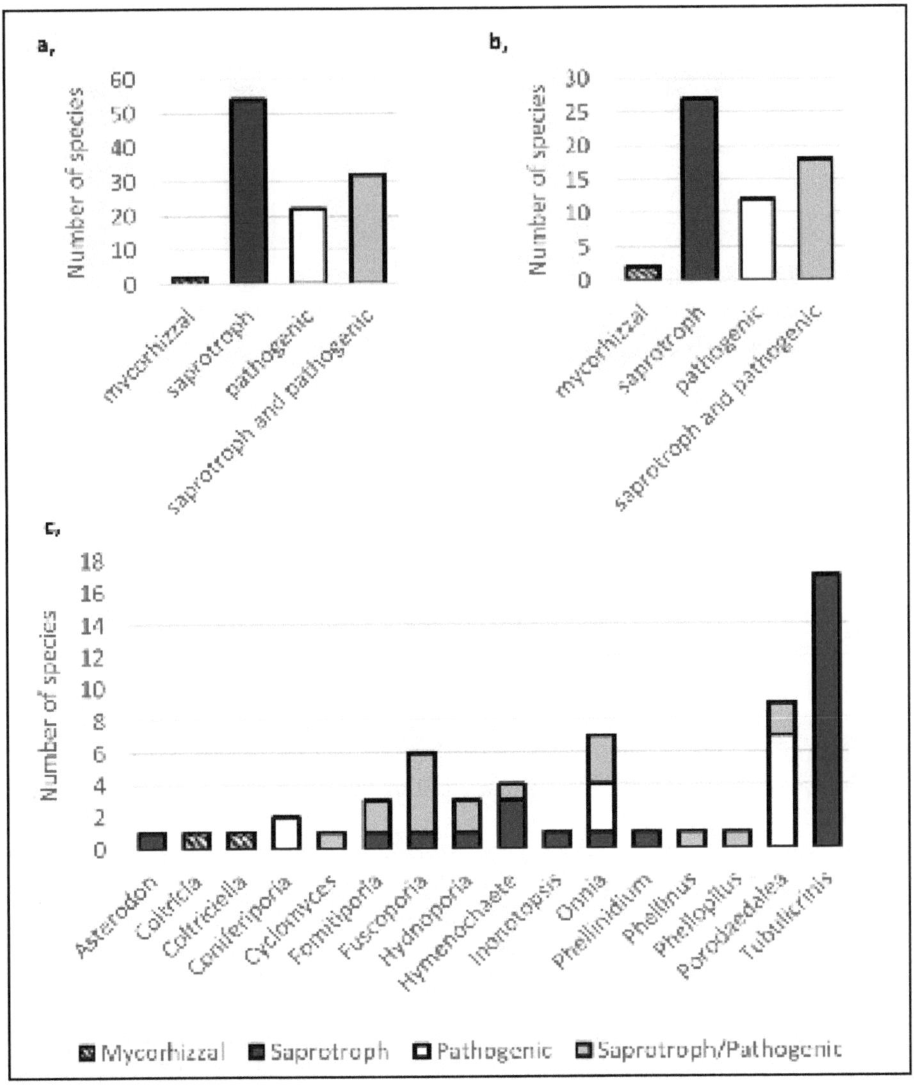

Fig. 4. Pooled frequencies of Hymenochaetaceae species on gymnosperms (a), on *Pinus* (b) and on *Pinus* separated by hymenochaetoid genera (c) in different ecological-nutritional groups.

Hymenochaete species regularly occur on smaller woody fractions of the host (i.e., branches, twigs, bark), while *Asterodon*, *Inonotopsis*, *Phellinidium* or *Porodaedalea* predominantly prefer substrata with higher diameter dimensions (trunks, stumps, logs). The gymnosperm specialist *Onnia* and *Coniferiporia* were frequently reported on basal parts of the hosts (ground, base, roots) and thus being able to cause butt-rot on the living substrata. In case of the basically terrestrial *Coltricia* genus, most of the observations refer to specimens found on the ground or on the roots of the host (Fig. 5).

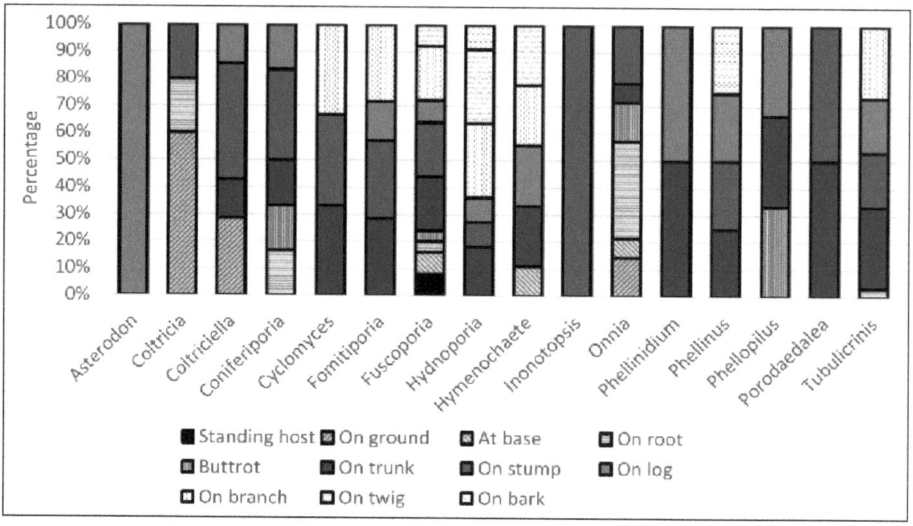

Fig. 5. Relations of fruiting body position and size of substrata in *Pinus*-related Hymenochaetaceae.

3.2 *Global Distribution and Habitat Preference of Pinus-Dwelling Hymenochaetaceae*

Hymenochaetaceae species that are able to colonize *Pinus* are mostly distributed in the Northern Hemisphere, following the main natural worldwide distribution of pines (Simpson, 2019). At the same time, predominantly generalist and terrestrial species also occur on angiosperm hosts south from the Equatorial (Fig. 6). The mostly distributed genera covering both hemispheres are the mainly generalist corticioid and resupinate species of *Tubulicrinis*, *Fuscoporia*, *Hydnoporia*; *Hymenochaete cinnamomea*, the pileate generalist *Phellinus nilgheriensis* and the basically terrestrial *Coltricia perennis*, *C. cinnamomea* and *Coltriciella dependens*. These taxa could utilize *Pinus* where it is naturally present and at the same time, their broad gymnosperm and angiosperm host spectrum let them to vastly establish themselves in the Paleotropical, Neotropical and Australian floral kingdoms as well. The idea of Hymenochaetaceae colonizing *Pinus* in its naturalized range in the Southern Hemisphere could also have its merits, as it was seen in case of other fungal taxa collected from *Pinus patula* and *P. taeda* plantations in Africa and in South America (Campi et al., 2015; Dejene et al., 2017), but to this day, no Hymenochaetaceae specimens have been collected from such exotic *Pinus* stands. Ecological measures such as the high IndVal fidelity indexes of *Tubulicrinis* species to *Pinus* in different forest vegetations are other proofs for the high potential of these genera to be widespread in a wide variety of sites containing the host species (Küffer et al., 2008). *Coltricia* and *Coltriciella* has also the potential to form ectomycorhizza with native and introduced species and as so, the ecological potential of such a nutritional form provides a much broader opportunity for these genera to be widely distributed (Tedersoo et al., 2007). Altogether 12, 19, 15 and 18 species have been observed in Central- and South America, Africa and Oceania, respectively, that are having a worldwide distribution (Fig. 6) (Table 2).

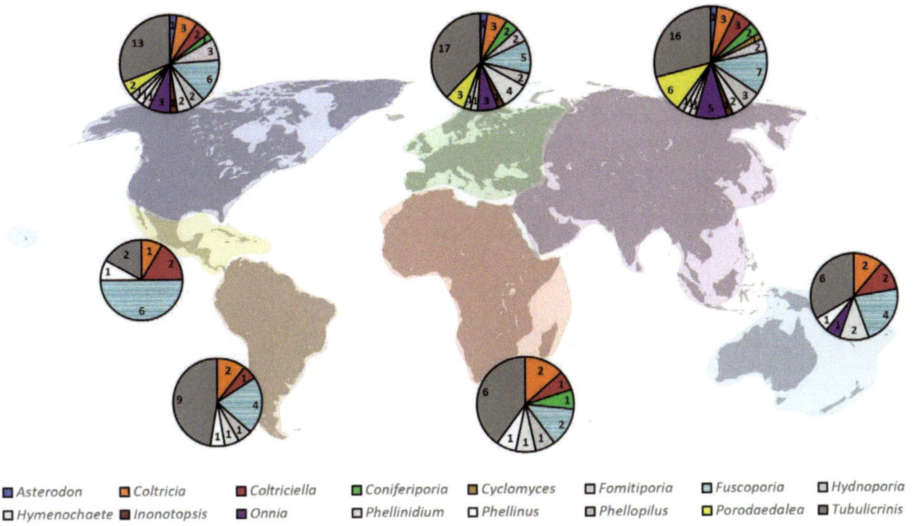

Fig. 6. Global distribution of *Pinus*-dwelling genera in Hymenochaetaceae.

Aside the circumglobally present taxa, the gymnosperm and *Pinus*-specialist species of *Onnia* and *Porodaedalea* are represented the most in the Northern Hemisphere. *Coniferiporia*, *Fomitiporia*, *Asterodon ferruginosus*, *Inonotopsis subiculosa*, *Phellinidium ferrugineofuscum* and *Phellopilus nigrolimitatus* are also widely distributed *Pinus*-related hymenochaetoids in the Northern Hemisphere. Asia harbors the highest amount of species (altogether 55), while in Europe and North America, hitherto 45 and 42 species have been observed, respectively (Fig. 6) (Table 2). The apparent natural restriction of *Pinus* to the Northern Hemisphere facilitated the northern distribution of these taxa in Hymenochaetaceae (Simpson, 2019).

3.2.1 Pinus-Related Hymenochaetaceae in North America

In North America, altogether 42 *Pinus*-related Hymenochaetaceae species are present, from which 25 are generalists, 13 are restricted to gymnosperms and three are angiosperm specialists, that could also occur on *Pinus*. Generalists include species mostly in *Asterodon* (1 sp.), *Coltricia* (2 spp.), *Coltriciella* (1 sp.), *Fomitiporia* (1 sp.), *Fuscoporia* (5 spp.), *Hydnoporia* (2 spp.), *Hymenochaete* (2 spp.), *Phellinus* (1 sp.) and *Tubulicrinis* (10 spp.), while gymnosperm specialists are represented in *Coniferiporia* (1 sp.), *Fomitiporia* (2 spp.), *Fuscoporia* (1 sp.), *Inonotopsis* (1 sp.), *Onnia* (2 spp.), *Phellinidium* (1 sp.), *Phellopilus* (1 sp.), *Porodaedalea* (2 spp.) and *Tubulicrinis* (2 spp.). *Onnia subtriquetra* are the only *Pinus*-specialist hymenochaetoid species in North America (Fig. 7) (Table 2). Among the continents of the Northern Hemisphere, North America possess the highest amount of *Pinus* species (altogether 66 spp.) that could provide substrata for these lignicolous fungi (Nobis et al., 2012).

Table 2. Checklist, geographical distribution, host and climatic preferences of Hymenochaetaceae species found on *Pinus* derived from the source literature. Abbreviations: GEN: generalist; A: angiosperm specialist; GYM: gymnosperm specialist; P: *Pinus* specialist; AF: Africa; NAM: North America; CAM: Central America; SAM: South America; EU: Europe; AS: Asia; OC: Australia and Oceania; COS/CIRG: Cosmopolitan/Circumglobal distribution; NH/CIRP: Northern Hemispere/Circumpolar distribution; TEMP: temperate; BOR: boreal; TROP: tropical; SUBTROP: subtropical; CONZON: conifer zone.

Species	**Host preference**	**Geographic distribution**									**Climatic/ecological zone**				
		AF	NAM	CAM	SAM	EU	AS	OC	COS/CIRG	NH/CIRP	TEMP	BOR	TROP	SUBTROP	CONZON
Asterodon ferruginosus	GEN		•			•	•			•	•	•			
Coltricia cinnamomea	GEN	•	•		•	•	•	•	•		•		•	•	•
Coltricia montagnei	A		•	•		•	•		•					•	
Coltricia perennis	GEN	•	•		•	•	•	•	•		•	•			•
Coltriciella dependens	GEN	•	•	•			•	•					•	•	
Coltriciella oblectabilis	A		•	•	•		•				•		•	•	
Coltriciella tasmanica	P						•	•					•	•	
Coniferiporia sulphurascens	GYM					•	•				•	•			
Coniferiporia weirii	GYM	•	•			•	•				•				
Cyclomyces xeranticus	A						•				•			•	
Fomitiporia hartigii	GYM		•			•	•			•	•	•			•
Fomitiporia punctata	GEN		•			•	•			•	•			•	
Fomitiporia repanda	GYM		•												
Fuscoporia chinensis	A						•				•		•	•	
Fuscoporia contigua	GEN	•	•	•	•	•	•	•			•	•	•	•	
Fuscoporia contiguiformis	P			•			•								
Fuscoporia coronadensis	GYM		•	•											
Fuscoporia ferrea	GEN		•	•	•	•	•	•	•		•			•	
Fuscoporia ferruginosa	GEN	•	•	•	•	•	•	•	•		•			•	
Fuscoporia torulosa	GEN		•			•	•	•					•	•	

Table 2 contd. ...

...Table 2 contd.

Species	**Host preference**	**Geographic distribution**									**Climatic/ecological zone**				
		AF	NAM	CAM	SAM	EU	AS	OC	COS/CIRG	NH/CIRP	TEMP	BOR	TROP	SUBTROP	CONZON
Fuscoporia viticola	GEN		•	•	•	•	•		•			•			
Hydnoporia corrugata	GEN		•		•	•	•		•						
Hydnoporia tabacina	GEN	•	•			•	•		•						
Hydnoporia yasudai	GEN						•				•	•			
Hymenochaete canescens	A					•									
Hymenochaete cinnamomea	GEN	•	•		•	•	•	•	•		•				
Hymenochaete fuliginosa	GEN		•			•	•	•				•			
Hymenochaete jaapii	GEN					•									
Inonotopsis subiculosa	GYM		•			•	•			•		•			•
Onnia kesiya	P						•								
Onnia leporina	GYM		•			•	•			•	•	•			•
Onnia microspora	P						•								
Onnia subtriquetra	P		•												
Onnia tibetica	P						•								
Onnia tomentosa	GYM		•			•	•	•		•	•	•			•
Onnia triquetra	P					•									
Phellinidium ferrugineofuscum	GYM		•			•	•			•	•	•			•
Phellinus nilgheriensis	GEN	•	•	•	•		•	•							
Phellopilus nigrolimitatus	GYM		•			•	•			•	•	•			•
Porodaedalea chinensis	P						•								
Porodaedalea chrysoloma	GYM					•					•	•			
Porodaedalea kesiyae	P						•								
Porodaedalea laricis	GYM					•	•			•	•	•			

Table 2 contd. ...

...Table 2 contd.

Species	**Host preference**	**Geographic distribution**									**Climatic/ecological zone**				
		AF	NAM	CAM	SAM	EU	AS	OC	COS/CIRG	NH/CIRP	TEMP	BOR	TROP	SUBTROP	CONZON
Porodaedalea mongolica	GYM						•								
Porodaedalea occidentiamericana	GYM		•												
Porodaedalea orientoamericana	GYM		•												
Porodaedalea pini	GYM					•	•				•				•
Porodaedalea yunnanensis	P						•								
Tubulicrinis accedens	GEN	•	•		•	•	•								
Tubulicrinis angustus	GEN		•			•	•			•		•			
Tubulicrinis borealis	GEN		•		•	•	•			•		•			
Tubulicrinis calothrix	GEN	•	•	•	•	•	•	•							
Tubulicrinis chaetophorus	GYM	•	•		•	•	•	•							
Tubulicrinis confusus	GEN		•		•	•	•								
Tubulicrinis effugiens	GYM		•			•		•							
Tubulicrinis globisporus	GYM					•									
Tubulicrinis gracillimus	GEN	•	•	•	•	•	•								
Tubulicrinis hamatus	GEN		•		•	•	•	•							
Tubulicrinis hirtellus	GYM					•	•					•			
Tubulicrinis inornatus	A		•		•	•	•								
Tubulicrinis medius	GEN		•			•	•					•			
Tubulicrinis propinquus	GYM					•	•	•							
Tubulicrinis sororius	GEN	•	•		•	•	•					•			
Tubulicrinis strangulatus	GYM					•	•								
Tubulicrinis subulatus	GEN	•	•			•	•	•				•			
Tubulicrinis xantha	P						•								

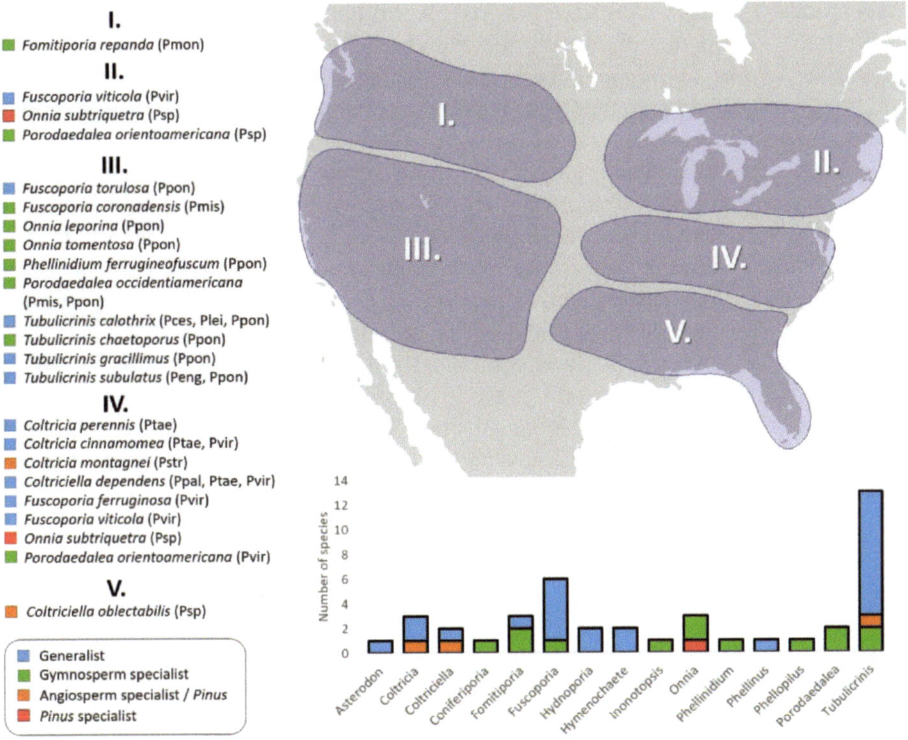

Fig. 7. Frequency of *Pinus*-related Hymenochaetaceae species in North America and occurrences on *Pinus* according to subregions. Abbreviations: Pces – *Pinus cembroides*, Peng – *Pinus engelmannii*, Plei – *Pinus leiophylla*, Pmis – *Pinus strobiformis*, Pmon – *Pinus monticola*, Ppal – *Pinus palustris*, Ppon – *Pinus ponderosa*, Psp – *Pinus* sp., Pstr – *Pinus strobus*, Ptae – *Pinus taeda*, Pvir – *Pinus virginiana*.

Eleven of the hymenochaetoid species were observed on *Pinus* in Western North America and West USA, while twelve species were collected from pines of Eastern North America and East USA. Western species include mostly gymnosperm-specialists such as *Fuscoporia coronadensis, Onnia leporina, O. tomentosa, Phellinidium ferrugineofuscum, Porodaedalea occidentiamericana* and *Tubulicrinis chaetophorus*, but generalist species such as *Fuscoporia torulosa, Tubulicrinis calothrix, T. gracillimus* and *T. subulatus* were also collected from *Pinus* in these regions (Fig. 7). These species distributions mostly overlap with the temperate conifer forests of the Rocky Mountains and its vicinity in the western mainland and coastal regions of USA and Canada (Wade et al., 2003). The coniferous biome subdivisions and vegetation of this area are serving as quite heterogeneous habitats for the mostly gymnosperm-specialist species (Brown et al., 1998; Rehfeldt et al., 2012). Cold temperate coniferous vegetation, that is mostly or partially characterized by *Pinus* species in Western North America include the Great Basin conifer woodlands with Pinyon-Juniper series (*Pinus edulis*); the Cascade-Sierran montane conifer forests with mixed conifer and yellow pine series (*Pinus ponderosa*); and the Rocky Mountain montane conifer forests with yellow pine series and their numerous associations (*Pinus ponderosa*, *P. flexilis*). Warm temperate vegetations with *Pinus* in Western USA include the California evergreen forests and woodlands with oak-pine series (*Pinus sabiniana*). The boreal-subalpine vegetations in Northwest USA and Southwest Canada are the Cascade-Sierran subalpine conifer forests with limber pine-lodgepole pine (*Pinus flexilis*, *P. contorta*) and whitebark pine (*Pinus albicaulis*) series; and the bristlecone pine-limber pine series (*Pinus flexilis* and mostly *P. balfouriana*, but also *P. aristata* and *P. longaeva*) of the Rocky Mountain subalpine conifer forests (Brown et al., 1998; Rehfeldt et al., 2012). *Pinus monticola* is also widely distributed in the west from British Columbia to the southern parts of Sierra Nevada, while *Pinus cembroides*, *P. engelmannii*, *P. leiophylla* and *P. strobiformis* are native in the southwestern United States and Mexico, distributed in a wide range of the local vegetation types and ecological settings (Debreczy and Rácz, 2000; Rehfeldt et al., 2012; Ávila-Flores et al., 2016; Shirk et al., 2018). All of these *Pinus* species are a vast source of potential substrata for *Pinus*-related Hymenochaetaceae in Western North America that is most probably utilized in higher frequencies shown here (Fig. 7).

In the east, generalists and angiosperm specialists that have been reported on *Pinus* are *Coltricia cinnamomea*, *C. perennis*, *C. montagnei*, *Coltriciella dependens*, *C. oblectabilis*, *Fuscoporia ferruginosa* and *F. viticola*. Gymnosperm specialists are represented by *Porodaedalea orientoamericana,* while the *Pinus* specialist *O. subtriquetra* was also collected here (Fig. 7; regions IV and V). Though temperate deciduous and evergreen forests are characteristic of the southeastern part of USA (mostly in the area of the Atlantic and Gulf Coastal Plain), the most of the eastern part of North America is predominantly occupied by deciduous forests (Brown et al., 1998; Rehfeldt et al., 2012). Mixed mesophytic series and evergreen oak vegetation of the southeastern deciduous and evergreen forests could have patches of subdominant pines such as *Pinus taeda*, *P. echinata*, *P. virginiana*, *P. rigida*, *P. strobus* and other xeric species (Hinkle et al., 1993; Brown et al., 1998). Such potential hosts are alternatively utilized by generalist or angiosperm specialist hymenochaetoids, that have an otherwise vast source of broadleaf substrata in these vegetation and are widely distributed in North America (i.e., *Coltriciella dependens*, *C. oblectabilis*, *Fuscoporia ferruginosa* or *Fuscoporia viticola*). Pine series, sand pine forests, mixed pine woodlands and other pine associations in the southeastern deciduous and evergreen forest biome also include *Pinus palustris*, *P. clausa*, *P. elliottii* and *P. serotina* (Brown et al., 1998; Debreczy and Rácz, 2000; Greller, 2003; Rehfeldt et al., 2012) as a wide range of potential hosts for the here present Hymenochaetaceae. *Porodaedalea orientoamericana* and *Onnia subtriquetra* have been reported mostly on *Pinus virginiana* and *Pinus* sp. in the central and northern parts of Eastern USA (Fig. 7; regions IV and II), corresponding with the locally present area of oak-pine and hemlock-white pine-mixed hardwood series of the cold temperate deciduous forests biome (Brown et al., 1998; Rehfeldt et al., 2012; Ji et al., 2017; Wu et al., 2022). Such vegetation types and other central-northeastern habitats could also include *Pinus taeda*, *P. pungens*, *P. resinosa*, *P. rigida* and *P. strobus*, which could also attract other generalist and angiosperm specialist hymenochaetoids shown above in Table 2 (Debreczy and Rácz, 2000; Decock et al., 2007).

3.2.2 *Pinus-Related Hymenochaetaceae in Europe*

In Europe, 45 *Pinus*-related species have occurrences. 24 species of *Asterodon* (1 sp.), *Coltricia* (2 spp.), *Fomitiporia* (1 sp.), *Fuscoporia* (5 spp.), *Hydnoporia* (2 spp.), *Hymenochaete* (3 spp.) and *Tubulicrinis* (10 spp.) represent generalists and 17 species are gymnosperm specialists like *Coniferiporia* (2 spp.), *Fomitiporia* (1 sp.), *Inonotopsis* (1 sp.), *Onnia* (2 spp.), *Phellinidium* (1 sp.), *Phellopilus* (1 sp.), *Porodaedalea* (3 spp.) and *Tubulicrinis* (6 spp.). The only *Pinus*-specialist hymenochaetoid fungi in Europe is *Onnia triquetra* (Fig. 8) (Table 2). In Europe, 12 *Pinus* species could serve as hosts for these fungi (Nobis et al., 2012).

In Southern Europe, mostly covered by the Mediterranean biogeographical region, generalist species reported on *Pinus* are *Coltricia perennis*, *Fuscoporia torulosa*, *Hymenochaete cinnamomea* and a wide range of *Tubulicrinis* species. Much less gymnosperm specialists have been collected from Mediterranean Europe, namely *Fomitiporia hartigii*, *Onnia tomentosa*, *Porodaedalea chrysoloma* and *P. pini* (Fig. 8; region V). As potential hosts in Southeast Europe, disjunct populations of *Pinus brutia*, *P. halapensis*, *P. sylvestris* and *P. nigra* are distributed in the Balkan Mountains, the mountain ranges of Western and Mid-Turkey, Crimea and in the Caucasus, integrated within the biome regions of both the Mediterranean and temperate mixed and conifer forests. From the Pindus Mountains and the Dinaric Alps through the

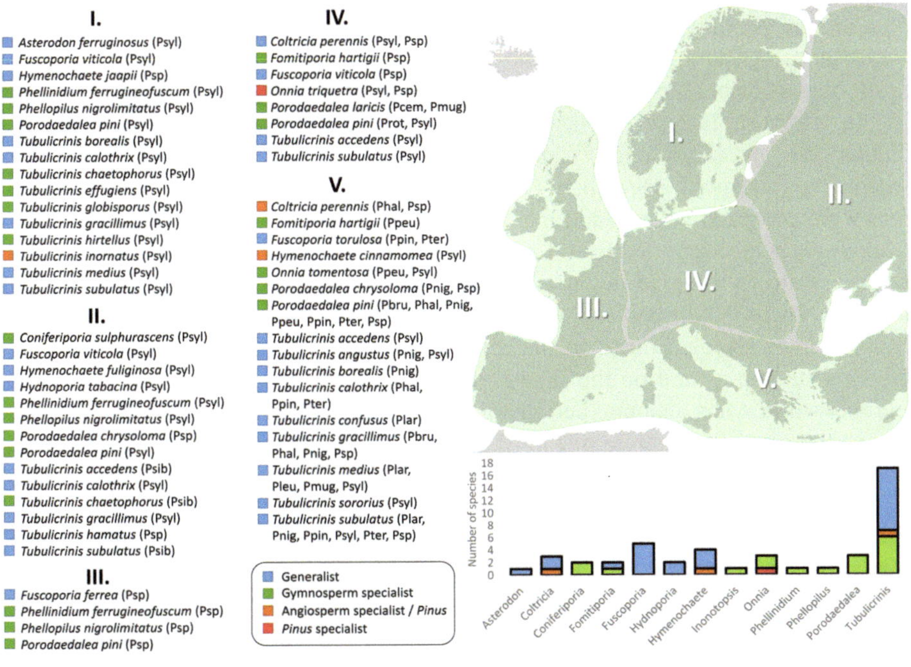

Fig. 8. Frequency of *Pinus*-related Hymenochaetaceae species in Europe and occurrences on *Pinus* according to subregions. Abbreviations: Pbru – *Pinus brutia*; Pcem – *Pinus cembra*; Phal – *Pinus halapensis*; Plar – *Pinus nigra* ssp. *laricio*; Pleu – *Pinus leucodermis*; Pmug – *Pinus mugo*; Pnig – *Pinus nigra*; Ppeu – *Pinus peuce*; Ppin – *Pinus pinea*; Prot – *Pinus rotundata*; Psib – *Pinus sibirica*; Psp – *Pinus* sp.; Psyl – *Pinus sylvestris*; Pter – *Pinus pinaster*.

Appenines to the mountain ranges in the Iberian Penninsula, *Pinus halapensis*, *P. nigra*, *P. sylvestris*, *P. peuce*, *P. pinea* and *P. pinaster* are also having disjunct, natural localities, serving also as potential hosts in a wide range of phytoassociations (e.g., *Abietion cephalonicae – Pinus nigra*, *Digitali viridiflorae-Pinetum peucis – Pinus peuce* and others). Mostly in the southwestern Mediterranean mountain ranges of the continent such as the Cantabrian Mountains, the Pyrenees, the Central-, Iberian- and the Beatic System of Spain, the Beira and Algarve region of Portugal and in the Massif Central of France, pines such as *Pinus sylvestris*, *P. mugo*, *P. pinaster*, *P. nigra*, *P. pinea* and *P. halapensis* are also naturally distributed, in characteristic Mediterranean vegetations such as the Mediterranean macchia or the Juniper-pine woods of southwest Europe (Tortić, 1987; Ortega and Lorite, 2000; Zervakis et al., 2002; Wade et al., 2003; Caudullo et al., 2017). Aside the upper presented observations on these host species (Fig. 8; region V), the angiosperm-specialist *Hymenochaete canescens* has been shown to be potentially present on *Pinus pinea* within its distribution in Italy and Spain (Corfixen and Parmasto, 2017).

Aside their natural distribution in Southwest Europe, *P. pinaster* and *P. nigra* also have naturalized occurrences in the mainland of France, while stands of *P. sylvestris* are distributed all the way to the northern corners of the British Isle and Ireland in the Atlantic biogeographical region (Caudullo et al., 2017). These naturalized forests in Western Europe provide a wide range of *Pinus* species as hosts for hymenochaetoid fungi, as gymnosperm specialist *Phellopilus nigrolimitatus* commonly and *Phellinidium ferruguineofuscum* rarely occur on *Pinus*, mostly in France and in the Netherlands. Generalist *Fuscoporia ferrea* occur predominantly on broadleaved hosts, but has been reported on *Pinus* as well in England (Fig. 8; region III).

In Central Europe, generalists *Coltricia perennis*, *Fuscoporia viticola*, *Tubulicrinis accedens* and *T. subulatus* were collected from *Pinus sylvestris*, whereas gymnosperm specialists *Fomitiporia hartigii*, *Porodaedalea laricis* and *P. pini* were observed on a wide range of *Pinus* species. The *Pinus*-specialist *Onnia triquetra* was also reported from Central Europe (Fig. 8; region IV). The midland of Europe is mostly covered by the coniferous mountain forests of the Ore Mountains, the Alps and the Carpathian Mountains. As a host, *Pinus sylvestris* is naturally and continuously distributed in the relatively warm and dry walleys of the Alps, the Ore Mountains and on the sandy soils of the northeastern lowlands of Central Europe, while it has disjunct, peripheral localities within the Carpathian Mountains. Above the continental strata of *Pinus sylvestris* in the Alps, Pine-larch forests with *P. cembra* could provide substrata for these hymenochaetoid species as well, which also has coherent natural distributions in the Southern- and the Eastern Carpathians. *Pinus mugo* is also naturally distributed in the alpine timber line of the Alps and in the Carpathians, while *P. uncinata* predominantly occur in the montane to subalpine belts of the Western Alps. Besides these, *P. nigra* and *P. sylvestris* also have huge naturalized distributions in the continental-alpine and Pannonian areas, respectively (Burley et al., 2004; Caudullo et al., 2017; Leuschner and Ellenberg, 2017). Among the gymnosperm-specialists, *Porodaedalea laricis* have been reported in the Šumava Mts. (Czech Republic) on *Pinus mugo* and inferred to be preferentially occupying localities above 1300 m.a.s.l. elevation, in contrast to *P. pini* in the adjacent areas, which mostly occur on *Pinus sylvestris* at elevations below 7–800 m.a.s.l. (Tomšovský et al., 2010; Ryvarden and Melo, 2014).

In East Europe, from the eastern and northern limits of the Carpathians and the Caucasus to the boreal forests and taiga of Eastern Scandinavia and the Ural Mountains, *Coniferiporia sulphurascens*, *Fuscoporia viticola*, *Hymenochaete fuliginosa*, *Hydnoporia tabacina*, *Phellinidium ferrugineofuscum*, *Phellopilus nigrolimitatus*, *Porodaedalea chrysoloma*, *P. pini* and six *Tubulicrinis* species have been observed on *Pinus sylvestris* and on *P. sibirica* (Fig. 8; region II). *Pinus sylvestris* is a common host from the Central Russian Upland to the Ural Mountains in the region, while spruce-fir-siberian stone pine forests of the Ural Mountains with *Pinus sibirica* are also a potential niche for these hymenochaetoids (Burley et al., 2004; Caudullo et al., 2017).

Among the possible *Pinus* hosts, only *P. sylvestris* is distributed naturally and dominantly in the boreal forest vegetation zones of Northern Europe (encompassing Scandinavia) and only *P. nigra* and *P. mugo* has naturalized occurrences restricted to Denmark (Caudullo et al., 2017). As a consequence, all of the registered occurencences of Hymenochaetaceae species are restricted to *Pinus sylvestris* in this region: *Asterodon ferruginosus*, *Fuscoporia viticola*, *Hymenochaete jaapii*, *Tubulicrinis borealis*, *T. calothrix*, *T. medius*, *T. subulatus* (generalists) and *Phellinidium ferrugineofuscum*, *Phellopilus nigrolimitatus*, *Porodaedalea pini*, *Tubulicrinis chaetophorus*, *T. effugiens*, *T. globisporus*, *T. hirtellus* (gymnosperm specialists) have been reported here on *Pinus sylvestris* (Fig. 8; region I). *Inonotopsis subiculosa* and *Porodaedalea laricis* are also distributed in Fennoscandia, but they prefer mainly *Picea* as host here. Still, they have the potential to alternatively utilize *Pinus sylvestris* as substrata (Dai, 2010; Tomšovský et al., 2010; Ryvarden and Melo, 2014).

Many of the gymnosperm-dwelling Hymenochaetaceae species are reported to be broadly present in most quarters of Europe, thus most probably occurring on *Pinus* in much broader ranges in the continent. These are mostly circumpolar and circumglobal species in *Coltricia* (*C. perennis*, *C. cinnamomea*), *Fomitiporia* (*F. punctata*), *Fuscoporia* (*F. ferrea*, *F. ferruginosa*, *F. torulosa*, *F. contigua*, *F. viticola*), *Hymenochaete* (*H. cinnamomea*, *H. fuliginosa*, *H. jaapii*), *Hydnoporia* (*H. corrugata*, *H. tabacina*), *Onnia* (*O. tomentosa*, *O. triquetra*), *Phellopilus* (*P. nigrolimitatus*), *Porodaedalea* (*P. chrysoloma, P. pini*) and most of the *Tubulicrinis* species. In the recent literature, the area of *Porodaedalea chrysoloma* have been indicated to be restricted to Europe (Wu et al., 2019) and its distribution were reported to cover the most of the continent as well with its certain northern limits in southern Sweden and Finland (Tomšovský et al., 2010; Ryvarden and Melo, 2014).

3.2.3 *Pinus-Related Hymenochaetaceae in Asia*

According to the available literature, the highest number of *Pinus*-related hymenochaetoid fungi could be found in Asia, altogether 55 species, from which 26 are generalists – including *Asterodon* (1 sp.), *Coltricia* (2 spp.), *Coltriciella* (1 sp.), *Fomitiporia* (1 sp.); *Fuscoporia* (5 spp.); *Hydnoporia* (3 spp.); *Hymenochaete* (2 spp.); *Phellinus* (1 sp.) and *Tubulicrinis* (10 spp.). Species of 8 genera are restricted to gymnosperms, such as *Coniferiporia* (2 spp.); *Fomitiporia* (1 sp.); *Inonotopsis* (1 sp.); *Onnia* (2 spp.); *Phellinidium* (1 sp.); *Phellopilus* (1 sp.); *Porodaedalea* (3 spp.) and *Tubulicrinis* (4 spp.). Three species of *Porodaedalea* (*P. chinensis*, *P. kesiyae*, *P. yunnanensis*), three species of *Onnia* (*O. kesiya*, *O. microspora*, *O. tibetica*), *Coltriciella tasmanica*, *Fuscoporia contiguiformis* and *Tubulicrinis xantha* have been reported to colonize *Pinus* exclusively. Thus, the highest number of gymnosperm- and genus-specialist *Pinus*-dwelling Hymenochaetaceae are occurring in Asia (Fig. 9) (Table 2) and 27 *Pinus* species are possible hosts in the continent (Nobis et al., 2012).

Mostly following the boreal forest and taiga belt in northern Asia, from the raised bogs in Western Siberia through Central- and Southern Siberian regions of the Yenisei River basin with boreal mountain and valley forests to northern- and middle boreal areas in the Tunguska River basin in East Siberia, a wide variety of Hymenochaetaceae were reported on *Pinus sylvestris* and *Pinus sibirica* (Wade et al., 2003; Filippova and Zmitrovich, 2013; Kotiranta and Shiryaev, 2015; Kotiranta et al., 2016; Park et al., 2020). These include generalists such as *Coltricia perennis*, *Fuscoporia contigua*, *Fuscoporia viticola*, *Hydnoporia tabacina* and a wide range of *Tubulicrinis* species, while gymnosperm specialists include *Coniferiporia weirii*, *Onnia leporina*, *Phellinidium ferrugineofuscum*, *Phellopilus nigrolimitatus*, *Porodaedalea laricis*, *P. pini* and four *Tubulicrinis* species. The *Pinus*-specialist *Fuscoporia contiguiformis* was also reported from Siberia (Fig. 9; regions I and II). In the eastern regions of North Asia, including the mostly boreal Khabarovsk Krai and the temperate Primorsky Krai regions of Russia (where *Pinus pumila* is naturally distributed) and

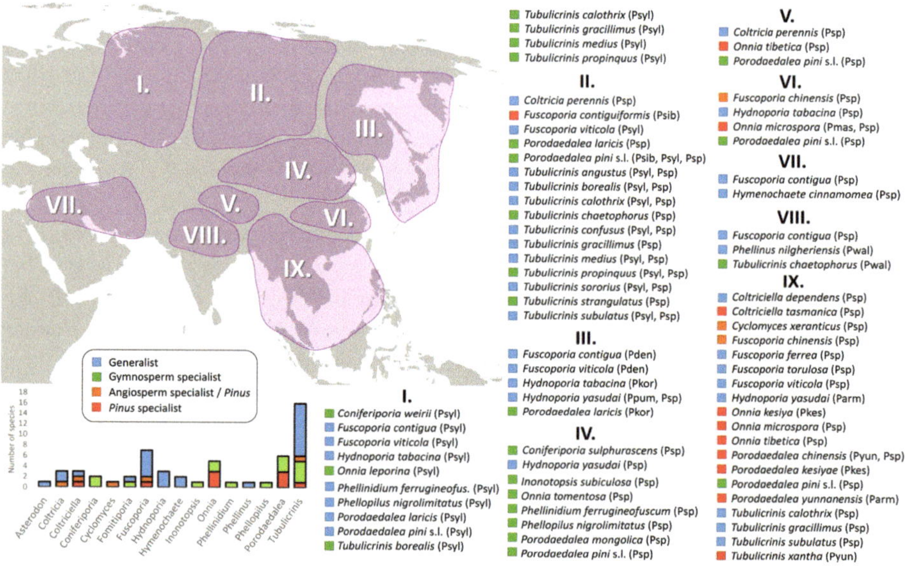

Fig. 9. Frequency of *Pinus*-related Hymenochaetaceae species in Asia and occurrences on *Pinus* according to subregions. Abbreviations: Parm – *Pinus armandii*; Pden – *Pinus densiflora*; Pkes – *Pinus kesiyae*; Pkor – *Pinus koraiensis*; Ppum – *Pinus pumila*; Psib – *Pinus sibirica*; Psp – *Pinus* sp.; Psyl – *Pinus sylvestris*; Pwal – *Pinus wallichiana*; Pyun – *Pinus yunnanensis*.

the temperate prefectures of Japan (Hattori, 1999; Wade et al., 2003; Tomšovský et al., 2010; Vasilyeva et al., 2018; Miettinen et al., 2019), species such as *Fuscoporia contigua*, *F. viticola*, *Hydnoporia tabacina*, *H. yasudai* and *Porodaedalea laricis* were reported, which all have specimens collected on pines such as *Pinus densiflora*, *P. koraiensis*, *P. pumila* and *Pinus* sp. (Fig. 9; region III). Aside these, species such as *Coniferiporia sulphurascens* and *Fuscoporia ferruginosa* have been indicated to be widely distributed in Northern Asia, that most probably have occurrences on *Pinus* in boreal Asia as well. It is worth to note, that the distribution of *Porodaedalea pini* have been indicated to be restricted to Europe, as phylogenetic studies inferred the distinct taxonomic position of Asian specimens that was previously described as *P. pini* (Tomšovský et al., 2010; Wu et al., 2019). Other records of *Porodaedalea* on *Pinus* in China have been also treated collectively as *Porodaedalea pini*, but probably not identical with the European species (Dai, 2010). Still, the Siberian and other Southern Asian records of *Porodaedalea pini* hasn't been revised to this day (Prasher and Lalita, 2013; Prasher, 2015; Park et al., 2020), thus we consider it to be present in these regions of Asia.

Most of the taxa are distributed in Southeast Asia. In Northern China, mostly gymnosperm specialist Hymenochaetaceae were reported on *Pinus*, such as *Coniferiporia sulphurascens*, *Inonotopsis subiculosa*, *Onnia tomentosa*, *Phellinidium ferrugineofuscum*, *Phellopilus nigrolimitatus*, *Porodaedalea mongolica* and *P. pini*, while the sole generalist found on *Pinus* in this region is *Hydnoporia yasudae* (Fig. 9; region IV). *Fomitiporia hartigii* and *Porodaedalea laricis* were also reported in Northern China, having the potential to colonize the here present *Pinus* species. Patches of temperate conifer forests in the Altai Mountains, the Loess Plateau and the Greater Khingan Range and the temperate broadleaf and mixed forests in Northeast China include more than 10 *Pinus* species, such as *P. armandii*, *P. bungeana*, *P. koraiensis*, *P. sylvestris var. mongolica* or *P. tabuliformis* as potential hosts (Debreczy and Rácz, 2000; Wade et al., 2003; Procheş et al., 2012).

The region of South-Southeast China and the Indochina Peninsula to Malaysia is also densely inhabited by a wide range of taxa, including generalists and angiosperm specialists such as *Coltriciella dependens*, *Cyclomyces xeranthicus*, *Fuscoporia chinensis*, *F. ferrea*, *F. torulosa*, *F. viticola*, *Hydnoporia yasudae* *Tubulicrinis calothrix*, *T. gracillimus*, *T. subulatus* and the gymnosperm specialist *Porodaedalea pini*, that have been collected from *Pinus*. Other gymnosperm specialists, such as *Coniferiporia sulphurascens*, *Fomitiporia hartigii*, *Onnia tomentosa* and *Phellopilus nigrolimitatus* are also supposedly inhabit pines in this region, as they are distributed here. This region incorporates the highest number of *Pinus*-specialist Hymenochaetaceae species: *Coltriciella tasmanica* (the host of the type and other specimens of this species from Australia haven't been specified), *Onnia kesiya*, *O. microspora*, *O. tibetica*, *Porodaedalea chinensis*, *P. kesiyae*, *P. yunnanensis* and *Tubulicrinis xantha* have been reported on *Pinus*, in some cases only on one species (i.e., *Onnia kesiya* and *Porodaedalea kesiyae* on *Pinus kesiya*, *Porodaedalea chinensis* and *Tubulicrinis xantha* on *Pinus yunnanensis* and *Porodaedalea yunnanensis* on *Pinus armandii*; Fig. 9; region IX). Most of the area of Southeast China and the Indochina Peninsula belongs to the Indo-Malay biogeographic realm and is covered by subtropical and tropical moist broadleaf forests. At the same time *Pinus kwangtungensis*,

P. massoniana, *P. kesiya*, *P. merkusii*, *P. dalatensis* and *P. krempfi* are characteristic potential hosts in this region, often forming stands with broadleaves or found in tropical pine forests. Amongst these, *P. merkusii* naturally occurs in Indonesia, thus being the only natural pine species of the Southern Hemisphere (Debreczy and Rácz, 2000; van Zonneveld et al., 2009). More to the west, the Hengduan Mountains and the eastern regions of the Tibetan Plateau is characterized by temperate conifer forests, where together with the areas of the Yunnan-Guizhou Plateau, *Pinus armandii*, *P. bungeana*, *P. kesiya*, *P. wallichiana* and *P. yunnanensis* are present as possible hosts (Debreczy and Rácz, 2000; Wade et al., 2003; van Zonneveld et al., 2009; Kanturski et al., 2017).

More northern hymenochaetoid occurrences on *Pinus* in East China include *Fuscoporia chinensis, Hydnoporia tabacina, Onnia microspora* and *Porodaedalea pini* (Fig. 9; region VI), where *Pinus hwangshanensis* is naturally distributed in the Huangshan Mountains (Debreczy and Rácz, 2000).

In the region of the Himalayas and the Plateau of Tibet of Southwest China, *Coltricia perennis*, *Onnia tibetica* and *Porodaedalea pini* have been reported on *Pinus* (Fig. 9; region V). In the Himalayas and in the southern regions of Tibet, *Pinus roxburghii* and *P. wallichiana* are present, while *P. tabuliformis* has its western distribution limit in Quinghai and Yunnan, mostly occurring in mixed forests on rocky mountainsides (*P. tabuliformis* var. *densata*) (Debreczy and Rácz, 2000; Nobis et al., 2012; Kanturski et al., 2017).

According to the reviewed literature, *Coltricia perennis*, *Coniferiporia sulphurascens*, *Cyclomyces xeranticus*, *Fomitiporia hartigii*, *Fuscoporia chinensis, F. contigua, F. ferrea, F. ferruginosa, Hymenochaete cinnamomea, Hydnoporia tabacina, Onnia tomentosa, Phellopilus nigrolimitatus, Porodaedalea laricis* and *P. pini* are rather characteristic in the entirety of China on a general level, which all most probably more frequently colonize *Pinus* in the region. In addition to specimens found in China, *Coltriciella dependens* and *Hydnoporia tabacina* were reported to be present in Malaysia, that could possibly colonize *Pinus merkusii* (van Zonneveld et al., 2009).

By and large, taxa in South Asia that were collected on *Pinus* mostly in the Nepalian and Indian Himalayas are comprised of *Fuscoporia contigua*, *Phellinus nilgheriensis* and *Tublicrinis chaetophorus* (Fig. 9; region VIII). At the same time, *Coltricia montagnei, C. perennis, Cyclomyces xeranthicus, Fuscoporia ferrea, F. torulosa, F. ferruginosa, Onnia tomentosa, Porodaedalea pini* and other species were also found here, as additional potential colonizers of *Pinus*. As a potential host, *P. wallichiana* occupies a wide range of habitats in the Himalayas and its vicinity: from the tropical-subtropical transition belt, it is associated with *P. roxburghii* in coniferous forests; in higher elevations, it is a composer of mixed forests; while at the upper treeline, it is part of the rhododendron-juniper scrub vegetation. In more moist areas, *P. wallichiana* appears together with *P. gerardiana* (Debreczy and Rácz, 2000; Sinha, 2002; Wade et al., 2003; Kanturski et al., 2017).

Hymenochaetaceae collected on *Pinus* in Southwest Asia are restricted to two generalist species, namely *Fuscoporia contigua* and *Hymenochaete cinnamomea*, that occurred in the Mediterranean forests of Israel, where *Pinus halapensis* and *P. pinea* are mixed with *Quercus calliprinos* (Fig. 9; region VII). At the same time, additional *Pinus*-related Hymenochaetaceae are also present here, mainly from *Fuscoporia*, which could also inhabit *Pinus* in the region (Ţura et al., 2008, 2010).

3.2.4 *Pinus*-Dwelling Hymenochaetaceae in Latin America, Africa and Oceania

Pinus-related species in Hymenochaetaceae are represented in an order of magnitude smaller near the Equator and in the Southern Hemisphere. Due to the lack of naturally distributed *Pinus* species in Oceania (including Australia, New Zealand, Micronesia and Polynesia), Africa and South America, no observations has been made on pines in these continents. Though other macrofungal taxa have been recorded to be established in naturalized and non-indigenous stands of *Pinus* in Africa and South America (Campi et al., 2015; Dejene et al., 2017), no Hymenochaetaceae specimens have been collected from exotic *Pinus* plantations in the Southern Hemisphere. *Pinus*-related Hymenochaetaceae occurring in these parts of the world are mostly generalist species that could inhabit the angiosperm flora of the Australasian, Afrotropic and Neotropic regions, or gymnosperm specialist that are able to colonize other naturally present gymnosperm genera than *Pinus* (i.e., *Dacrydium* in Oceania) colonized by *Tubulicrinis cinctus* or *Austrocedrus* in South America (colonized by *Tubulicrinis chaetophorus* and the generalist *T. ellipsoideus*). The host of the type and other collected specimens of *Coltriciella tasmanica* from Oceania haven't been specified and it only has been reported on *Pinus* in Southeast Asia, thus it is the only species south from the Equator, that has been related solely to *Pinus*. The most characteristic generalist and angiosperm-specialist Hymenochaetaceae taxa of the Southern Hemisphere are from the genera *Coltricia*, *Coltriciella*, *Fuscoporia*, *Hydnoporia*, *Hymenochaete*, *Phellinus* and *Tubulicrinis*, which predominantly are having cosmopolitan distributions (Table 2) (Fig. 10).

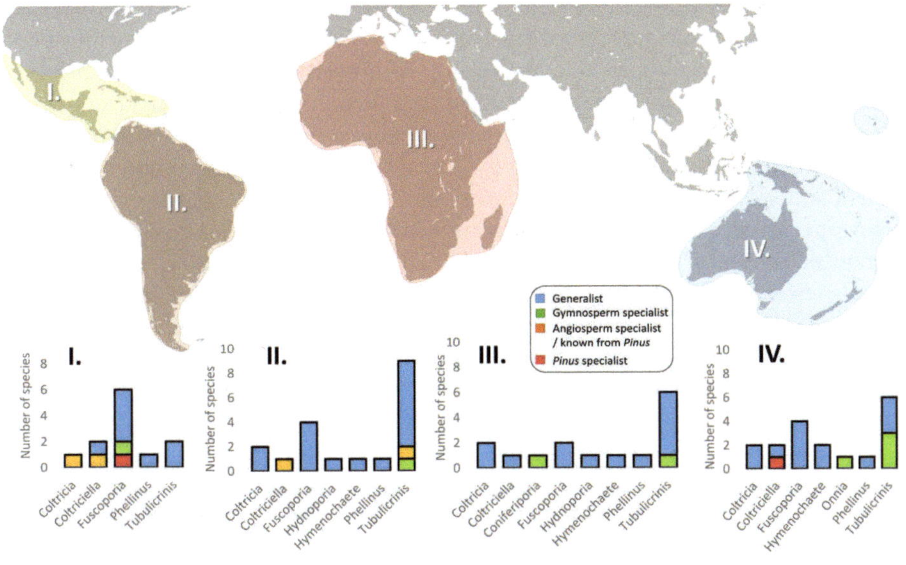

Fig. 10. Frequency of *Pinus*-related Hymenochaetaceae species in Central America, South America, Africa and Oceania.

In contrary to the Southern Hemisphere, Central America possess the most diverse and highest amounts of pine species, compared to other similar sized regions in the world – almost around 50 species. A few examples are *Pinus arizonica*, *P. cembroides*, *P. cooperi*, *P. durangensis*, *P. engelmannii*, *P. leiophylla*, *P. lumholtzii* and *P. strobiformis* that are the most common pine inhabitants of the Mexican Sierra Madre Occidental with the Madrean-Transvolcanic Pine-Oak Woodlands and the Madrean Montane Conifer Forests (Brown et al., 1998; Rehfeldt et al., 2012; Ávila-Flores et al., 2016; Shirk et al., 2018). Consequently, the relevant literature reported occurrences of the generalist *Fuscoporia viticola* and *Tubulicrinis calothrix*, the gymnosperm specialist *Fuscoporia coronadensis* and the *Pinus*-specialist *F. contiguiformis* on *Pinus durangensis*, *P. lumholtzii* and on *Pinus* sp. (Fig. 10; region I). As additional species are present in Central America, that are able to colonize *Pinus* (Table 2), the list of Hymenochaetaceae colonizing pines in this region is presumably longer.

Conclusion

The host preference of *Pinus*-dwelling Hymenochaetaceae as we know today is affected by the series of evolutional and phylogeographical developments that are closely intertwined with the evolutional history of the host species. White rotting lignicolous fungi such as Hymenochaetaceae diversified its original gymnosperm host range in the direction of angiosperm facilitation, leading to generalist and angiosperm specialist hymenochaetoid species, that are likewise able to colonize *Pinus*. These generalist species are the most widespread on Earth among *Pinus*-related Hymenochaetaceae. At the same time, the prevailingly Northern Hemisphere presence of *Pinus*-related Hymenochaetaceae species mirror the natural circumboreal distribution of the *Pinus* genera. In this core area, the extent of geographical dispersion varies between and inside the ecological groups: while the area of some gymnosperm specialist species could span all over the three continents of the Northern Hemisphere (North America, Europe, Asia) or could be restricted to one geographical subregion, the strictly *Pinus*-specialist species are most commonly confined to a limited area. In the same time, the nutritional strategy is also characteristic of the hymenochaetoid genera and species in question: though most *Pinus*-related species are saprotrophs, many species also or solely utilize living hosts, the latter being typical in case of gymnosperm- and *Pinus* specialist genera. The patterns of basidiocarp positioning reveals terrestrial, basal, high-diameter and low-diameter substrate preferences that are in some cases characteristic to certain genera. Altogether, 32 *Pinus* species have been reported as hosts in the reviewed literature.

References

Akamatsu, Y., Ma, D.B., Higuchi, T. and Shimada, M. (1990). A novel enzymatic decarboxylation of oxalic acid by the lignin peroxidase system of white-rot fungus Phanerochaete chrysosporium. FEBS Lett., 269(1): 261–263.

Armenise, L., Simeone, M.C., Piredda, R. and Schirone, B. (2012). Validation of DNA barcoding as an efficient tool for taxon identification and detection of species diversity in Italian conifers. Eur. J. For. Res., 131(5): 1337–1353.

Aust, S.D. (1995). Mechanisms of degradation by white rot fungi. Environ. Health Persp., 103: 59.

Ávila-Flores, I.J., Hernández-Díaz, J.C., González-Elizondo, M.S., Prieto-Ruíz, J.Á. and Wehenkel, C. (2016). *Pinus engelmannii* Carr. in Northwestern Mexico: A review. Pak. J. Bot., 48(5): 2159–2166.

Baltazar, J.M. and Gibertoni, T.B. (2009). A checklist of the aphyllophoroid fungi (Basidiomycota) recorded from the Brazilian Atlantic Forest. Mycotaxon, 109(1): 439–442.

Baltazar, J.M., Ryvarden, L. and Gibertoni, T.B. (2010). The genus Coltricia in Brazil: New records and two new species. Mycologia, 102(6): 1253–1262.

Bernicchia, A., Savino, E. and Gorjón, S.P. (2007). Aphyllophoraceous wood-inhabiting fungi on *Pinus* spp. in Italy. Mycotaxon, 101: 5–8.

Běťák, J., Holec, J., Beran, M. and Riebesehl, J. (2021). Ecology and distribution of *Kneiffiella curvispora* (Hymenochaetales, Basidiomycota) in Central Europe and its phylogenetic placement. Nova Hedw., 113(1-2): 161–189.

Bolshakov, S.Y., Potapov, K.O., Ezhov, O.N., Volobuev, S.V., Khimich, Y.R. and Zmitrovich, I. V. (2016). New species for regional mycobiotas of Russia. 1. Report 2016. Mikologiya I Fitopatologiya, 50(5): 275–286.

Borgen, T., Elborne, S.A. and Henning, K. (2006). A checklist of the Greenland basidiomycetes. pp. 37–59. *In*: Boertmann, D. and Knudsen, H. (eds.). Meddelelser om Grønland, Monographs on Greenland: Bioscience 56. Commision for Scientific Research in Greenland, Copenhagen, Denmark.

Brazee, N.J. (2013). Phylogenetic and differentiation analysis of the trunk rot pathogen *Fomitiporia tsugina* in North America. For. Pathol., 43(5): 407−414.

Brown, D.E., Reichenbacher, F. and Franson, S.E. (1998). A Classification of North American Biotic Communities. University of Utah Press, Salt Lake City, USA.

Buchanan, P.K. and Ryvarden, L. (2000). An annotated checklist of polypore and polypore-like fungi recorded from New Zealand. NZ J. Bot., 38(2): 265–323.

Burley, J., Evans, J. and Youngquist, J.A. (2004). Encyclopedia of Forest Sciences. Elsevier, Oxford, Great Britain.

Campi, M.G., Maubet, Y.E. and Britos, L. (2015). Mycorrhizal fungi associated with plantations of *Pinus taeda* L. from the National University of Asunción, Paraguay. Mycosphere, 6(4): 486–492.

Cardoso, J., Melo, I. and Tellería, M.T. (1992). Aphyllophorales of Peneda-Gerês national Park (Portugal). Crypt. Bot., 2(4): 395–404.

Caudullo, G., Welk, E. and San-Miguel-Ayanz, J. (2017). Chorological maps for the main European woody species. Data in Brief, 12: 662–666.

Chen, Q., Du, P., Vlasák, J., Wu, F. and Dai, Y.C. (2020). Global diversity and phylogeny of *Fuscoporia* (Hymenochaetales, Basidiomycota). Mycosphere, 11(1): 1477–1513.

Cloete, M., Fischer, M., Du Plessis, I.L., Mostert, L. and Halleen, F. (2016). A new species of Phellinus sensu stricto associated with esca on grapevine in South Africa. Mycol. Prog., 15(3): 25.

Corfixen, P. and Parmasto, E. (2017). *Hymenochaete* and *Hymenochaetopsis* (Basidiomycota) in Europe. Karstenia, 58(1): 49–80.

Cowling, E.B. (1961). Comparative Biochemistry of the Decay of Sweetgum Sapwood by White-Rot and Brown-Rot Fungi. Technical Bulletin No. 1258. Washington, D.C., USA.

Dai, Y.C. (2010). Hymenochaetaceae (Basidiomycota) in China. Fungal Diversity, 45(1): 131–343.

Dai, Y.C. (2012). Polypore diversity in China with an annotated checklist of Chinese polypores. Mycoscience, 53(1): 49–80.

Dai, Y.C., Cui, B.K., Yuan, H.S. and Li, B.D. (2007). Pathogenic wood-decaying fungi in China. For. Pathol., 37(2): 105–120.

Dai, Y.-C., Cui, B.-K., Yuan, H.-S., He, S.-H., Wei, Y.-L. et al. (2011). Wood-inhabiting fungi in southern China. 4. Polypores from Hainan Province. Ann. Bot. Fenn., 48(3): 219–231.

Dai, Y.C., Wei, Y.L. and Wang, Z. (2004). Wood-inhabiting fungi in southern China 2. Polypores from Sichuan Province. Ann. Bot. Fenn., 41(5): 319–329.

Daniel, G. (1994). Use of electron microscopy for aiding our understanding of wood biodegradation. FEMS Microbiol. Rev., 13(2-3): 199–233.

Daniel, G. (2016). Fungal degradation of wood cell walls. pp. 131–167. *In*: Kim, Y.S., Funada, R. and Singh, A.P. (eds.). Secondary Xylem Biology. Elsevier.

Debreczy, Z. and Rácz, I. (2000). Fenyők a Föld Körül. Dendrológiai Alapítvány, Budapest, Hungary.

Decock, C., Figueroa, S.H., Robledo, G. and Castillo, G. (2007). *Fomitiporia punctata* (Basidiomycota, Hymenochaetales) and its presumed taxonomic synonyms in America: Taxonomy and phylogeny of some species from tropical/subtropical areas. Mycologia, 99(5): 733–752.

de la Fuente, J.I., García-Jiménez, J., López, C.Y., Oros-Ortega, I., Vela-Hernández, R.Y. et al. (2020). An annotated checklist of the macrofungi (Ascomycota, Basidiomycota and Glomeromycota) from Quintana Roo, Mexico. Check List, 16(3): 627–648.

Dejene, T., Oria-de-Rueda, J.A. and Martín-Pinto, P. (2017). Fungal diversity and succession following stand development in *Pinus patula* Schiede ex Schltdl. & Cham. plantations in Ethiopia. For. Ecol. Manage., 395: 9–18.

Doğan, H.H. and Kurt, F. (2016). New macrofungi records from Turkey and macrofungal diversity of Pozantı-Adana. Tur. J. Bot., 40(2): 209–217.

Doğan, H.H., Aktaş, S., Öztürk, C. and Kaşik, G. (2012). Macrofungi distribution of Cocakdere valley (Arslanköy, Mersin). Tur. J. Bot., 36(1): 83–94.

Doğan, H.H., Karadelev, M. and Işiloğlu, M. (2011). Macrofungal diversity associated with the scale-leaf juniper trees, *Juniperus excelsa* and *J. foetidissima*, distributed in Turkey. Tur. J. Bot., 35(2): 219–237.

Drechsler-Santos, E.R., Santos, P.J.P., Gibertoni, T.B. and Cavalcanti, M.A.Q. (2010). Ecological aspects of Hymenochaetaceae in an area of Caatinga (semi-arid) in Northeast Brazil. Fungal Diversity, 42(1): 71–78.

Elo, M., Halme, P., Toivanen, T. and Kotiaho, J.S. (2019). Species richness of polypores can be increased by supplementing dead wood resource into a boreal forest landscape. J. Appl. Ecol., 56(5): 1267–1277.

Eriksson, K.-E.L., Blanchette, R.A. and Ander, P. (1990). Microbial and Enzymatic Degradation of Wood and Wood Components. Springer Berlin Heidelberg, Berlin, Heidelberg, Germany.

Ezhov, O.N. and Zmitrovich, I.V. (2015). Checklist of aphyllophoroid fungi (Agaricomycetes, Basidiomycota) in boreal forests of Pinega Reserve, north-east European Russia. Check List, 11(1): 1495.

Farjón, A. (2010). A Handbook of the World Conifers. Brill Press, Leiden, The Netherlands.

Filippova, N.V. and Zmitrovich, I.V. (2013). Wood decay community of raised bogs in West Siberia. Environ. Dyna. Global Clim. Chan., 4(1): 1–16.

Fischer, M. and Binder, M. (2004). Species recognition, geographic distribution and host-pathogen relationships: a case study in a group of lignicolous basidiomycetes, *Phellinus* s.l. Mycologia, 96(4): 799–811.

Ghobad-Nejhad, M. (2011). Updated checklist of corticioid and poroid basidiomycetes of the Caucasus region. Mycotaxon, 117: 508.

Ghobad-Nejhad, M. and Dai, Y.-C. (2010). Diplomitoporus rimosus is found in Asia and belongs to the Hymenochaetales. Mycologia, 102(6): 1510–1517.

Ghobad-Nejhad, M. and Hallenberg, N. (2012). Checklist of Iranian non-gilled/non-gasteroid hymenomycetes (Agaricomycotina). Mycotaxon, 119(1): 493–494.

Ghobad-Nejhad, M., Hallenberg, N. and Kotiranta, H. (2008). Additions to the corticioids of the Caucasus from NW Iran. Mycotaxon, 105: 269–293.

Gibertoni, T.B., Santos, P.J.P. and Cavalcanti, M.A.Q. (2007). Ecological aspects of Aphyllophorales in the Atlantic rain forest in northeast Brazil. Fungal Diversity, 25: 49–67.

Gilbertson, R.L. (1980). Wood-rotting fungi of North America. Mycologia, 72(1): 1–49.

Gilbertson, R.L., Martin, K.J. and Lindsey, J.P. (1974). Annotated Check List and Host Index for Arizona Wood-Rotting Fungi. College of Agriculture, University of Arizona, Tucson, Arizona, USA.

Gilbertson, R.L. and Ryvarden, L. (1986). North American Polypores 1. Fungiflora, Oslo, Norway.

Gilbertson, R.L. and Ryvarden, L. (1987). North American Polypores 2. Fungiflora, Oslo, Norway.

Gorjón, S.P., Greslebin, A.G. and Rajchenberg, M. (2013). *Uncobasidium roseocremeum* sp. nov. and other corticioid basidiomycetes from the Patagonian Andes of Argentina. Mycotaxon, 121(1): 349–364.

Gorjón, S.P., Hallenberg, N. and Bernicchia, A. (2009). A survey of the corticioid fungi from the Biosphere Reserve of Las Batuecas-Sierra de Francia (Spain). Mycotaxon, 109(1): 161–164.

Grand, L.F. and Vernia, C.S. (2005). Biogeography and hosts of poroid wood decay fungi in North Carolina: species of *Coltricia*, *Coltriciella* and *Inonotus*. Mycotaxon, 91: 35–38.

Green, F. and Highley, T.L. (1997). Mechanism of brown-rot decay: Paradigm or paradox. Int. Biodeter. Biodegrad., 39(2-3): 113–124.

Greller, A.M. (2003). A review of the temperate broad-leaved evergreen forest zone of Southeastern North America: Floristic affinities and arborescent vegetation types. The Bot. Rev., 69(3): 269–299.

Greslebin, A.G. and Rajchenberg, M. (2003). Diversity of Corticiaceae sens. lat. in Patagonia, Southern Argentina. NZ J. Bot., 41(3): 437–446.

Halme, P., Kunttu, P., Niemelä, T. and Kulju, M. (2008). New records of *Inonotopsis subiculosa* and other rare basidiomycetes in Central Finland. Memoranda Societatis pro Fauna et Flora Fennica, 84(3): 102–107.

Hatakka, A. and Hammel, K.E. (2011). Fungal biodegradation of lignocelluloses. pp. 319–340. *In*: Osiewacz, H.D. (ed.). The Mycota: Vol. X. Industrial Applications. Springer, Berlin, Germany.

Hattori, T. (1999). *Phellinus setifer* sp. nov. and *P. acontextus*, two noteworthy polypores from temperate areas of Japan, with notes on their allies. Mycoscience, 40(6): 483–490.

He, M.-Q., Zhao, R.-L., Hyde, K.D., Begerow, D., Kemler, M. et al. (2019). Notes, outline and divergence times of Basidiomycota. Fungal Diversity, 99(1): 105–367.

He, X., Chen, J.Z. and Zhao, C.L. (2021). Diversity of wood – decaying fungi in Haikou Forestry Farm, Yunnan Province, P.R. China. Stud. Fungi, 6(1): 365–377.

He, X., Shi, Z.-J. and Zhao, C.-L. (2020). Morphological and molecular identification of two new species of *Tubulicrinis* (Hymenochaetaceae, Hymenochaetales) from southern China. Mycoscience, 61(4): 184–189.

Hibbett, D.S. and Donoghue, M.J. (2001). Analysis of character correlations among wood decay mechanisms, mating systems and substrate ranges in homobasidiomycetes. Syst. Biol., 50(2): 215–242.

Hinkle, C.R., McComb, W.C., Safley, J.M. and Schmalzer, P.A. (1993). Mixed mesophytic forests. pp. 203–253. *In*: Martin, W.H., Boyce, S.G. and Echternacht, A.C. (eds.). Biodiversity of the Southeastern United State: Upland Terrestrial Communiites. John Wiley & Sons, New York, USA.

Høiland, K. and Bendiksen, E. (1996). Biodiversity of wood-inhabiting fungi in a boreal coniferous forest in Sør-Trøndelag County, Central Norway. Nor. J. Bot., 16(6): 643–659.

Huang, S., Li, S., Wang, Z., Lin, S. and Deng, J. (2021). Enzyme degradation mechanism of white rot fungi and its research progress on Refractory Wastewater. E3S Web of Conferences, 237: 01002.

Hubregtse, J. (2019). Fungi in Australia, Rev. 2.2. Blackburn: Field Naturalists Club of Victoria Inc.

Ji, X.-H., He, S.-H., Chen, J.-J., Si, J., Wu, F. et al. (2017). Global diversity and phylogeny of Onnia (Hymenochaetaceae) species on gymnosperms. Mycologia, 109(1): 27–34.

Käärik, A. (1983). Decomposition of wood. Nederlands Bosbouw Tijdschrift, 55(2/3): 43–50.

Kanturski, M., Akbar, S.A. and Favret, C. (2017). The Bhutan Pine Aphid *Pseudessigella brachychaeta* Hille Ris Lambers (Hemiptera: Aphididae: Lachninae) From India Reveals the Hitherto Unknown Oviparous Female and Dwarfish Male. Zool. Stud., 56: e12.

Karadelev, M., Rusevska, K., Mitić-Kopanja, D. and Lambevska, A. (2014). Ecology and Distribution of lignicolous Fungi in Albania. pp. 633–636. *In*: Dursun, S., Mankolli, H., Zuchetti, M. and Kongoli, C. (eds.). 4th International Conference of Ecosystems (ICE2014). Health and Environment Association, Tirana, Albania.

Kinge, T.R., Goldman, G., Jacobs, A., Ndiritu, G.G. and Gryzenhout, M. (2020). A first checklist of macrofungi for South Africa. MycoKeys, 63: 1–48.

Kirk, P.M., Cannon, P.F., Minter, D.W. and Stalpers, J.A. (2008). Ainsworth & Bisby's Dictionary of the Fungi (10th Edition). CABI, Wallingford, Great Britain.

Kirk, T.K. and Adler, E. (1970). Methoxy-deficient structural elements in lignin of sweetgum decayed by a brown-rot fungus. Acta Chem. Scandin., 24: 3379–3390.

Kohler, A., Kuo, A., Nagy, L.G., Morin, E., Barry, K.W. et al. (2015). Convergent losses of decay mechanisms and rapid turnover of symbiosis genes in mycorrhizal mutualists. Nat. Gen., 47(4): 410–415.

Korotkin, H.B., Swenie, R.A., Miettinen, O., Budke, J.M., Chen, K.-H., Lutzoni, F., Smith, M.E. and Matheny, P.B. (2018). Stable isotope analyses reveal previously unknown trophic mode diversity in the Hymenochaetales. American Journal of Botany, 105(11): 1869–1887.

Kotiranta, H. and Shiryaev, A.G. (2015). Aphyllophoroid fungi (Basidiomycota) in Tunguska River basin, central East Siberia, Russia. Karstenia, 55(1-2): 25–42.

Kotiranta, H., Shiryaev, A.G. and Spirin, V. (2016). Aphyllophoroid fungi (Basidiomycota) of Tuva Republic, southern Siberia, Russia. Folia Cryptogamica Estonica, 53: 51–64.

Krah, F.-S., Bässler, C., Heibl, C., Soghigian, J., Schaefer, H. and Hibbett, D.S. (2018). Evolutionary dynamics of host specialization in wood-decay fungi. BMC Evol. Biol., 18(1): 119.

Krah, F.-S., Seibold, S., Brandl, R., Baldrian, P., Müller, J. and Bässler, C. (2018). Independent effects of host and environment on the diversity of wood-inhabiting fungi. J. Ecol., 106(4): 1428–1442.

Küffer, N. and Senn-Irlet, B. (2005). Diversity and ecology of wood-inhabiting aphyllophoroid basidiomycetes on fallen woody debris in various forest types in Switzerland. Mycol. Prog., 4(1): 77–86.

Küffer, N., Gillet, F., Senn-Irlet, B., Job, D. and Aragno, M. (2008). Ecological determinants of fungal diversity on dead wood in European forests. Fungal Diversity, 30: 83–95.

Kunttu, P. (2016). Extensions of known geographic distribution of aphyllophoroid fungi (Basidiomycota) in Finland. Mycosphere, 7(3): 333–357.

Kunttu, P., Helo, T., Kulju, M., Julkunen, J., Pennanen, J. et al. (2019). Aphyllophoroid funga (Basidiomycota) of Finland: Range extensions and records of nationally new and rare species. Acta Mycol., 54(2): 1–66.

Kunttu, P., Helo, T., Kulju, M., Veteli, P., Julkunen, J. et al. (2021). Diversity and distribution of Finnish aphyllophoroid and heterobasidioid fungi (Basidiomycota): An update. Pl. Fungal Syst., 66(1): 79–105.

Larsen, M.J. (2000). Phellinus gilbertsonii sp. nov. from western North America causing heart-rot of coastal Douglas-fir. Folia Cryptogamica Estonica, 37: 51–54.

Larsen, M.J., Lombard, F.F. and Aho, P.E. (1979). A new variety of *Phellinus pini* associated with cankers and decay in white firs in southwestern Oregon and northern California. Can. J. For. Res., 9(1): 31–38.

Lee, S.S., Alias, S.A., Jones, E.G.B., Zainuddin, N. and Chan, H.T. (2012). Checklist of Fungi of Malaysia. Forest Research Institute, Kepong, Malaysia.

Leuschner, C. and Ellenberg, H. (2017). Ecology of Central European Forests. Vegetation Ecology of Central Europe, Volume I. Springer International Publishing, Cham, Switzerland.

Lin, W.-C., Deng, J.-S., Huang, S.-S., Wu, S.-H., Chen, C.-C. et al. (2017). Anti-inflammatory activity of *Sanghuangporus sanghuang* Mycelium. Int. J. Mol. Sci., 18(2): 347.

Lizoň, P. and Kautmanová, I. (2004). Fungi collected during the 21st European *Cortinarius* Foray. Cathatelasma, 5: 23–35.

Lombard, F.F., Davidson, R.W. and Gilbertson, R.L. (1972). Studies of two species of *Phellinus* in Western North America. Mycopathol. Mycol. Appl., 46(4): 351–365.

Maekawa, N. (2021). Taxonomy of corticioid fungi in Japan: Present status and future prospects. Mycoscience, 62(6): 345–355.

May, T.W., Milne, J., Shingles, S. and Jones, R.H. (2003). Fungi of Australia. Volume 2B: Catalogue and Bibliography of Australian Fungi 2, Basidiomycota p.p. & Myxomycota p.p. ABRS/CSIRO Publishing, Melbourne, Australia.

Miettinen, O., Larsson, K.-H. and Spirin, V. (2019). *Hydnoporia*, an older name for Pseudochaete and *Hymenochaetopsis* and typification of the genus *Hymenochaete* (Hymenochaetales, Basidiomycota). Fungal Syst. Evol., 4: 77–96.

Miyauchi, S., Hage, H., Drula, E., Lesage-Meessen, L., Berrin, J.-G. et al. (2020). Conserved white-rot enzymatic mechanism for wood decay in the Basidiomycota genus *Pycnoporus*. DNA Res., 27(2): 1–14.

Nilsson, T., Obst, J.R. and Daniel, G. (1988). The possible significance of the lignin content and lignin type on the performance of CCA-treatedd timber in ground contact. The Intrenational Research Group on Wood Preservation (IRG/WP No. 1357).

Nobis, M.P., Traiser, C. and Roth-Nebelsick, A. (2012). Latitudinal variation in morphological traits of the genus *Pinus* and its relation to environmental and phylogenetic signals. Pl. Ecol. Divers., 5(1): 1–11.

Ortega, A. and Lorite, J. (2000). A floristic and ecological catalogue of lignicolous Aphyllophorales s.l. (Basidiomycota, Macrofungi) from southern Spain (Andalusia). Cryptog. Mycol., 21(1): 35–48.

Park, J.-H., Pavlov, I.N., Kim, M.-J., Park, M.S., Oh, S.-Y. et al. (2020). Investigating wood decaying fungi diversity in central siberia, russia using ITS sequence analysis and interaction with host trees. Sustainability, 12(6): 2535.

Piepenbring, M., Maciá-Vicente, J.G., Codjia, J.E.I., Glatthorn, C., Kirk, P. et al. (2020). Mapping mycological ignorance – checklists and diversity patterns of fungi known for West Africa. IMA Fungus, 11(1): 13.

Piirto, D.D., Parmeter, J.R.J. and Wilcox, W.W. (1984). California Forestry and Forest Products. University of California, Berkeley, Forestry and Forest Products, 55: 4.

Prasher, I.B. (2015). Wood-rotting non-gilled Agaricomycetes of Himalayas. Springer Netherlands, Dordrecht, The Netherlands.

Prasher, I.B. and Ashok, D. (2013). A checklist of wood rotting fungi (non-gilled Agaricomycotina) of Himachal Pradesh. J. New Biol. Rep., 2(2): 71–98.

Prasher, I.B. and Lalita. (2013). A checklist of wood rotting fungi (non-gilled Agaricomycotina) of Uttarakhand. J. New Biol. Rep., 2(2): 108–123.

Presley, G.N. and Schilling, J.S. (2017). Distinct growth and secretome strategies for two taxonomically divergent brown rot fungi. Appl. Environ. Microbiol., 83(7): e02987–16.

Procheş, Ş., Wilson, J.R.U., Richardson, D.M. and Rejmánek, M. (2012). Native and naturalized range size in *Pinus*: Relative importance of biogeography, introduction effort and species traits. Glob. Ecol. Biogeogr., 21(5): 513–523.

Ranadive, K.R. (2013). An overview of Aphyllophorales (wood rotting fungi) from India. Int. J. Curr. Microbiol. Appl. Sci., 2(12): 112–139.

Raymundo, T., Valenzuela, R., Bautista-Hernández, S., Esqueda, M., Cifuentes, J. and Pacheco, L. (2013). El género *Fuscoporia* (Hymenochaetales, Basidiomycota) en México. Revista Mexicana de Biodiversidad, 84(SUPPL.): S50–S69.

Rehfeldt, G.E., Crookston, N.L., Sáenz-Romero, C. and Campbell, E.M. (2012). North American vegetation model for land-use planning in a changing climate: a solution to large classification problems. Ecol. Appl., 22(1): 119–141.

Renvall, P. (1995). Community structure and dynamics of wood-rotting Basidiomycetes on decomposing conifer trunks in northern Finland. Karstenia, 35(1): 1–51.

Rizzo, D.M., Gieser, P.T. and Burdsall, H.H. (2003). *Phellinus coronadensis*: A new species from southern Arizona, USA. Mycologia, 95(1): 74–79.

Robledo, G.L. and Rajchenberg, M. (2007). South American polypores: First annotated checklist from Argentinean Yungas. Mycotaxon, 100: 5–9.

Ryvarden, L. (2004). Neotropical polypores: Part 1: Introduction, Ganodermataceae & Hymenochaetaceae. Synopsis Fungorum. Fungiflora, Oslo, Norway.

Ryvarden, L. (2007). Synopsis Fungorum 22. Synopsis Fungorum. Fungiflora, Oslo, Norway.

Ryvarden, L. and Melo, I. (2014). Poroid fungi of Europe. Synopsis Fungorum, 31. Fungiflora, Oslo, Norway.

Saitta, A. and Losi, C. (2016). New records of corticioid fungi from Sicily. Check List, 12(5): 1972.

Santisuk, T. (1997). Geographical and ecological distributions of the two tropical pines, *Pinus kesiya* and *Pinus merkusii*, in Southeast Asia. Thai For. Bull. (Bot.), 25: 102–123.

Sell, I. and Kotiranta, H. (2011). Diversity and distribution of aphyllophoroid fungi growing on Common Juniper (*Juniperus communis* L.) in Estonia. Folia Cryptogamica Estonica, 48: 73–84.

Shevchenko, M.V. (2018). Noteworthy records of corticioid fungi from Ichnia National Nature Park. Ukr. Bot. J., 75(1): 77–83.

Shirk, A.J., Cushman, S.A., Waring, K.M., Wehenkel, C.A., Leal-Sáenz, A. et al. (2018). Southwestern white pine (*Pinus strobiformis*) species distribution models project a large range shift and contraction due to regional climatic changes. For. Ecol. Manage., 411: 176–186.

Simpson, M.G. (2019). Evolution and diversity of woody and seed plants. pp. 131–165. *In*: Simpson, M. (ed.). Plant Systematics (Third Edition). Academic Press, Cambridge, Massachusetts, USA.

Sinha, B. (2002). Pines in the Himalayas: Past, present and future scenario. Ener. Environ., 13(6): 873–881.

Tedersoo, L., Suvi, T., Beaver, K. and Saar, I. (2007). Ectomycorrhizas of *Coltricia* and *Coltriciella* (Hymenochaetales, Basidiomycota) on Caesalpiniaceae, Dipterocarpaceae and Myrtaceae in Seychelles. Mycol. Prog., 6(2): 101–107.

Thind, K.S. and Rattan, S.S. (1970). The thelephoraceae of India—III. Proc. Ind. Acad. Sci., 71(3): 118–131.

Tomšovský, M., Sedlák, P. and Jankovský, L. (2010). Species recognition and phylogenetic relationships of European *Porodaedalea* (Basidiomycota, Hymenochaetales). Mycol. Prog., 9(2): 225–233.

Tortić, M. (1987). Main Characters of the Mycoflora in Forests of *Pinus* peuce Griseb. Acta Bot. Croat., 46(1): 145–151.

Ţura, D., Zmitrovich, I.V., Wasser, S.P. and Nevo, E. (2008). Species diversity of heterobasidiomycetous and non-gilled hymenomycetous (Aphyllophorales s.l.) fungi in Israel. Isr. J. Pl. Sci., 56(4): 349–359.

Ţura, D., Zmitrovich, I.V., Wasser, S.P. and Nevo, E. (2010). Checklist of hymenomycetes (Aphyllophorales s.l.) and heterobasidiomycetes in Israel. Mycobiology, 38(4): 256–273.

Urbizu, M., Siqueiros, M.E., Abrego, N. and Salcedo, I. (2014). New records of aphyllophoroid fungi from Aguascalientes, Mexico and an approach to their ecological preferences. Rev. Mex. Biodiver., 85(4): 1007–1018.

van Zonneveld, M., Koskela, J., Vinceti, B. and Jarvis, A. (2009). Impact of climate change on the distribution of tropical pines in Southeast Asia. Unasylva, 60(231/232): 24–29.

Vasco Palacios, A. and Franco Molano, A. (2013). Diversity of Colombian macrofungi (Ascomycota - Basidiomycota). Mycotaxon, 121(1): 499–500.

Vasilyeva, G., Vavilova, V., Ustyantsev, K., Sukhikh, I., Blinov, A. et al. (2018). Genetic diversity of *Pinus sibirica, P. pumila* and their natural hybrids based on non-linked nuclear loci. Dendrobiology, 79: 168–173.

Viner, I. (2015). Polyporoid and corticioid Basidiomycetes in pristine forests of the Pechora-Ilych Nature Reserve, Komi Republic, Russia. Folia Cryptogamica Estonica, 52: 81–88.

Viner, I.A., Schigel, D.S. and Kotiranta, H. (2016). New occurrences of aphyllophoroid fungi (Agaricomycetes, Basidiomycota) in the Central Forest State Biosphere Nature Reserve, Tver Region, Russia. Folia Cryptogamica Estonica, 53: 81–91.

Vlasák, J., Kout, J., Vlasák, J. and Ryvarden, L. (2011). New records of polypores from southern Florida. Mycotaxon, 118(1): 159–176.

Volk, T.J., Burdsall, H.H. and Reynolds, K. (1994). Checklist and host index of wood-inhabiting fungi of Alaska. Mycotaxon, 52(1): 1–46.

Wade, T.G., Riitters, K., Wickham, J.D. and Jones, K.B. (2003). Distribution and causes of global forest fragmentation. Conserv. Ecol., 7(2): 7.

Wagner, T. and Fischer, M. (2001). Natural groups and a revised system for the European poroid Hymenochaetales (Basidiomycota) supported by nLSU rDNA sequence data. Mycol. Res., 105(7): 773–782.

Wagner, T. and Fischer, M. (2002). Proceedings towards a natural classification of the worldwide taxa *Phellinus* s.l. and *Inonotus* s.l. and phylogenetic relationships of allied genera. Mycologia, 94(6): 998–1016.

Wang, X.-Q. and Ran, J.-H. (2014). Evolution and biogeography of gymnosperms. Mol. Phyl. Evol., 75(1): 24–40.

Wang, X.-W., May, T.W., Liu, S.-L. and Zhou, L.-W. (2021). Towards a natural classification of *Hyphodontia* Sensu Lato and the trait evolution of Basidiocarps within Hymenochaetales (Basidiomycota). Journal of Fungi, 7: 478.

Wijayawardene, N. (2020). Outline of Fungi and fungus-like taxa. Mycosphere, 11(1): 1060–1456.

Wojewoda, W. (2003). Morphology of some rare and threatened Polish Basidiomycota. Acta Mycol., 38(1-2): 3–20.

Wright, J.E. and Wright, A.M. (2005). Checklist of the Mycobiota of Iguazú National Park (Misiones, Argentina). Boletin de la Sociedad Argentina de Botanica, 40(1-2): 1–22.

Wu, F., Dai, S.-J., Vlasák, J., Spirin, V. and Dai, Y.-C. (2019). Phylogeny and global diversity of *Porodaedalea*, a genus of gymnosperm pathogens in the Hymenochaetales. Mycologia, 111(1): 40–53.

Wu, F., Zhou, L.-W., Vlasák, J. and Dai, Y.-C. (2022). Global diversity and systematics of Hymenochaetaceae with poroid hymenophore. Fungal Diversity, 113(1): 1–192.

Yuan, H.-S., Lu, X., Dai, Y.-C., Hyde, K.D., Kan, Y.-H. et al. (2020). Fungal diversity notes 1277–1386: Taxonomic and phylogenetic contributions to fungal taxa. Fungal Diversity, 104(1): 1–266.

Yurchenko, E.O. (2003). Annotated list of non-poroid Aphyllophorales of Belarus. Mycotaxon, 86: 37–66.

Zervakis, G., Dimou, D. and Balis, C. (1998). A check-list of the greek macrofungi, including hosts and biogeographic distribution: I. Basidiomycotina. Mycotaxon, 66: 273–336.

Zervakis, G.I., Dimou, D.M., Polemis, E. and Karadelev, M. (2002). Mycodiversity studies in selected ecosystems of Greece: II. Macrofungi associated with conifers in the Taygetos Mountain (Peloponnese). Mycotaxon, 83: 97–126.

Zhou, L.-W., Nakasone, K.K., Burdsall, H.H., Ginns, J., Vlasák, J. et al. (2016). Polypore diversity in North America with an annotated checklist. Mycol. Prog., 15(7): 771–790.

Zhou, L.-W., Vlasák, J. and Dai, Y.-C. (2016). Taxonomy and phylogeny of *Phellinidium* (Hymenochaetales, Basidiomycota): A redefinition and the segregation of *Coniferiporia* gen. nov. for forest pathogens. Fungal Biol., 120(8): 988–1001.

Zhou, L.-W., Vlasák, J., Qin, W.-M. and Dai, Y.-C. (2016). Global diversity and phylogeny of the *Phellinus igniarius* complex (Hymenochaetales, Basidiomycota) with the description of five new species. Mycologia, 108(1): 192–204.

Zhou, M. and Wu, F. (2018). A new species of *Onnia* (Hymenochaetales, Basidiomycota) from Vietnam. Phytotaxa, 349(1): 73–78.

11

Occurrence and Adaptive Potential of Indoor Macrofungi

Dmitry Yu Vlasov,[1,*] *Nadezhda V Psurtseva,*[1] *Ivan V Zmitrovich,*[1] *Katerina V Sazanova,*[1] *Oleg N Ezhov*[2] and *Margarita A Bondartseva*[1]

1. Introduction

Among basidiomycetes that destroy timber, so-called indoor macrofungi (house fungi, domestic fungi or indoor wood-destroyers) attract the most attention. They are capable to wood damage in dwellings, non-residential constructions, mines and others (Schmidt, 2007). It has been estimated that the cost of repairing fungal damage to timber in building in 1977 amounted to £3 million/week in Britain (Rayner, Boddy, 1988). In many cases, the house fungi were found in historical buildings (Burova, 1986; Serova, 2015), which significantly impacted wood preservation. This group includes representatives of the genera *Amyloporia*, *Coniophora*, *Gloeophyllum*, *Neolentinus*, *Serpula*, *Tapinella* and others. Many species have received descriptive names that characterize the external signs of such fungi on a wood substrate: *Serpula lacrymans* (Lacrymose Domestic Dry-Rot Fungus), *Coniophora puteana* (Chaffy Domestic Dry-Rot Fungus), *Gloeophyllum sepiarium* (Palisade Dry-Rot Fungus), *Neolentinus lepideus* (Pile Dry-Rot Fungus), *Amyloporia sinuosa* (White

[1] V.L. Komarov Botanical Institute, Russian Academy of Sciences, 2 Professor Popov Street, Saint Petersburg, 197376, Russia.

[2] N. Laverov Federal Center for Integrated Arctic Research of the Ural Branch of the Russian Academy of Sciences, 23 North Dvina Enbankment, Arkhangelsk, 163000, Russia.

* Corresponding author: dmitry.vlasov@mail.ru

Domestic Dry-Rot Fungus) (Bondartsev, 1956). Such species as *Serpula lacrymans* is considered the most dangerous wood destroyer in Europe (Schmidt, 2007). In addition to true house fungi, many other xylotrophic basidiomycetes can settle on commercial wood outside the houses. It is also known that microscopic fungi can have a noticeable effect on the destruction of industrial wood, damaging the surface layer of the substrate (Serova, 2015). Thus, in the anthropogenic environment, fungal communities of complex composition and structure are often formed on wood structures, in which house fungi play the most prominent role.

The rate of destruction of commercial wood, proceeding as a brown-rot (destructive), is significant. According to Bondartsev (1956), species such as *Amyloporia sinuosa, Coniophora arida, Fibroporia vaillantii, Gloeophyllum sepiarium, Neolentinus lepideus, Serpula lacrymans, Tapinella panuoides* are capable of destroying the wood building elements in a year and a half under favorable conditions. The high rate of destructive processes is due to the activity of cellulolytic enzymes of house fungi released into the substrate (Ripachek, 1967; Rabinovich et al., 2001; Baldrian and Valaskova, 2008). When damaged by brown-rot basidiomycetes, the wood takes on a brown color, which is associated with an increase in the lignin contents. The wood becomes brittle, easily breaks and crumbles, cracks, disintegrates into prismatic fragments, and loses its volume and weight (Rabinovich et al., 2001).

Wood-destroying basidiomycetes have different mechanisms of wood decomposition associated with a set of enzymes and secondary metabolites that they are able to produce during their biosynthetic activity. It is known, for example, that *Serpula lacrymans* have the ability to affect wood, besides its enzyme complex, also with secondary phenol-type metabolites that accumulate in the substrate mycelium (Watkinson and Eastwood, 2012). According to many authors, wood-destroying basidiomycetes, especially brown rotters, are capable of producing a large amount of oxalic acid (Espejo and Agosin, 1991; Dutton et al., 1993; Munir et al., 2005; Makela et al., 2009; Barinova et al., 2010; Graz and Jarosz-Wilkołazka, 2011). In these fungi, the production of oxalic acid is most likely associated with its function of reducing iron Fe (III) to Fe (II) with subsequent participation of the latter in Fenton reactions leading to cellulose depolymerization.

The mechanism relies, at least in the incipient stage of decay, on the oxidative cleavage of glycosidic bonds in cellulose and hemicellulose and the oxidative modification and arrangement of lignin upon attack by highly destructive oxygen reactive species such as the hydroxyl radical generated non-enzymatically via Fenton chemistry ($Fe^{3+} + H_2O_2 \rightarrow Fe^{2+} + \cdot OH + {}^{-}OH$). Modifications in the lignocellulose macrocomponents associated with this non-enzymatic attack are believed to aid in the selective, near-complete removal of polysaccharides by an incomplete cellulase suite and without causing substantial lignin removal (Arantes et al., 2012).

According to some authors, the activity of basidiomycetous xylotrophs correlates with the content of low molecular weight substances in the mycelium (Watkinson and Eastwood, 2012). However, this issue remains insufficiently researched. Despite rather a long study of house fungi, the revealing of their new biological features is still of considerable interest. One of the key questions remains what determines the ability of these fungi to adapt to the anthropogenic environment and cause intensive destruction of commercial wood in various environmental conditions? It is obvious that studies of morphological and biochemical characteristics of key species strains can contribute to knowledge accumulation on the indoor macrofungi and their adaptive potential.

2. Biodiversity

On timber outside the buildings, we can find many species that also grow on fallen logs in the surrounding forests. Sometimes these species are called "indoor macrofungi" too, however, strictly speaking, this is not entirely true and a more adequate term here is "timber ware indoor macrofungi", since their increased occurrence is noted in timber warehouses with decorticated wood and boards (the most famous species are *Cylindrobasidium laeve* and *Phlebiopsis gigantea*). Actually, true indoor macrofungi develop inside year-round operated buildings where the temperature regime is stable throughout the year and there is no winter peak with negative temperatures, characteristic of buildings in the temperate and cold zones of the planet. The humidity of the air, in comparison with hygrophilous boreal forests and unexploited buildings, on the contrary, is lowered. Accordingly, the group of true indoor macrofungi is the result of the selection of species from the surrounding forests that are able to actively moisten the substrate in the absence of external sources of moisture and show a tendency to active exploratory growth when the mycelium quickly overgrows non-degradable substrate areas, and develops well above the zone of the boreal temperature optimum. Another important feature of indoor macrofungi is a weakened dependence on circadian rhythms and the ability to form more or less regularly functioning sporulation under daylight conditions. As we can see, there are many requirements for fungi to occupy this specific niche, so the set of true house fungal species is quite limited.

Considering most timber buildings and floors in stone buildings are made of coniferous wood, then indoor macrofungi are often adapted specifically to coniferous substrates, and given the moisture component, often to pine and larch wood, which can develop better in a natural xerophilic conditions. Since most conifericolous basidiomycetes cause brown-rot with decomposition (oxidation) of cellulose according to the Fenton mechanism, the key species of indoor macrofungi are brown-rot fungi. Since the anthropogenic conditions for indoor macrofungi development are unified on a global scale, geographically we observe among indoor macrofungi (1) "pine" cosmopolitan species, (2) species of the boreal ("spruce") belt of the planet, (3) initially subtropical species that have expanded into buildings, located in the temperate zones of the planet. Thus, *Serpula lacrymans* originally comes from South Asia, and its closest relatives from the genus *Austropaxillus* are common in South America. In different regions of the world, the species composition of indoor macrofungi somewhat varies, but its core remains unchanged, including such species as *Coniophora puteana*, *Hydnomerulius pinastri*, *Serpula lacrymans*, *Meruliporia pulverulenta*, *Tapinella panuoides* (Boletales), *Amyloporia xantha*, *A. sinuosa* (Polyporales), *Gloeophyllum sepiarium*, and *Neolentinus lepideus* (Gloeophyllales). Pictures of representative indoor macrofungi have been presented in Fig. 1.

Fig. 1. Key species of the indoor macrofungi: 1 – *Amyloporia xantha* (*a* – lateral form with nodulose pilei, *b* – young hymenophore, *c* – mature citrine-yellow hymenophore); 2 – *Coniophora puteana*; 3 – *Gloeophyllum sepiarium*; 4 – *Neolentinus lepideus* (*a* – fertile basidiome, *b* – sterile bodies f. ceratoides, *c* – affected wood); 5 – *Serpula himantioides*; 6 – *S. lacrymans*; 7 – *Tapinella panuoides*.

Each of these species is described in several descriptions of varying degrees of detail (Falck, 1912; Bondartsev, 1956; Cooke, 1956; Ginns, 1976, 1982; Ryvarden and Gilbertson, 1986; Pegler, 1983; Bondartseva et al., 2016; Vlasenko et al., 2017; Zmitrovich et al., 2019), but it would be helpful to present them in a standardized form and in the outline of the indoor macrofungi issue systematization. Because the quick and inexpensive identification of indoor macrofungi is possible by fruiting bodies morphology (basidiomata as a rule regularly appear indoors), below we provide morphological descriptions of the key species of indoor macrofungi.

Order Boletales E.-J. Gilbert, Les Livres du Mycologue Tome I–IV, Tom. III: Les Bolets: 83, 1931).

Agaricomycetes with perforated parentosomata, chiastic basidia and thick-walled (often Melzer's positive), cyanophilous basidiospores. Basidiomata monomitic, boletoid, agaricoid, pleurotoid, resupinate. Redox-active pigments secreted in many species, wood-destroying fungi produce brown-rot, ectomycorrhizal, wood and litter saprotrophic.

Family Coniophoraceae Ulbr., Krypt.-Fl. Anfäng. 1(3): 120, 1928.

Wood-inhabiting, athelioid or merulioid, hyphae simple-septate or nodose-septate, basidia utriculate, basidiospores indextrinoid or weakly dextrinoid.

Genus *Coniophora* DC. in DC. et Lamarck, Fl. Franç. Edn 3, 6: 34, 1815.

Basidiomata annual, totally effused, adhering the substrate, pellicular, with byssoid margin and subiculum; hymenophore as a concrete pellicular layer, even to warted-tuberculate, cinnamon or olivaceous-brown when mature; hyphal system monomitic, hyphae simple-septate or furnished by verticillate clamp connections; cystidia absent or present as cylindrical septocystidia protruding the hymenium; basidia utriculate, elongate, 2–4-spored; basidiospores ellipsoid to fusoid, with the smooth two-layered brownish wall, CB+, IKI–; on the wood of various conditions, causes a brown rot.

Type: *Coniophora membranacea* DC., Fl. Franç. Edn 3, 5/6: 34, 1815.

Species boreales: *C. arida* (Fr.) P. Karst., Not. Sällsk. Fauna et Fl. Fenn. Förh. 9: 370, 1868; *C. bimacrospora* Decock, Bitew et G. Castillo in Carlier, Bitew, Castillo et Decock, Cryptog. Mycol. 25(3): 264, 2004; *C. fusispora* (Cooke et Ellis) Cooke in Sacc., Syll. Fung. 6: 650, 1888; *C. olivacea* (Fr.) P. Karst., Hattsvampar: 162, 1879.

2.1 *Chaffy Domestic Dry-Rot Fungus*

Coniophora puteana (Schumach.) P. Karst., Not. Sällsk. Fauna et Fl. Fenn. Förh. 9: 370, 1868.

≡ *Thelephora puteana* Schumach., Enum. Pl. 2: 397, 1803.

= *Th. cerebella* Pers., Syn. Meth. Fung. 2: 580, 1801.

= *Fibrillaria ramosissima* Sowerby, Col. Fig. Engl. Fung. Mushr. 3(27): tab. 387, f. 2, 1803.

= *Coniophora membranacea* DC., Fl. Franç. 5/6: 34, 1815.

= *C. cellaris* Pers., Mycol. Eur. 1: 154, 1822.

= *C. cuticularis* Pers., Mycol. Eur. 1: 154, 1822.

= *Thelephora laxa* Fr., Elench. fung. 1: 196, 1828.

= *Th. luteocincta* Berk., J. Linn. Soc. Bot. 13: 168, 1873.

= *Corticium puteanum* var. *areolatum* Fr., Hymenomyc. Eur.: 658, 1874.

= *Coniophora lurida* P. Karst., Meddn Soc. Fauna Flora Fenn. 6: 12, 1881.

= *Tomentella brunnea* J. Schröt. in Cohn, Krypt.-Fl. Schlesien 3.1(25–32): 419, 1889.

= *Coniophora incrustans* Massee, J. Linn. Soc., Bot. 25(170): 132, 1889.

= *Merulius polychromus* Petch, Ann. R. Bot. Gdns Peradeniya 6(3): 204, 1917.

= *Coniophora incrustata* P.H.B. Talbot, Bothalia 7: 139, 1958.

= *C. piceae* Černý, Lesnictví 22: 122, 1976.

Basidiomata annual (or into the houses expanding several years), corticioid to merulioid, widely effused, loosely attached to the substrate, pellicular with byssoid margin and subiculum, thickening, of byssoid-fleshy consistency. Hymenophore as a concrete pellicular layer, even to warted-tuberculate, in some areas rugose and sub-merulioid, cream-rufous, then cinnamon or olivaceous-brown. Margin and subiculum white to cream. Sterile mycelium cords and films white to cinnamon, initially arachnoid, then solid, of tough consistency.

Hyphal system is monomitic in the basidiomata. Subicular hyphae 2–20 μm in diam., hyaline, thin- to prominently-walled, mostly simple-septate, also some with single, double or multiple clamps, sometimes strongly ampullate up to 20 μm. Subhymenial hyphae 2–8 μm in diam., regularly branching. Within hyphal cords, the differentiation into wessel hyphae and fibrohyphae is allocable. Wessel hyphae up to 10–35 μm wide, thick-walled, with inflated segments, hyaline to yellowish-brown. Fibrohyphae 2–4 μm wide, subsolid, hyaline. Cystidia absent, but harrow hyphoid sterile elements abundant among the basidia, often wavy or contorted, branched or lobed, thin-walled, 2–4 μm in diam, up to 80 μm long. Basidia 30–120 × 7–11 μm, narrowly clavate to utriform with a slightly swollen base, 4-spored, simple-septate at the base. Basidiospores 8–15 × 5–10 μm, narrowly to broadly ellipsoid, sometimes with a certain fusoid tendency, pale yellowish-brown in KOH, smooth, thick-walled with a shallow germ pore at the apical end, negative to weakly dextrinoid in Melzer's reagent, cyanophilous.

Associated with a cubical brown-rot. Cracks are quite frequent both in transverse and longitudinal directions. Most often, the rot is dry and associated with the woodworms of Anobiidae, but sometimes, especially in the central parts of the overlap, it is wet. The color varies from brown to dark rusty-brown, without green or red hues.

The species is especially common inside buildings. It affects partitions and ceilings, outside the buildings also sleepers, bridges, fences and piles, lashing timber, vegetable stores, cellars, etc. In nature, it occurs in autumn through the boreal and temperate zones on stumps and dead trunks of conifers, rarely deciduous (birch) trees.

Family Paxillaceae Lotsy, Vortr. Bot. Stammesgesch. 1: 706, 1907.

Wood-inhabiting, litter-inhabiting or ectomycorrhizal; lepistoid, phylloporoid, merulioid, or gasteroid; hyphae with oculate clamp connections, basidia clavate to pedunculate, basidiospores dextrinoid.

Genus *Hydnomerulius* Jarosch et Besl, Pl. Biol. 3(4): 447, 2001.

Basidiomata annual, resupinate, orbicular and adhering to the substrate, with byssoid cordonic margin and subiculum; within the subiculum and a sterile films the blackish sclerotia can be scattered; hymenophore as ceraceous hydnoid-merulioid field, greenish-citric to olivaceous when mature; hyphal system monomitic, hyphae with rare clamp connections; leptocystidia absent or of fusoid appearance; basidia clavate, 4-spored, with a basal clamp connection; basidiospores short cylindric, with the smooth one-layered prominent wall, CB+, weakly dextrinoid; on coniferous wood of various conditions or forest litter, cause a brown rot.

Type: *Hydnum pinastri* Fr., Novit. fl. svec. 2: 38, 1814. ≡ *Hydnomerulius pinastri* (Fr.) Jarosch et Besl, Pl. Biol. 3(3): 448, 2001. Monotypic genus.

2.2 *Mine Dry-Rot Fungus*

Hydnomerulius pinastri (Fr.) Jarosch et Besl, Pl. Biol. 3(3): 448, 2001.

≡ *Hydnum pinastri* Fr., Novit. Fl. Svec. 2: 38, 1814.

= *Merulius sclerotiorum* Falck, Hausschwamm-Forsch. 1: 93, 1907.

Basidiomata annual, merulioid, resupinate, orbicular and adhering to the substrate, with byssoid cordonic margin and subiculum, a separate patch reaches 15 cm in diam. Within whitish subiculum and sterile films, the blackish sclerotia can be scattered. Hymenophore as ceraceous hydnoid-merulioid field up to 5 mm thick, initially clay-cream, then greenish-citric to olivaceous.

Hyphal system is monomitic. Hyphae 2–10 μm wide, with clamps, sub-hymenial ones thin-walled, 2–5 μm wide, more or less parallelly arranged, mostly indistinct and immersed in a gelatinous matrix, subicular hyphae ones 4–10 μm wide, packed together forming hyphal strands, with sparse somewhat oculate clamps, thin- to slightly thick-walled. Cystidia are absent except for fusoid leptocystida-like elements. Basidia 20–30 × 5–7.5 μm, clavate-utriform, 4-spored, clamped at the base. Basidiospores 5–6.5 × 3.5–4.5 μm, ellipsoid or short-cylindrical, smooth, thick-walled (wall visually one-layered), yellowish to olivaceous-brown, weakly dextrinoid, cyanophilous.

Associated with a cubical brown rot. The rot is predominantly dry, with more pronounced longitudinal cracks.

The species occurs occasionally as a destructor of floors and load-bearing timber of the building basement and often affecting wood contact with the ground. Found also in greenhouses. In a nature, it is frequent in boreal zone on stumps, fallen logs and small coniferous debris.

Family Serpulaceae Jarosch et Bresinsky in Jarosch, Biblthca Mycol. 191: 90, 2001.

Wood-inhabiting, litter-inhabiting or ectomycorrhizal; clitocy-boid, merulioid, or gasteroid; hyphae with oculate clamp connections, basidia clavate to utriculate, basidiospores weakly dextrinoid to indextrinoid.

Genus *Meruliporia* Murrill, Mycologia 34(5): 596, 1942.

Basidiomata annual, pileate to resupinate with a dorsal attachment or totally effused, porioid, with cordonic and rhizomorphic margin and subiculum; hymenophore as ceraceous merulioid field to true poroid, cinnamon or ochraceous, almost black when old; hyphal system monomitic to dimitic in rhizomorphs; generative hyphae with oculate clamp connections, fibrohyphae subsolid, CB+; cystidia absent; basidia utriculate, 4-spored, with a basal clamp connection; basidiospores ellipsoid-ovoid, with 1–2-layered brownish wall, CB+, IKI–; on the wood of various states, causes a brown rot.

Type: *Merulius incrassatus* Berk. et M.A. Curtis, Hooker's J. Bot. Kew Gard. Misc. 1: 234, 1849.

2.3 American Domestic Dry-Rot Fungus

Meruliporia incrassata (Berk. et M.A. Curtis) Murrill, Mycologia 34(5): 596, 1942.

≡ *Merulius incrassatus* Berk. et M.A. Curtis, Hooker's J. Bot. Kew Gard. Misc. 1: 234, 1849.

Basidiomata annual, merulioid-porioid, resupinate, effused up to 15 cm diam. one patch, soft and easily separable when fresh, with white to pale rhizomorphic margin and rufous to black hymenophore. Hymenophore as strongly merulioid and appears porioid layer of fleshy-ceraceous consistency, cream to light buff or ochraceous at first, becoming darker grayish brown to blackish on drying; the pores irregular, 2–3 mm. Subiculum up to 5 mm thick, soft-fibrous, white to pale-buff.

Hyphal system monomitic in basidiomata, but dimitic in mycelium cordons. Subicular hyphae 3–9 μm in diam., hyaline, thin-walled, with abundant single clamps and some simple septa, tramal hyphae similar, 3–5 μm in diam.; wessel hyphae in cordons 7–20 μm in diam., thick-walled and inflated as sarcoskeletals; cordonic fibrohyphae 2–5 μm in diam., sub-solid, hyaline. Cystidia absent. Basidia 30–50 × 8–9.5 μm clavate-utriform, 4-spored, with a basal clamp. Basidiospores 9.5–14 × 5–8 μm, broadly to narrowly ellipsoid, some slightly curved, pale brown in KOH, smooth, thick-walled, inamyloid, cyanophilous.

Associated with brown-rot. The rot is usually cubic (the wood cracks into large polygons), dry to wet, and the color varies from light to dark brown.

The most common species of indoor macrofungi on the North American continent, occupying a niche of the Eurasian *Serpula lacrymans*. It affects load-bearing timber and ceilings and boardwalks. It actively spreads throughout the building due to rhizomorphs and films that have a characteristic pale shade. In nature, the fungus is common in North America on pine logs, capturing a warm belt. It usually enters Europe in ships with contaminated wood materials, but it does not get much distribution in the buildings of the Old World due to competitive restrictions.

2.4 Minor Domestic Dry-Rot Fungus

Meruliporia pulverulenta (Sowerby) Zmitr., Kalinovskaya et Myasnikov, Folia Cryptogamica Petropolitana 7: 15, 2019.

≡ *Auricularia pulverulenta* Sowerby, Col. Fig. Engl. Fung. Mushr. 2(17): tab. 214, 1799.

= *Merulius lacrymans* var. *minor* Rea, Brit. Basidiomyc.: 622, 1922.

= *M. minor* Falck, Hausschwamm-Forsch. 6: 53, 1912.

= *M. tignicola* Harmsen, Friesia 4(4–5): 245, 1953.

Basidiomata annual, merulioid, pileate to resupinate with a dorsal attachment or totally effused, up to 10 cm diam. one patch, with cordonic margin and subiculum. Hymenophore as fleshy-ceraceous merulioid field up to 0.7 cm thick, initially lemon yellow, then isabelline to ochraceous-brown. Margin white with citrine tints, then pale-cream.

Hyphal system is monomitic. Subicular hyphae 2.5–10 μm wide, with oculate clamp connections, tramal ones similar, up to 7 μm wide, gelatinized, wessel hyphae as sarcoskeletals up to 15 μm diam. Cystidia absent. Basidia 20–35 × 5–7.5 μm clavate-utriform, 4-spored, with a basal clamp. Basidiospores 5–7 × 3–5 μm, broadly to narrowly ellipsoid, golden-brown in KOH, smooth, thick-walled, inamyloid, cyanophilous.

Associated with brown-rot. The rot is dry, light brown or red, mostly longitudinally fissured.

The species is found mainly in ceilings above basements and in cellars; it is a rather dangerous wood destroyer, as it spreads through the building in the form of mycelium films. Mycelium films are white, with a characteristic lemon-yellow tint.

Genus *Serpula* (Pers.) Gray, Nat. Arr. Brit. Pl. 1: 637, 1821.

Basidiomata annual, pileate to resupinate with a dorsal attachment or totally effused, growing as large patches adhering the substrate or pelliculose films, with cordonic and rhizomorphic margin and subiculum; abhymenial layer a trichoderm tending to transformation into obscure cutis; hymenophore as sub-gelatinose merulioid field, hyaline with isabelline, rufous, ferrugineous, ochraceous tints, hymenophoral growth ageotropic; hyphal system monomitic to dimitic in rhizomorphs; generative hyphae with oculate clamp connections, fibrohyphae sub-solid, CB+; cystidia absent; basidia utriculate, 4-spored, with a basal clamp connection; basidiospores ellipsoid-ovoid, with 1–2-layered brownish wall, CB+, IKI–; on wood of various states, causes a brown rot.

Type: *Merulius destruens* Pers., Syn. Meth. Fung. 2: 496, 1801.

Species borealis: *Serpula himantioides* (Fr.) P. Karst., Meddn Soc. Fauna Flora fenn. 11: 137, 1884.

2.5 *Lacrymose Domestic Dry-Rot Fungus*

Serpula lacrymans (Wulfen) J. Schröt. in Cohn, Krypt.-Fl. Schlesien 3(25-32): 466, 1888.

≡ *Boletus lacrymans* Wulfen in Jacquin, Miscell. Austriac. 2: 111, 1781.

= *Merulius vastator* Tode, Abh. Naturforsch. Ges. Halle 1: 351, 1783.

= *Boletus obliquus* Bolton, Hist. Fung. Halifax 2: 74, tab. 74, 1788.

= *Merulius destruens* Pers., Syn. Meth. Fung. 2: 496, 1801.

= *Sistotrema cellare* Pers., Syn. Meth. Fung. 2: 554, 1801.

= *Xylophagus destruens* Link, Mag. Gesell. Naturf. Freunde 3: 38, 1809.

= *Merulius lacrymabundus* Link, Handbuch zur Erkennung der natuzbarsten und am häufigsten vorkommenden Gewächse 3: 289, 1833.

= *M. giganteus* Saut., Hedwigia 16: 72, 1877.

= *M. guillemotii* Boud. (ut '*guillemoti*'), Bull. Soc. Mycol. France 10(1): 63, 1894.

= *M. domesticus* Falck, Hausschwamm-Forsch. 6: 53, 1912.

= *M. terrestris* Burt, Ann. Mo. bot. Gdn 4: 346, 1917.

= *M. carbonarius* Lloyd, Mycol. Writ. 6(63): 963, 1920.

Basidiomata annual (but expanding into the houses several years), pileate to resupinate and form prostrate patches up to 50 cm diam., usually 5–20 mm thick. Abhymenial layer a trichoderm tending to transformation into obscure cutis, white, then pale-rufous to tan. Hymenophore merulioid, of fleshy-gelatinous consistency, yellowish to orange when young to isabelline or dark brown when mature, margin distinct, white, cottony, with rhizomorphs.

Hyphal system dimitic. Generative hyphae with oculate clamps, in sub-hymenium 2–4 μm wide, immersed in a gelatinous matrix; in subiculum 3–15 μm wide, thin-walled; wessel hyphae 10–35 μm wide, thick-walled as sarcoskeletals;

fibrohyphae 3–5 μm wide, sub-solid, hyaline. Cystidia present as hypoid leptocustidia up to 8 μm wide. Basidia 30–60 × 5–8 μm, clavate-utriform, 4-spored, with a basal clamp. Basidiospores are narrowly ellipsoid, 9–15 × 5–8 μm, smooth, thick-walled, yellowish-brown, indextrinoid, cyanophilous.

Associated with brown-rot. The rot is active, varies from dry to wet, tan to dark-brown, usually with the wood cracking into cubic polygons. Often the wood pieces are perforated by dirty white rhizomorphs of the fungus; the mycelium films with veins of rhizomorphs are also found.

The species occurs mainly in various ceilings and in more or less closed structures of residential or factory buildings as well as in vegetable stores, basements, greenhouses and warehouse buildings. In many cases, the fungus appears in new houses built from raw timber without well-functioning ventilation in the ceilings. The natural range of the species is confined to the Himalayas and other regions of temperate and southern Asia, however, its anthropogenic distribution captures all the natural zones of Eurasia.

2.6 *Wild Domestic Dry-Rot Fungus*

Serpula himantioides (Fr.) P. Karst., Meddn Soc. Fauna Flora Fenn. 11: 137, 1885.

≡ *Merulius himantioides* Fr., Observ. Mycol. 2: 238, 1818.

= *Boletus arboreus* Sowerby, Col. Fig. Engl. Fung. Mushr. 3(24): tab. 346, 1802.

= *Xylomyzon versicolor* Pers., Mycol. Eur. 2: 30, 1825.

= *Merulius papyraceus* Fr., Elench. Fung. 1: 61, 1828.

= *M. squalidus* Fr., Elench. Fung. 1: 62, 1828.

= *M. lacrymans* var. *tenuissimus* Berk. et Ravenel, N. Amer. Fung.: no. 134, 1875.

= *M. silvester* Falck in Møller, Hausschwamm-Forsch. 6: 53, 1912.

= *M. americanus* Burt, Ann. Mo. Bot. Gdn 4: 345, 1917.

= *M. gelatinosus* Lloyd, Mycol. Writ. 7(67): 1158, 1922.

= *Coniophora dimitiella* S.S. Rattan, Biblthca Mycol. 60: 75, 1977.

Basidiomata annual, merulioid, resupinate, effused or, when dry, with an inrolled margin, of membranaceous consistency, up to 2 mm thick and 15 cm diam. in one patch. Hymenophore at first more or less smooth hymenial field with prominent yellowish, then folded, merulioid, of fleshy-gelatinous consistency, yellow-brown, grayish-isabelline to dark brown. Margin distinctly white, fimbriate or rhizomorphic.

Hyphal system is monomitic in basidiomata. Generative hyphae, with large clamps, in the sub-hymenium 2–6 μm wide, thin-walled, densely packed, in the subiculum up to 10–12 μm wide, with ampullate clamps, swelling up; wessel hyphae as sarcoskeletals, up to 30 μm wide; fibrohyphae in the subiculum and rhizomorphs, 2–5 μm wide, sub-solid. Cystidia absent. Basidia 35–60 × 6–9 μm, clavate-utriform, 4-spored, clamped at the base. Basidiospores 8.5–13 × 5–6.5 μm, narrowly ellipsoid, smooth, thick-walled (double wall), yellowish to brownish, indextrinoid or only weakly dextrinoid, cyanophilous.

Associated with brown-rot. The rot is quite active, with moistening of the substrate. Cracks appear late, mostly longitudinal. Color brown to dark brown.

As a house fungus, it is most often found in basements on crowns in contact with the ground. It is also common in abandoned houses in the lower crowns, ceilings and mauerlats. In nature, it is quite often found all over the world on fallen coniferous trees. In the boreal zone, often on spruce, to the south – on pine.

Family Tapinellaceae C. Hahn, Sendtnera 6: 122, 1999.

Wood-inhabiting; pleurotoid, tyromycetoid or merulioid; hyphae with medallion or oculate clamp connections, basidia clavate sub-pedunculate, basidiospores dextrinoid.

Genus *Tapinella* E.-J. Gilbert, Les Livres du Mycologue Tome I–IV, Tom. III: Les Bolets: 67, 1931.

Basidiomata annual, pleurotoid, pileate to sub-resupinate with a dorsal attachment or differentiated into the pileus and strong hispid lateral stem; abhymenial surface a trichodermis transforming into the cutis; hymenophore lamellate with forked lamellae (lamellulae of 2–3 ranks) or merulioid fields near the base, creamish, citric-yellow, golden-yellow to olivaceous-brown when old; hyphal system sarcomonomitic; generative hyphae with oculate clamp connections; leptocystidia as pleurocystidia of fusoid or hyphoid appearance; basidia clavate, sub-pedunculate, 4-spored, with a basal clamp connection; basidiospores ellipsoid-cylindric, with a prominent golden wall, CB+, dextrinoid; on fallen logs and stumps of coniferous, rarely deciduous trees, causes a brown rot.

Type: *Agaricus panuoides* Fr., Observ. Mycol. 2: 227, 1818.

Species boreales: *Tapinella atrotomentosa* (Batsch) Šutara, Česká Mykol. 46(1–2): 50, 1992; *T. polychroa* (Singer) C. Hahn, Mycol. Bavarica 13: 64, 2012

2.7 Mine Wet-Rot Fungus, Gilled Domestic Fungus

Tapinella panuoides (Fr.) E.-J. Gilbert, Les Livres du Mycologue Tome I–IV, Tom. III: Les Bolets: 68, 1931.

≡ *Agaricus panuoides* Fr., Observ. Mycol. 2: 227, 1818.

= *Agaricus acheruntius* Humb., Fl. Friberg. Spec.: 73, 1793.

= *Paxillus fagi* Berk. et Broome, Ann. Mag. nat. Hist., Ser. 5 9: 181, 1882.

Basidiomata annual, pleurotoid – pileate, sessile or resupinate, dimidiate to flabelliform, up to 8 × 5 × 1.5 cm, solitary or confluent in imbricate clusters. Pileus upperside a trichoderm tending to transformation into obscure cutis, yellowish, clay-buff to pale cinnamon or olivaceous with greenish or violaceous tints, smooth, finely tomentose to strigose at the base. Stipe absent or obscurely expressed as a sterile basal (or, in pendent pilei, dorsal) knob. Hymenophore lamellate to sub-merulioid, as sub-ceraceous layer up to 7 mm thick; gills of 2–3 ranks, wavy, forked, anastomosing, edges entire, color yellow, apricot, pinkish-buff to clay-olivaceous. Context clay-buff, fleshy (spongious when dry), up to 1 cm thick, without prominent odor. Exploratory growth is carried out by greenish-white mycelium films, permeated with the white rhizomorph concretions.

Hyphal system is monomitic. Hyphae hyaline, 2–15 μm in diam., with inflated or medallion clamps; in sub-hymenium thin-walled, gelatinized; tramal ones thin- to thick-walled (CB+), inflated up to 15 μm in diam. Cordons consisted with inflated hyphae identical to tramal ones, but intermixed with uninflated clamped hyphae imitating fibrohyphae. Cystidia absent. Basidia 25–37 × 5–5.5 μm, clavate-utriform, 4-spored, with a basal clamp. Basidiospores 4–6 × 2.5–4.5 μm, ellipsoid, pale-yellowish in KOH, smooth, dextrinoid, cyanophilous.

Associated with brown-rot. Causes an active cubic (rather wet in the center) rot with a brownish-green color of the wood closer to the surface and a reddish-red color inside the rotting log. A very dangerous floor destructor.

It is found in the ceilings of buildings, often also in mines. In nature, it has a very wide distribution in the temperate zone, being associated with conifers, less often deciduous trees.

Order Polyporales Gäum., Vergl. Morph. Pilze: 503, 1926.

Agaricomycetes with perforated parentosomata, chiastic basidia and thin-walled, rarely thick-walled basidiospores, basidia as a rule up to 35 μm long. Basidiomata monomitic, dimitic or trimitic, lentinoid, polyporoid, tyromycetoid, merulioid, porioid, stereoid, corticioid. Mostly wood and litter saprotrophic.

Family Fomitopsidaceae Jülich, Biblthca Mycol. 85: 367, 1982.

Wood-inhabiting, tyromycetoid, trametoid to fomitoid, monomitic or dimitic, generative hyphae clamped, basidia clavate, 4-spored, basidiospores thin-walled, IKI–, CB+, cause brown-rot.

Genus *Amyloporia* Singer, Mycologia 36(1): 67, 1944.

Basidiomata annual to perennial, porioid; hymenophore poroid, one-two-layered; hyphal system dimitic; generative hyphae with clamp connections; skeletal hyphae fibrous, CB+, sometimes IKI+; leptocystidia of fusoid or hyphoid appearance; basidia clavate, 4-spored, with a basal clamp connection; basidiospores cylindric to allantoid, thin-walled, CB–, IKI–; on fallen logs and stumps of coniferous, rarely deciduous trees, causes a brown rot.

Type: *Polyporus vulgaris* var. *calceus* Fr., Syst. Mycol. 1: 381, 1821.

Examples: *Amyloporia subxantha* (Y.C. Dai et X.S. He) B.K. Cui et Y.C. Dai, Antonie van Leeuwenhoek 104(5): 825, 2013.

2.8 White Domestic Dry-Rot Fungus

Amyloporia sinuosa (Fr.) Rajchenb., Gorjón et Pildain, Aust. Syst. Bot. 24(2): 117, 2011.

≡ *Polyporus sinuosus* Fr., Syst. Mycol. 1: 381, 1821.

= *P. holoporus* Pers., Mycol. Eur. 2: 107, 1825.

= *Poria sinuosa* f. *ptychopora* Bourdot et Galzin, Bull. Trimest. Soc. Mycol. France 41(2): 232, 1925.

= *P. sinuosa* f. *vaporaria* Bourdot et Galzin, Bull., *ibid.*

Basidiomata annual, porioid, becoming effused up to 40 cm diam. × 0.8 cm thick one patch, adherent, margin abrupt to thinning out, then narrowly sterile up to 1.5 mm, cream to light buff. Hymenophore as a single tube layer up to 5 mm thick, fibrous; pores 1–5/1 mm, variable in shape and size, initially round, then angular, elongated or sinuose, sometimes dulited into irpicoid or raduloid plates, white-cream to buffy-brown, often with grayish tint (stays white indoor). Context tough-fibrous, white to cream. Mycelium films and rhizomorphs are white andabundant.

Hyphal system dimitic. Generative hyphae 2.5–3.5 µm wide, clamped, hyaline, thin-walled. Skeletal hyphae 2.5–4 µm wide, hyaline, thick-walled, non-septate, rarely branched; wessel hyphae present in rhizomorphs, up to 20 µm wide, clamped, with inflated segments. Cystidia as fusoid leptocystidial elements 9–20 × 4–5 µm. Basidia 9–17 × 4–5 µm clavate, 4-spored, with a basal clamp. Basidiospores 4.5–6 × 1.5–2.5 µm, cylindric to sub-allantoid, hyaline, smooth, inamyloid, acyanopilous.

Associated with brown-rot. The rot is quite moist, coarsely cracked and abundantly permeated with sterile mycelial formations of the fungus, at first rufous-cinnamon, then reddish-brown.

The species is found regularly on the floors of residential buildings, autonomous from external sources of moisture. In nature, it is common in the mountain and lowland forests of the Holarctic, where it prefers pine wood.

2.9 *Yellow* Amyloporia

Amyloporia xantha (Fr.) Bondartsev et Singer, Ann. Mycol. 39(1): 50, 1941.

≡ *Polyporus xanthus* Fr., Observ. Mycol. 1: 128, 1815.

= *Physisporus flavus* P. Karst., Sydvestra Finlands Polyporeer, Disp. Praes. Akademisk Afhandling: 40, 1859.

= *Polyporus selectus* P. Karst., Not. Sällsk. Fauna et Fl. Fenn. Förh. 9: 360, 1868.

= *P. sulphurellus* Peck, Ann. Rep. N.Y. St. Mus. 42: 123, 1889.

= *Poria greschikii* Bres., Ann. Mycol. 18(1/3): 38, 1920.

= *P. selecta* Rodway et Cleland, Pap. Proc. R. Soc. Tasm.: 15, 1930.

= *P. calcea* subf. *stratosa* Pilát, Bull. Trimest. Soc. Mycol. France 48(2): 180, 1932.

= *P. greschikii* var. *subiculosa* Pilát, Bull. Trimest. Soc. Mycol. France 48(1): 33, 1932.

= *Amyloporiella flava* A. David et Tortič, Trans. Br. Mycol. Soc. 83(4): 662, 1984.

= *Antrodia flava* Teixeira, Rev. Brasil. Botân. 15(2): 125, 1992.

Basidiomata annual to perennial into the houses, porioid, resupinate or forming numerous stalagmioid pseudopilei (f. *pachymeres* J. Erikss.), often widely effused (one patch up to 50 cm diam.), up to 10 mm thick, adnate, soft when fresh, crumbly and chalky when dry, bitter in taste, margin narrow and white. Hymenophore a single (rarely double) tube layer, surface initially white-cream, then citric to sulphurous, canary, or golden yellow, fading on drying and storing to almost pure white or pale cream, smooth when young, when older characteristically cracking into square pieces 5–15 mm long and wide, pores circular or sometimes elongated, 5–7/1 mm. Subiculum thin and white, chalky, whereas the tube layer pale yellowish-cream to white, up to 5 mm thick.

Hyphal system dimitic. Generative hyphae 2–5 µm wide, clamped, thin-walled, hyaline; skeletal hyphae predominant, 3–6 µm in diam., fibrous, straight to slightly sinuous, non-septate, unbranched to occasionally dichotomously branched, thick-walled, weakly amyloid. Cystidia as fusoid leptocystidial elements 10–14 × 3–4 μm. Basidia 10–17 × 4–6 μm, clavate, 4-spored, with a basal clamp. Basidiospores 4–5 × 1–1.5 μm, allantoid, hyaline, smooth, thin-walled, inamyloid, indextrinoid.

Associated with brown-rot. The rot is dry, usually inactive, but engulfs alarge volume of the wood, at first yellow, then light brownish, at first without cracks, then with large longitudinal cracks, and finally with numerous transverse cracks appearing. Usually in association with Anobiidae.

The species is regularly found in attic ceilings on the mauerlats and the ends of the rafters adjacent to them, from where it passes to the beams. Geographically, it is a cosmopolitan, found in the Southern Hemisphere mainly on pines, and in the Northern Hemisphere on a wide range of conifers, occasionally affecting the hardwoods.

Order Gloeophyllales Thorn in Hibbett et al., Mycol. Res. 111(5): 540, 2007.

Agaricomycetes with perforated parentosomata, chiastic basidia and thin-walled basidiospores, basidia as a rule exceed 35 μm long. Basidiomata dimitic with pseudocystidia, lentinoid, trametoid, or stereoid. Basidiospores thin-walled (sometimes pigmented), IKI–, CB+. Wood saprotrophic. Causes brown-rot.

Family Gloeophyllaceae Jülich, Biblthca Mycol. 85: 368, 1982.

Type family, see above.

2.10 *Palisade Dry-Rot Fungus*

Gloeophyllum sepiarium (Wulfen) P. Karst., Bidr. Känn. Finl. Nat. Folk 37: 79, 1882.

≡ *Agaricus sepiarius* Wulfen in Jacquin, Collnea Bot. 1(2): 339, 1787.

= *A. asserculorum* Batsch, Elench. Fung.: 95, 1783.

= *A. undulatus* Hoffm., Veg. Herc. Subterr. 2: 7, 1797.

= *A. boletiformis* Sowerby, Col. Fig. Engl. Fung. Mushr., Suppl. 30: tab. 418, 1814.

= *Lenzites argentinus* Speg. Anal. Mus. Nac. Hist. Nat. B. Aires 6: 114, 1898.

= *Daedalea ungulata* Lloyd, Mycol. Writ. 4(60): 15, 1915.

Basidiomata annual to perennial, trametoid (lenzitoid), pileate, broadly sessile, dimidiate or rozette-shaped, often imbricate in clusters with a common base, or fused laterally to, up to 10 cm wide, 12 cm long and 1 cm thick at the base, tough and flexible (coriaceous), margin initially bolster-like, then sharp and slightly wavy. Upperside at first pale or bright yellowish-brown, then rusty brown and finally grayish to black, when young and along the margin clearly tomentose, in age the hyphae agglutinate and the surface becomes tufted and hispid, scrupose with coarse protuberances, finally more or less smooth in zones mixed with narrow, more persistent hispid bands, narrowly to broadly zonate reflecting different stages of growth and thus, the zones from the margin to base in old specimens are often differently colored. Hymenophore false gilled (lenzitoid) with dense anastomosing lamellae up to 7 mm deep and 15–20/cm behind the margin, more rarely mixed with poroid areas with rounded to irregular, sinuous, radially elongated pores, about 0.5–2/mm; edges of lamellae light golden brown in active growth, later umber brown, side surface of lamellae ochraceous to pale brown, usually distinctly lighter than the context and trama. Context dark brown, denser next to the lamellae than towards the upper surface, up to 5 mm thick, black in KOH.

Hyphal system dimitic. Generative hyphae 2.5–4 μm in diam., clamped, thin- to thick-walled, hyaline. Skeletal hyphae predominate in the basidiome, up to 6 μm in diam., golden-brown, straight, thick-walled, non-septate, fibrous or simpodially branched. Pseudocystidia abundant in the hymenium, subulate to obtuse, thick-walled in age, some extremely elongated, not or only slightly projecting, 25–100 × 3–7.5 μm, usually smooth, more rarely with a small crown of crystals. Basidia narrowly clavate, 20–45 × 4.5–7 μm (some elongated up to 110 μm), 4-spored, with a basal clamp. Basidiospores 8–12 × 3–5.5 μm, cylindrical, hyaline, smooth, inamyloid, cyanophilous.

Associated with brown-rot. The rot is active, dry, with predominant longitudinal fissuring, at first yellowish, then brown.

It occurs on various types of timber wood (logs, poles, piles), including in non-residential buildings (sheds, fences, bridges) as well as residential buildings, where it is confined to attics and basements. In nature, it occurs all over the world mainly on coniferous wood, although in the boreal zone it is not uncommon to find it on stumps of hardwoods, especially in clearings, where this xerophilous species gains a competitive advantage.

2.11 *Pile Dry-Rot Fungus*

Neolentinus lepideus (Fr.) Redhead et Ginns, Trans. Mycol. Soc. Japan 26(3): 357, 1985.

≡ *Agaricus lepideus* Fr., Observ. Mycol. 1: 21, 1815.
= *A. tubaeformis* Schaeff., Fung. Bavar. Palat. Nasc. 4: 65, 1774.
= *A. cyprinus* Batsch, Elench. Fung.: 57, 1783.
= *A. serpentiformis* Batsch, Elench. Fung.: 89, 1783.
= *Ramaria ceratoides* Holmsk., Beata Ruris Otia Fungis Dan. 1: 101, 1799.
= *Agaricus suffrutescens* Brot., Fl. Lusit. 2: 466, 1805.
= *A. polymorphus* Pers., Mycol. Eur. 3: 52, 1828.
= *Lentinus sitaneus* Fr., Syn. Generis Lentinorum: 8, 1836.
= *L. cryptarum* Fuckel, Jb. Nassau. Ver. Naturk. 23–24: 15, 1870.
= *L. contiguus* Fr., Hymenomyc. Eur.: 482, 1874.
= *L. maximus* A.E. Johnson, Bull. Minn. Acad. Nat. Sci. 1: 338, 1878.
= *L. gallicus* Quél., C. R. Assoc. Franç. Avancem. Sci. 13: 280, 885.
= *L. domesticus* P. Karst., Revue Mycol. 9(33): 9, 1887.
= *L. magnus* Peck, Bull. Torrey Bot. Club 23(10): 413, 1896.
= *L. platensis* Speg., Anal. Mus. Nac. Hist. Nat. B. Aires 6: 113, 1899.
= *L. spretus* Peck, Bull. N.Y. St. Mus. 105: 24, 1906.

Basidiomata annual, lentinoid, differentiated into the convex, plane, or depressed pileus up to 18 cm diam. and central or eccentric stipe up to 2.5 cm diam. (length can vary from 3 to 20 cm), of tough consistency. Upperside glabrous, white-cream, cracking into numerous brownish-grayish scales. Hymenophore gilled, more or less decurrent; gills of 2–3 ranks, whitish to cream, often with citrine tints (become rusty to reddish brown in age), initially entire, then with saw-toothed edges. Stipe tough and thick, often with a rough ring, and the lower portions bear small brownish scales, initially whitish, then reddish brown. Context up to 1.2 cm thick, tough-fleshy, with a fragrant, sometimes anise-like odor.

Sterile bodies as clusters of 2–4 unbranched or apically-branched stipes devoid of pilei, appearing from dense mycelial mats as orthotropic finger-like formations. Individual sprouts reaching 20 cm long, 0.4–0.9 cm diam., flexuose, anisodiametric, but apically thinning out, in some cases one time branched at upper 1/3 (branches slightly curved). Surface initially sub-tomentose, then matted, near basal parts rusty brown to cinnamon, in apical parts pale to cream, covered with thin whitish pruina.

The hyphal system is dimitic with the presence of pseudoskeletal and skeletal hyphae. Generative hyphae 2.5–4 μm in diam., thin-walled, regularly branched. Pseudoskeletal hyphae 2–6 μm in diam., thick-walled, regularly branched, giving up near the surface of the clavate to utriform pseudocystidia 8–10 μm wide. Skeletal hyphae 8–9 μm in diam., unbranched, fibrous, thick-walled to sub-solid, predominate in superficial tissues. Pseudocystidia clavate to utriform, 30–50 × 8–13 μm wide, with the narrow lumen and brownish or bearing chrysescent content wall. Basidia 25–50 × 8–15 μm, clavate-cylindraceous, 4-spored, with a basal clamp. Basidiospores 9–15 × 3–5 μm, fusoid-cylindraceous, thin-walled, smooth, inamyloid, acyanophilous.

Associated with brown-rot. The rot is active, dry (cracks appear locally and go both in longitudinal and transverse directions, later they join), at first brown, then dark brown.

Like a previous species, *N. lepideus* is distributed in a wide range of residential buildings and non-residential structures, it is found on sleepers (including those impregnated with creosote), poles and inside buildings it is not uncommon in attics, where it destroys both ceilings and boardwalks. Common in clearings on stumps. In nature, it shows bipolar distribution tendencies, being especially common within the Holarctic. It occurs mainly in pine forests on pine stumps and deadwood, although it can also affect other conifers as well as deciduous species.

3. Techniques for Assessing the Physiological and Biosynthetic Activity

The materials for the experimental study of indoor macrofungi can be pure cultures isolated from fresh basidiomata of these fungi or strains maintained in culture collection.

The study of the cellulolytic activity of indoor macrofungi strains can be carried out in Petri dishes by two methods. In the first case, fungi are cultivated at a temperature of 25°C on a modified glucose-peptone agar medium (glucose, 10; peptone, 3; $MgSO_4$, 0.5; $CaCl_2$, 0.05; $ZnSO_4 \times 7\ H_2O$, 0.01; 0.6, K_2HPO_4 – 0.4, agar – 20 g/l), in which glucose was replaced by cellulose (10 g/l). Measurement of the cellulose dissolution zone is carried out weekly during the entire duration of the experiment (4 weeks). In the second case, a substrate consisting of microcrystalline cellulose (Chemapol: 1 g/100 ml distilled water) and agar (Difco: 1 g/100 ml distilled water) is used. Disks with the inoculum of indoor macrofungi are placed on the surface of the substrate. The study of activity was carried out at 1, 2, 3 and 4 weeks of growth of strains. The dissolution zone of cellulose around the discs is determined after 48 hours of incubation at 25°C.

An estimation of the activity of proteolytic enzymes of indoor macrofungi can be carried out on a gelatin substrate (4 g of edible gelatin is dissolved in 100 ml of water by bringing it to 90°C, cooled to 40°C and poured into Petri dishes). The manifestation of gelatinase activity (GA) is recorded in 2-week cultures by the characteristic formation of zones of softening of the medium around the mycelium disks after incubation at 25°C for 48 and 72 hours. The activity is assessed by the diameter of the formed zones (mm), and the degree of softening of the medium is also taken into account in the gelatin lysis zone (from "–" to "++").

Identification of the acidifying activity of indoor macrofungi can be carried out by their ability to dissolve calcium carbonate added to the medium. Previously, this method was successfully used to detect and quantify the release of acids by microfungi. The area of the dissolution zone is considered to be directly proportional to the number of acids released into the nutrient medium (Barinova et al., 2010). Determination of acidifying activity (AA) of fungi can be carried out on four agar nutrient media: BWA; GPA with a glucose content of 10 g/l); malts-extract agar (MEA, Oxoid); potato dextrose agar (PDA, Biokar Diagnostics). The 2% $CaCO_3$ solution of is added to each nutrient medium. The size of the zone of dissolution of calcium carbonate around the fungal colony (clarification of the medium) is determined as the difference between the diameter of the zone of clarification of the medium and the diameter of the colony. Activity measurements are carried out every week for a month of cultivation at a temperature of 25°C.

All experiments to assess the biochemical activity of strains of indoor macrofungi are recommended to be carried out in 3-fold biological replication. For statistical processing of measurement results, the Microsoft Excel program can be used for the calculation of average values and standard deviation.

For metabolomic analysis, the mycelium at the 3rd week of growth of strains, which is extracted with methanol, can be used. The resulting extract is evaporated at 40°C, the dry residue is dissolved in pyridine. Further, using N,O-bis-(trimethylsilyl) trifluoroacetamide (BSTFA) gets TMS(trimethylsilyl)-derivatives. The analysis is carried out by gas chromatography-mass spectrometry (GC-MS), for example, on an Agilent instrument with a 5975C mass-selective detector (USA), HP-5MS column.

4. Evaluation of Cellulolytic Activity of Key Species

The cellulolytic activity was studied for the following key species of indoor macrofungi: *Amyloporia xantha*, *Coniophora puteana*, *Gloeophyllum sepiarium*, *Neolentinus lepideus*, *Serpula lacrymans*, and *S. himantioides* (Kolker et al., 2018).

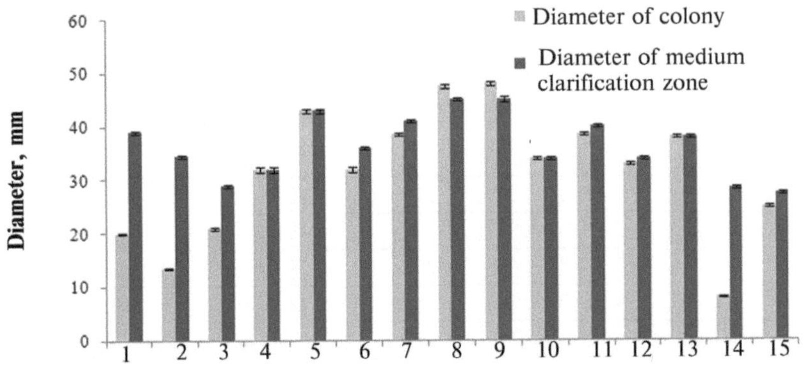

Fig. 2. Cellulolytic activity of indoor macrofungi strains from the LE-BIN collection (7th day of cultivation): 1 – *Amyloporia xantha* LE-BIN 1029; 2 – *Coniophora puteana* LE-BIN 001; 3 – *C. puteana* LE-BIN 006; 4 – *C. puteana* LE-BIN 1370; 5 – *Gloeophyllum sepiarium* LE-BIN 0158; 6 – *G. sepiarium* LE-BIN 2058; 7 – *G. sepiarium* LE-BIN 2059; 8 – *G. sepiarium* LE-BIN 3412; 9 – *G. sepiarium* LE-BIN 3667; 10 – *Neolentinus lepideus* LE-BIN 0525; 11 – *N. lepideus* LE-BIN 0848; 12 – *N. lepideus* LE-BIN 0963; 13 – *N. lepideus* LE-BIN 2278; 14 – *Serpula lacrymans* LE-BIN 1192; 15 – *S. himantioides* LE-BIN 1368.

The study was carried out using strains maintained in the Komarov Botanical Institute Basidiomycetes Culture Collection (LE-BIN). When evaluating the activity by the first method (during the growth of strains), the results on the 7th day of cultivation were the most informative (Fig. 2).

The largest zones of medium clarification (cellulose decomposition) were noted for *Coniophora puteana* (LE-BIN 001) and *Serpula lacrymans* (LE-BIN 1192), which are characterized by the lowest growth rate on all nutrient media. At the same time, for well-growing strains of *Gloeophyllum sepiarium* and *Neolentinus lepideus*, the diameter of the cellulose decomposition zone was close to or equal to the size of colonies, and in some cases turned out to be even smaller than the diameter of fungal colonies. Probably, this may be due both to the lower cellulolytic activity of these fungi under the conditions of a particular experiment, and to a higher growth rate of their colonies (in comparison with *Coniophora puteana* and *Serpula lacrymans*), which, in our opinion, makes it difficult to objectively assess the cellulolytic activity of these fungi. Such widely wild-growing species as *S. himantioides* (LE-BIN 1368), characterized by a high growth rate, showed significantly lower cellulolytic activity in comparison with the true house fungus *S. lacrymans* (LE-BIN 1192). The second method (application of a disk cut from a mycelium colony onto a cellulose substrate) turned out to be more applicable to basidiomycete strains with different growth rates. In this series of experiments, the cellulolytic activity was studied depending on the age of house fungus strains at 1, 2, 3, and 4 weeks of growth (Table 1).

It was found that for most of the studied strains, the highest cellulolytic activity appears on the 2nd growth week. With an increase in the age of the colony, a slight decrease in this type of activity was noted. For five strains of indoor macrofungi, the largest zone of substrate clarification was observed after 1 week of growth, and the tendency for an activity to decrease with culture age persisted. It is interesting to note that for *S. lacrymans* (LE-BIN 1192), due to the low growth rate, the determination

Table 1. The cellulolytic activity of different-aged cultures of indoor macrofungi.

Species	Strain number	The diameter of the zone of clarification of the medium around the mycelium discs (mm)			
		Colony age (days)			
		7	14	21	28
Amyloporia xantha	LE-BIN 1029	24.0 ± 0.1	22.0 ± 0.2	20.3 ± 0.2	20.4 ± 0.2
Coniophora puteana	LE-BIN 001	–	31.5 ± 0.3	31.0 ± 0.1	28.8 ± 0.2
	LE-BIN 006	26.0 ± 0.1	28.1 ± 0.2	26.7 ± 0.2	24.0 ± 0.1
	LE-BIN 1370	26.0 ± 0.1	22.5 ± 0.3	18.0 ± 0.1	18.3 ± 0.2
Gloeophyllum sepiarium	LE-BIN 0158	25.8 ± 0.2	27.0 ± 0.4	23.0 ± 0.3	22.5 ± 0.3
	LE-BIN 2058	25.0 ± 0.1	24.8 ± 0.2	22.8 ± 0.2	21.0 ± 0.1
	LE-BIN 2059	25.0 ± 0.1	26.5 ± 0.1	23.0 ± 0.1	21.2 ± 0.2
	LE-BIN 3412	27.0 ± 0.1	28.0 ± 0.1	24.0 ± 0.1	20.5 ± 0.1
	LE-BIN 3667	26.0 ± 0.1	23.0 ± 0.1	20.3 ± 0.2	17.8 ± 0.2
Neolentinus lepideus	LE-BIN 0525	21.5 ± 0.1	22.0 ± 0.1	20.5 ± 0.2	18.8 ± 0.2
	LE-BIN 0848	21.8 ± 0.2	22.5 ± 0.1	21.0 ± 0.1	20.8 ± 0.2
	LE-BIN 0963	22.0 ± 0.1	20.8 ± 0.2	20.8 ± 0.2	20.0 ± 0.3
	LE-BIN 2278	23.2 ± 0.2	25.5 ± 0.2	20.8 ± 0.2	19.5 ± 0.1
Serpula lacrymans	LE-BIN 1192	–	–	23.0 ± 0.1	23.3 ± 0.2
S. himantioides	LE-BIN 1368	23.0 ± 0.1	23.8 ± 0.2	22.2 ± 0.2	19.3 ± 0.2

Note: The diameter of the zone of clarification of the medium (mm) was measured for each strain after 48 h of incubation; "–", sowing was not carried out due to the low growth rate of colonies.

of cellulolytic activity was possible only from the 3rd week of growth. At the same time, the formation of a clearly visible zone of substrate clarification was noted, while in *S. himantioides* (LE-BIN 1368) the cellulolytic activity was more pronounced in the first two weeks. The most noticeable differences for strains of the same species were recorded in *Coniophora puteana*.

5. Evaluation of the Proteolytic Activity of Key Species

The results of experiments to assess the proteolytic activity (by gelatin) of collection strains of indoor macrofungi indicate that the higher gelatinase activity 72 hours after inoculation (Table 2) possessed all strains of *Coniophora puteana* as well as *Gloeophyllum sepiarium* (LE-BIN 2059, LE-BIN 3412) and *Serpula lacrymans*.

Comparing the *Serpula lacrymans* – *S. himantioides* pair, one can note a much higher proteolytic activity of *S. lacrymans*. Only two strains of *Neolentinus lepideus* (LE-BIN 0963 and LE-BIN 2278) showed practically no gelatinase. Thus, in most of the studied strains of indoor macrofungi, a manifestation of proteolytic activity was noted, which may be of additional importance in the development of indoor macrofungi in the anthropogenic environment.

Table 2. Proteolytic (gelatinase) activity of indoor macrofungi from the LE-BIN collection.

Species	strain number	Lysis zone density	Gelatinase activity (c.u.)	
			48 h	72 h
Antrodia xantha	LE-BIN 1029	±	14.2 ± 0.1	18.0 ± 0.2
Coniophora puteana	LE-BIN 001	+	15.8 ± 0.2	21.3 ± 0.2
	LE-BIN 006	+	19.5 ± 0.2	21.3 ± 0.2
	LE-BIN 1370	++	26.0 ± 0.1	28.0 ± 0.3
Gloeophyllum sepiarium	LE-BIN 0158	–	no	14.0 ± 0.1
	LE-BIN 2058	+	15.8 ± 0.2	19.0 ± 0.1
	LE-BIN 2059	+	22.0 ± 0.1	25.9 ± 0.1
	LE-BIN 3412	+	20.8 ± 0.2	23.5 ± 0.2
	LE-BIN 3667	+	15.8 ± 0	17.8 ± 0.2
Neolentinus lepideus	LE-BIN 0525	–	trace amounts	12.3 ± 0.2
	LE-BIN 0848	–	15.7 ± 0.2	18.4 ± 0.1
	LE-BIN 0963	–	no	no
	LE-BIN 2278	–	no	no
Serpula lacrymans	LE-BIN 1192	+	21.3 ± 0.2	23.2 ± 0.2
S. himantioides	LE-BIN 1368	+	20.5 ± 0.1	21.8 ± 0.2

Note: "++" is the maximum manifestation of activity (liquid consistency lysis zone), "+" is a noticeable manifestation of activity (softening of the medium in the lysis zone), "±" is a weak manifestation of activity (weak softening of the medium in the lysis zone), –, the density of the medium has not changed, "no", no activity.

6. Production of Organic Acids by Key Species

The results of assessing the total production of acids by strains of indoor macrofungi indicate that the highest acidifying activity is manifested under culture conditions in *Coniophora puteana*, *Serpula lacrymans*, and *Amyloporia xantha*, which are known as the most dangerous decomposers of commercial wood (Fig. 3).

Numerous bipyramidal-prismatic crystals of calcium oxalate were formed in the cultures of these fungi. Their appearance can be explained by the release of oxalic acid by fungi into the nutrient medium.

Species such as *Gloeophyllum sepiarium* and *Neolentinus lepideus* practically didn't release organic acids during their growth. A weak manifestation of total acid production was noted only in two strains of *Gloeophyllum sepiarium* (LE-BIN 0158 and LE-BIN 2058). Moreover, the formation of zones of clarification of the medium was observed in them only on the 6–8th day of cultivation, and the diameter of these zones turned out to be significantly smaller than the diameter of the colonies themselves. A similar situation was observed for strains of *Neolentinus lepideus*. For most of the studied strains of *Coniophora*, *Serpula*, and *Amyloporia*, the largest clarification zones were observed on commercial media MEA and PDA with the addition of calcium carbonate.

In general, the degree and dynamics of the acidifying activity of strains under culture conditions varied depending on the composition of the nutrient medium and

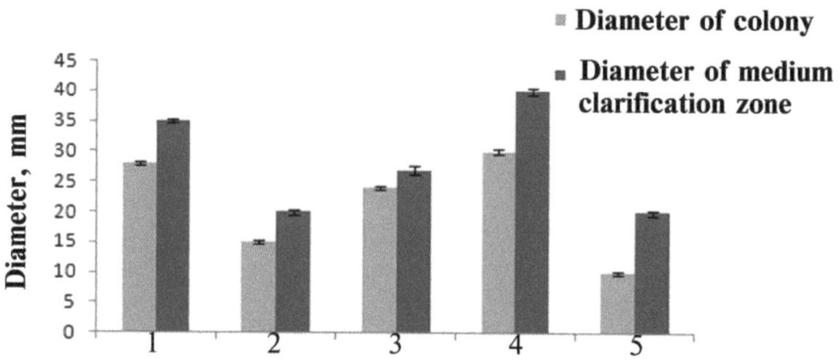

Fig. 3. Acid production by some strains of indoor macrofungi (6th day of cultivation): 1 – *Amyloporia xantha* LE-BIN 1029; 2 – *Coniophora puteana* LE-BIN 001; 3 – *C. puteana* LE-BIN 006; 4 – *C. puteana* LE-BIN 1370; 5 – *Serpula lacrymans* LE-BIN 1192.

the properties of a particular strain. It is interesting to note that strains of *Amyloporia xantha* (LE-BIN 1029), *Coniophora puteana* (LE-BIN 001), and *Serpula lacrymans* (LE-BIN 1192), which are characterized by high acidifying activity, the highest cellulolytic activity was also observed, which may indicate a positive correlation of these types of activities in the most aggressive species of indoor macrofungi.

7. Metabolomic Profiling of Key Species

As a result of metabolomic profiling, more than 100 different compounds were found in the mycelium of all collection strains of indoor macrofungi, among which about 60 were identified. The data obtained indicate a fairly high similarity of the metabolomic profiles of the studied strains in the qualitative composition of the identified compounds, although some specific metabolites were found in the mycelium of several species (Table 3).

Thus, cyclohexene-1-carboxylic acid and thiophene were found in *Coniophora puteana*. The mycelium of *Amyloporia xantha* contained 2-pyrrolidone-5-carboxylic acid and cyclic sugar alcohol pinitol, which was also found in the mycelium of *Serpula lacrymans* and *Gloeophyllum sepiarium*. The mycelium of *Serpula himantioides* contained the largest amount of organic acids, especially oxalate and citrate. The same species was characterized by high acidifying activity levels.

Tartaric acid was found only in the mycelium of *Neolentinus lepideus*. The mycelium of *N. lepideus* also contained ornithine and erythropenthol. *Gloeophyllum sepiarium* and *Amyloporia xantha* were distinguished by a high content of di- and trisaccharides compared to other species. According to the composition of unidentified compounds, the greatest differences were characteristic of the metabolomic profile of *Neolentinus lepideus* strains. Among unidentified substances in the mycelium of this species, more than 20 unique compounds were found that are not characteristic of other species. However, their content did not exceed trace amounts, which made it difficult to further identify and study their functional role.

When comparing the total metabolomic profile for different strains of the same species, significant differences were noted for the strain *N. lepideus* LE-BIN 2278 in comparison with other strains of this species. The metabolomic profile of this strain

Table 3. Review of the metabolomic profile of indoor macrofungi strains from the LE-BIN basidiomycete collection.

Compounds	Characteristic of all strains	Specific
Amino acids	glycine, valine, lysine, serine, threonine, phenylalanine, asparagine, leucine, proline, oxyproline, tyrosine, aspartic acid, glutamine	ornitine (*Neolentinus lepideus* LE-BIN 0525, LE-BIN 0848, LE-BIN 0963, LE-BIN 2278)
Carboxylic acids of the aliphatic series	pyruvic, hydroxypropionic, oxalic, amber, fumaric, threonic, malic, 2-ketoglutaric, citric, gluconic; fatty acids: oleic, stearic, linoleic, linolenic, palmitic	tartaric (*Neolentinus lepideus* LE-BIN 0525, LE-BIN 0848, LE-BIN 0963, LE-BIN 2278)
Polyols	glycerol, erythritol, hiro-inositol, mio-inositol, allo-inositol, arabitol, mannitol, sorbitol, pinitol	erythropentitol (*Neolentinus lepideus* LE-BIN 0525, LE-BIN 0848, LE-BIN 0963, LE-BIN 2278)
Cyclic compounds	benzoic, 4-hydroxybenzoic acid	2-pyrrolidone-5-carboxylic acid (*Amyloporia xantha* LE-BIN 1029), cyclohexene-1-carboxylic acid, thiophene (*Coniophora puteana* LE-BIN 001, LE-BIN 006, LE-BIN 1370), pinitol (*Amyloporia xantha* LE-BIN 1029, *Serpula lacrymans* LE-BIN 1192 and *Gloeophyllum sepiarium* LE-BIN 0158, LE-BIN 2058, LE-BIN 2059, LE-BIN 3412, LE-BIN 3667)
Sterols	ergosterol, sitosterol, lanosterol	–
Sugars	glucose, fructose, galactose, mannose, sucrose, trehalose	–
Other compounds	phosphate, glycerol-3-P, glycerol-2-P, uridine, methyl uridine, adenosine	–

was dominated by arabitol at relatively low concentrations of mono-, disaccharides and organic acids, while the strains of *N. lepideus* LE-BIN 0525, LE-BIN 0848, LE-BIN 0963 didn't have a clear predominance of arabitol. Strains LE-BIN 0525 and LE-BIN 0848 contained the highest amounts of glycerol-2-P and glycerol-3-P, compared with strains LE-BIN 0963 and LE-BIN 2278. The strain age correlated with the accumulation of glycerol-3-P in the mycelium (Fig. 4).

A similar pattern was observed for *Coniophora puteana* strains. Strains LE-BIN 001 and LE-BIN 006 accumulated more glycerol-2-P and glycerol-3-P compared to strain LE-BIN 1370. In addition, the mycelium of the *C. puteana* strain LE-BIN 001 contained significantly more disaccharides than strains LE-BIN 006 and LE-BIN 1370. The metabolomic profiles of the strains *Gloeophyllum sepiarium* LE-BIN 2058, LE-BIN 2059, LE-BIN 3667 were almost identical, and the profile of the strain *G. sepiarium* LE-BIN 0158 was significantly different. The mycelium of this strain was characterized by a very high content of glycerol-2-P and glycerol-3-P compared to the strains of this species LE-BIN 2058 and LE-BIN 2059 and a low content of arabitol compared to the same strains and LE-BIN 3667. It can be assumed

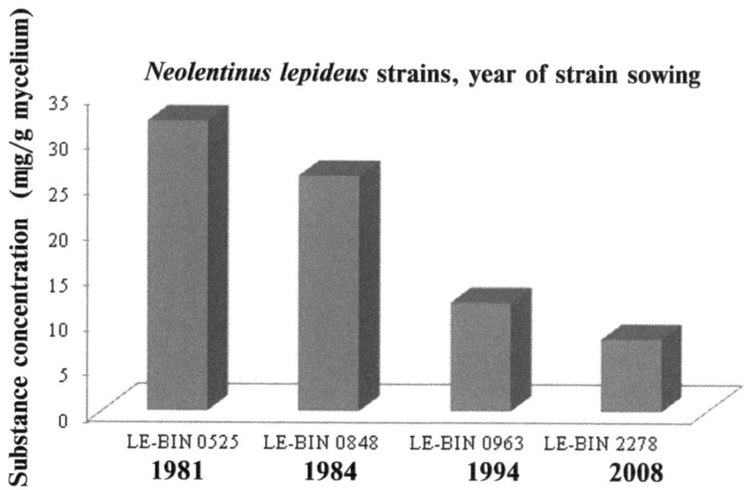

Fig. 4. Strain variability of *Neolentinus lepideus* in the contents of glycerol-3 phosphate.

that the accumulation of glycerol-2-P and glycerol-3-P (glycolysis intermediates) is characteristic of older strains with a lower growth rate, and may be associated with a slowdown in the work of respiratory enzymes. The revealed differences in the content of glycolysis products in different strains of the same species may be related to the age of the strains and due to changes in the metabolism of fungi during storage in the subculture.

The data of metabolomic profiling indicate that the strains of the studied species have certain differences in biochemical properties, however, these differences do not correlate with the manifestation of cellulolytic, gelatinase and acidifying activity of different strains. The patterns of biosynthesis of a number of metabolites associated with the age of the culture (the period of maintenance of strains in the collection fund) were noted, but the relationship between the biochemical properties of strains and their geographical origin could not be identified. The data obtained indicate the specificity of the studied types of activity for different species and strains of indoor macrofungi from the LE-BIN collection. Probably, the studied fungi can differ both in the degree of impact on wood and in the mechanisms involved in their trophic specialization. The greatest cellulolytic and acidifying activity is characteristic of three strains of *Amyloporia xantha* (LE-BIN 1029), *Coniophora puteana* (LE-BIN 001), and *Serpula lacrymans* (LE-BIN 1192), which allows us to recommend these strains for testing wood as well as protective compounds for resistance to the effects of indoor macrofungi.

8. Control Strategies

There has been a tremendous interest to explore and understanding deep physiological mechanisms of how wood-decaying fungi act, and to discover efficient methods against their damaging effects. However, as previously stated, prevention measures must be adopted against timber decaying fungi. Information on how frequently

Table 4. Moisture requirements of dry-rot and -rot (Viitanen and Paajanen, 1988).

Species	Temperature °C/(F)			Timber moisture content (%)		
	Minimum	Optimum	Maximum	Minimum	Optimum	Maximum
Coniophora puteana	0 to +5 (32–41)	20–25 (68–77)	40–46 (104–114.8)	15–25	15–25	60–80
Serpula lacrymans	–5 to +5 (23–41)	15–22 (59–71.6)	30–40 (86–104)	17–25	20–55	55–90

occurring wood-destroying fungi act on particular types of wood, storing measures, decay influencing factors, and solid wood decay control strategies are essential. Moreover, wood can be treated with a preservative that improves service life. Such preserved wood is used most often for outdoor wood products such as houses, bridges, fences, decks, railroad ties and so on. Some of the most important decay influencing factors are wood moisture and temperature (Table 4). Wood samples with low moisture content can be degraded and high temperatures as an alternative control measure do not kill mycelia (Schmidt, 2007).

In terms of control, various chemical fungicides were found to eradicate indoor wood-decaying fungi. Most of these chemicals are water soluble and their application to the timber implies spraying or injection methods. In the latter method, wood walls need to be drilled first and then injected with chemical fungicides. However, such methods were found to be more or less effective and depending on a complex of factors such as: wood size, the penetration rate of the chemical within wood, moisture levels in the wood, temperature and so on copper-based wood preservatives were commonly used for pressure treatment of wood used for building construction. Each copper-based preservative type proved to be effective against one or several species of decaying fungi, but in many cases, some fungi seemed to be tolerant to such chemicals. For example, in the study of Hastrup et al. (2005), 11 of 12 isolates of *Serpula lacrymans* were shown to be tolerant towards copper citrate. The report of Clausen and Yang (2007) suggests the use of biocides intended for indoor applications that must be non-toxic, non-volatile, odorless, hypoallergenic, and should be able to provide long-term protection under conditions of high humidity. Such biocides (borate-based multi-component biocide) were found to be effective against moulds, wood-decaying fungi (*Postia placenta*, *Gloeophyllum trabeum*, and *Trametes versicolor*) and termites. According to Ridout (2000), preservatives based on organic solvents perform well and do not have the disadvantages of some water-based systems. They do not appear to affect the strength characteristics of the timber, are non-deliquescent, do not cause metal corrosion and are not leached by water movement. Other preservative types are based on liquid petroleum gases.

In terms of biological control, species of *Trichoderma* were reported to eradicate the dry-rot fungus *Serpula lacrymans* (Score and Palfreyman, 1994), but its usage is limited because *Trichoderma* has a negative impact on human health. Limiting fungal spread in buildings without the use of toxic pesticides may be done also by preventing amino acid translocation through mycelial cords. According to Watkinson and Tlalka (2008), the amino acid analogue AIB (α-aminoisobutyric acid) competitively excludes utilizable amino acids from the mycelium, and from the nutrient supply network of mycelia cords that enables basidiomycetes fungi including *Serpula lacrymans*, *Coniophora puteana*, *Gloeophyllum trabeum*, *Neolentinus lepideus* to spread through buildings (Ţura et al., 2016).

However, the most effective is a complex of engineering measures that hinder the development of indoor macrofungi indoors. To reach this, it is necessary to plan such a construction of buildings that favors the best ventilation and drying and prevents soil and atmospheric moisture from entering from the outside. This is especially important if healthy, but raw wood got into the construction. As with any construction, and during repairs, it is necessary to ensure that there is always a well-ventilated air layer in the interstitial ceilings, which is fully achieved by using slotted baseboards (Bondartsev, 1956).

Ground moisturizing plays an important role in the development of indoor macrofungi. If the wood house is located on a flat plot, it needs contouring drainage with a depth of at least 2 m. If the house is located on a hill, but pressure water is coming close, this is also a "lost place", to which indoor macrofungi are constantly returning. The building in the boreal zone constructed according to the "Veps type" (with high sub-floors or on stilts) solves the problem of initiating a destructive process in the contact between wood with the moist ground. The contact of wood with the ground should be generally excluded due to the construction of stone and concrete basements with a well-established passive ventilation system (Zmitrovich, 2010–2021).

If a building has been damaged, then the start of work should also be associated with the elimination of the causes for the accumulation of external or internal moisture (ventilation, repair of the roof, walls, etc.). Secondly, radical removal of all affected and contorted wood from the affected wood with a margin of 50 cm along the wood fibers and 20 cm across the fibers with thorough drying is required hearth. Only after that, it makes sense to do wood antiseptization, although it is not always possible to carry out radical repairs without damaging the supporting structure of the building and sometimes it is necessary to completely rebuild the house. Wood residues affected by rot should be carefully removed from the building and burned (Bondartsieva and Parmasto, 1986).

Conclusion

1. The indoor macrofungi represent a clearly distinct group of xylotrophic basidiomycetes which originated as the result of certain anthropogenic-associated selection. This includes the species adapted to the temperature and humidity conditions of residential buildings.
2. A set of adaptations of various indoor macrofungi is rather wide and cannot be unified, however, the common to all indoor macrofungi are (1) the ability to active exploratory growth with the formation of mycelial films, cordons and rhizomorphs, (2) the destruction of wood by the Fenton mechanism, (3) the moistening of the substrate with metabolic moisture, and (4) the active acid production.

3. The indoor macrofungi should not to be confused with timber wareindoor macrofungi (the species number of the latter is wider), while the degree of their synanthropization is such that they continue to hold their niche in the wild, mainly in coniferous forests of the Northern and Southern hemispheres. However, in buildings, their occurrence is much higher, and some species, e.g., *Serpula lacrymans*, irradiate across the widest areas from a center of their speciation.
4. The core of the species composition of indoor macrofungi consists of such species as *Coniophora puteana*, *Hydnomerulius pinastri*, *Serpula lacrymans*, *S. himantioides*, *Meruliporia pulverulenta*, *Tapinella panuoides* (Boletales), *Amyloporia sinuosa*, *A. xantha* (Polyporales), *Gloeophyllum sepiarium*, and *Neolentinus lepideus* (Gloeophyllales).
5. Within a key species of indoor macrofungi, the strains of *Coniophora puteana*, *Serpula lacrymans*, and *Amyloporia xantha* are characterized by the greatest cellulolytic and acidifying activities.
6. The control strategies for indoor macrofungi include chemical and biological control measures, however the main condition for suppressing these organisms is engineering solutions aimed at eliminating the accumulation of condensate moisture inside the building as well as its intake from outside. An important area of work is the installation of a fungal-resistant basement and drainage of basement ground.
7. A further study of the biology of indoor macrofungi is an urgent task. An interesting little-studied issue of their biology is a violation of circadian rhythms which allows the development of normal sporulation in shaded conditions, as well as the phenomena of ageotropism accompanying the differentiation of aerial mycelium. In practical terms, it is important to assess the stress resistance of various species.

Acknowledgements

The work of D. Yu. Vlasov, N.V. Psurtseva, I.V. Zmitrovich, K.V. Sazanova, and M.A. Bondartseva was completed according to the state task No. 122011900033-4. The work of O.N. Ezhov was completed according to task No. 122011400384-2 (“Study of patterns of spatio-temporal changes in forest ecosystems in the subarctic territories of the European North of Russia”), topic number FUUW-2022-0057.

References

Arantes, V., Jellison, J. and Goodell, B. (2012). Peculiarities of brown-rot fungi and biochemical Fenton reaction with regard to their potential as a model for bioprocessing biomass. Appl. Microbiol. Biotech., 94: 323–338. 10.1007/s00253-012-3954-y.

Baldrian, P. and Valaškova, V. (2008). Degradation of cellulose by basidiomycetous fungi. FEMS Microbiol. Rev., 32: 501–521. 10.1111/j.1574-6976.2008.00106.x.

Barinova, K.V., Vlasov, D.Yu. and Schiparev, S.M. (2010). Organic acids of micromycetes – biodestructors. Ecological significance, metabolism, dependence on environmental factors. Lambert Academic Publishing, Saarbrücken (in Russian).

Bondartsev, A.S. (1956). Manual for the determination of indoor fungi. Izdatelstvo AN SSSR, Moscow, Leningrad (in Russian).

Bondartseva, M.A. and Parmasto, E. (1986). Definitorium Fungorum USSR. Aphyllophorales. Issue 1. Nauka, Leningrad (in Russian).

Bondartseva, M.A., Zmitrovich, I.V. and Zarudnaya, G.I. (2016). New combination for the sterile form of *Neolentinus lepideus* (Gloeophyllales, Agaricomycetes). Mikologiya i fitopatologiya, 50(3): 195–197.

Burova, L.G. (1986). Ecology of Fungi – Macromycetes. Nauka, Moscow (in Russian).

Clausen, C.A. and Yang, V. (2007). Protecting wood from mould, decay, and termites with multicomponent biocide systems. Int. Biodet. Biodeg., 59(1): 20–24. 10.1016/j.ibiod.2005.07.005.

Cooke, W.B. (1957). The genera *Serpula* and *Meruliporia*. Mycologia, 49(2): 197–225.

Dutton, M.V., Evans, C.S., Atkey, P.T. and Wood, D.A. (1993). Oxalate production by Basidiomycetes, including the white-rot species *Coriolus versicolor* and *Phanerochaete chrysosporium*. Appl. Microbiol. Biotechnol., 39: 5–10. 10.1007/BF00166839.

Espejo, E. and Agosin, E. (2011). Production and degradation of oxalic acid by brown rot fungi. Appl. Environm. Microbiol., 57(7): 1980–1986. 10.1128/aem.57.7.1980-1986.1991.

Falck, R. (1912). Die Merliusfäule der Baumholzes. *In*: Möller, A. (ed.). Hausschwammforschungen. 6. Jena, p. 403.

Gilbertson, R.L. and Ryvarden, L. (1986). North American polypores. Vol. 1. Lubrecht and Cramer, Port Jervis, p. 431.

Gilbertson, R.L. and Ryvarden, L. (1987). North American polypores. Vol. 2. Lubrecht and Cramer, Port Jervis, p. 886.

Ginns, J. (1978). *Leucogyrophana* (Aphyllophorales): Identification of species. Can. J. Bot., 56(16): 1953–1973.

Ginns, J. (1982). A monograph of the genus *Coniophora* (Aphyllophorales, Basidiomycetes). Opera Bot., 61: 1–61.

Graz, M. and Jarosz-Wilkołazka, A. (2011). Oxalic acid, versatile peroxidase secretion and chelating ability of *Bjerkandera fumosa* in rich and limited culture conditions. J. Microbiol. Biotechnol., 27: 1885–91. 10.1007/s11274-010-0647-5.

Hastrup, A.C.S., Green, F.I., Clausen, C.C. and Jensen, B. (2005). Tolerance of *Serpula lacrymans* to copper-wood preservatives. Int. Biodet. and Biodeg., 56(3): 173–77.

Kolker, T.L., Psurtseva, N.V., Sazanova, K.V. and Vlasov, D.Yu. (2018). Biochemical features of house fungi maintained in the LE-BIN collection of basidiomycete cultures. Mikologiya i fitopatologiya, 52(6): 398–407.

Makela, M.R., Hilden, K., Hatakka, A. and Lundell, T.K. (2009). Oxalate decarboxylase of the white-rot fungus *Dichomitus squalens* demonstrates a novel enzyme primary structure and non-induced expression on wood and in liquid cultures. Microbiology, 155: 2726–38. 10.1099/mic.0.028860-0.

Munir, E., Hattori, T. and Shimada, M. (2005). Role for oxalate acid biosynthesis in growth of copper tolerant wood-rotting and pathogenic fungi under environmental stress. 55th Meeting of the Japan Wood Research Society. Tokyo.

Pegler, D.N. (1983). The genus *Lentinus*: A world monograph. Kew Bulletin additional series X. Her Majesty's stationary office, London, p. 281.

Rabinovich, M.L., Bolobova, A.V. and Kondrashchenko, V.I. (2001). Theoretical foundations of biotechnology of wood composites. Book 1. Wood and its destroying fungi. Nauka, Moscow (in Russian).

Rayner, A.D.M. and Boddy, L. (1988). Fungal Decomposition of Wood. John Wiley, Chichester, p. 587.

Ridout, B. (2000). Timber Decay in Buildings: The Conservation Approach to Treatment. Taylor and Francis, p. 260.

Ripachek, V. (1967). Biology of Wood-destroying Fungi. Publishing House of the Forest Industry, Moscow (in Russian).

Schmidt, O. (2007). Indoor wood-decay basidiomycetes: Damage, causal fungi, physiology, identification and characterization, prevention and control. Mycol. Progr., 6: 261–279. 10.1007/s11557-007-0534-0.

Score, A.J. and Palfreyman, J.W. (1994). Biological control of the dry rot fungus *Serpula lacrymans* by *Trichoderma* species. Int. Biodet. Biodeg., 33(2): 115–128. 10.1016/0964-8305(94)90031-0.

Serova, T.A. (2015). Mycobiota of wooden constructions of historical buildings of St. Petersburg. PhD Thesis. All-Russian Institute of Plant Protection, St Petersburg (in Russian).

Ţura, D., Zmitrovich, I.V. and Wasser, S.P. (2016). Wood-inhabiting fungi: Applied aspects. *In*: Fungi: Applications and Management Strategies. CRC Press, London.

Viitanen, H. and Paajanen, L. (1988). The critical moisture and temperature conditions for the growth of some mould fungi and the brown-rot fungus *Coniophora puteana* on wood. International Research Group on Wood Preservation, Stockholm. Document # IRG/WP 1369.

Vlasenko, V.A., Vlasenko, A.V. and Zmitrovich, I.V. (2017). First record of *Neolentinus lepideus* f. *ceratoides* (Gloeophyllales, Basidiomycota) in Novosibirsk Region. Current Res. Environm. Appl. Mycol., 7(3): 187–192. 10.5943/CREAM/7/3/5.

Watkinson, S.C. and Eastwood, D.C. (2012). *Serpula lacrymans*, wood and buildings. Adv. Appl. Microbiol., 78: 121–149. 10.1016/B978-0-12-394805-2.00005-1.

Zmitrovich, I.V. (2010–2021). Reports on Wood State. Komarov Botanical Institute, St. Petersburg.

Zmitrovich, I.V., Kalinovskaya, N.I. and Myasnikov, A.G. (2019). Funga photographica. Boletales I: Coniophoraceae, Hygrophoropsidaceae, Paxillaceae, Serpulaceae, Tapinellaceae boreales. Folia Cryptogamica Petropolitana, 7: 1–60.

Polymorphism in Macrofungi

12

Genetic and Morphological Polymorphism in Wood-Decaying Fungi

Badalyan SM,[1,*] *Vladykina VD*,[2] *Zhuykova EV*,[3] *Diyarova DK*[3] and *Mukhin VA*[2,3,*]

1. Introduction

Xylotrophic or wood-decaying fungi (phylum Basidiomycota, class Agaricomycetes) represent an ecologically significant group of organisms that play an important role in the carbon cycle of forest ecosystems. As the only known group of organisms capable of biochemical conversion of the lignocellulose complex of wood, they determine the main parameters of biological decomposition of wood debris and CO_2 and O_2 exchange. This makes wood-decaying fungi significant as regulators of the atmospheric gas composition, the change in which as a result of CO_2 accumulation is one of the main causes of climate change (Mukhin et al., 2021). According to available expert, the biodiversity of xylotrophic Agaricomycetes in Eurasia and North America is estimated around 1500 species (Mukhin, 1978; Stepanova and Mukhin, 1979; Gilbertson, 1980). Among these fungi, one of the most important groups belongs

[1] Laboratory of Fungal Biology and Biotechnology, Institute of Biology, Yerevan State University, 1 A. Manoogian St, 0025 Yerevan, Armenia.
[2] Department of Biodiversity and Bioecology, Institute of Natural Sciences and Mathematics, Ural Federal University named after the first President of Russia B.N. Yeltsin, 19 Mira Street, 620003 Ekaterinburg, Russia.
[3] Institute of Plant and Animal Ecology, Ural Branch of Russian Academy of Sciences, 202, 8 Marta Street, 620144, Ekaterinburg, Russia.
* Corresponding authors: s.badalyan@ysu.am/victor.mukhin@ipae.uran.ru

to polypore fungi (order Polyporales) with tubular, less often, lamellar or daedaloid hymenophore. The polypores are widespread organisms – about 200 species of 552 known in Europe and North America (38%) are found in both continents (Gilbertson and Ryvarden, 1986, 1987; Ryvarden and Gilbertson, 1993, 1994).

Among all existing concepts of biological species (biological, phylogenetic, ecological, polythetic) morphological concept is predominated in mycology (Taylor et al., 2006; Kirk et al., 2008; Stengel et al., 2022). One of its central elements is the position on the type (reference specimen) or holotype, in relation to which other specimens of a given species are compared by morphological features. According to Mayr (1970), the species described according to this concept are monotypic morphological species existing outside of time and space – the same any time in any geographical point of distribution. The vast majority of existing species taxa of fungi (about 70,000) is monotypic morphological species (Taylor et al., 2006). It is partly explained by the fact, that the simple morphology of basidiocarps and a limited number of morphological characters make the study of intraspecific variability difficult (Júdová et al., 2012). The monotypic type of species contradicts to the wide geographic distribution of wood-decaying fungi. A wide geographic distribution implies geographic variability in environmental conditions, the inevitable consequence of which is geographic intraspecific variability of fungi because each species must correspond to the environmental conditions where it grows.

The situation changed due to wide use of molecular genetic methods in taxonomy and ecology of fungi expanding the study of intraspecific variability of these organisms. The most popular locus in sequence-based mycological research is the internal transcribed spacer (ITS) of the nuclear ribosomal RNA repeat unit (Horton and Bruns, 2001; Bridge et al., 2005; Kõljalg et al., 2013). This is the basis of an ITS-based taxon delimitation method, according to which the similarity level 97–99% of sequences is allowed to assume "species hypothesis", that they are belong to the one species (Nilsson et al., 2008; Kõljalg et al., 2013; Taylor and Hibbett, 2014). However, it has been shown that the ITS region has intraspecific and interspecific variability. For the phylum Basidiomycota, intraspecific variability varies from 0 to 17.3%, and averages 3.33% (Nilsson et al., 2008). The presence of intraspecific variability is a sign of polytypic species consisting of many distinct populations (Mayr, 1970).

Indeed, using the ITS-based taxon delimitation method, it has been shown that many monotypic morphological species are heterogeneous, consisting of phylogenetic groups comparable in their characteristics to biological and phylogenetic species (Rogers et al., 1999; Kirk et al., 2008; Júdová et al., 2012; Pristaš et al., 2013; Grienke et al., 2014; Dresch et al., 2015; Gáper et al., 2016; Mukhin et al., 2018; Peintner et al., 2019; Náplavova et al., 2020; Badalyan et al., 2022). Previous studies have shown that closely related morphological species may not have differences in the species level in the ITS region, and they are morphologically distinct parts of one phylogenetic species (Ko and Jung, 1999; Bernicchia et al., 2005; Koukol et al., 2014; Mentrida et al., 2015; Galović et al., 2018; Mukhin et al., 2020). This probably indicates the beginning of the transition from the monotypic morphological concept of species to the biological polytypic one.

This process was initiated not only based on recent achievements in fungal phylogenetic taxonomy, but also by practical usage of fungi in modern biotechnology and biomedicine. It is known that many species, including polypores discussed in the current chapter, are producers of valuable bioactive metabolites. However, for their practical usage, it is necessary to know the boundaries of producer species, how homogeneous/heterogeneous they are in genetic, morphological and ecological terms, and how it is related to their chemical composition (Dresch et al., 2015; Gáper et al., 2016; Kües and Badalyan, 2017; Badalyan et al., 2019; Peintner et al., 2019). Therefore, the study of monotypic/polytypic character of fungal species, with all evidence, is not only a fundamental, but also a practical problem. The transition from a morphological understanding of species to a biologically defined polytypic species occurred gradually with respect to different groups of organisms (Mayr, 1970). Regarding fungi, it is only at the very early stage. However, we suppose that available data indicates the existence of two types of polytypic species in wood-decaying fungi consisting of (a) genetically different, but morphologically indistinguishable (cryptic) phylogenetic lineages and (b) phenotypically distinct groups. The first one is presented by monomorphic cryptic polytypic species, whereas the second by polymorphic polytypic species.

2. Monomorphic Cryptic Polytypic Species

The representative of this group of polypore fungi is *Fomes fomentarius* (L.) Fr. *sensu lato* or tinder fungus, one of the most common bracket species in the forests of Eurasia and North America (Gilbertson and Ryvarden, 1986; Ryvarden and Gilbertson, 1993). In Eurasia, it is predominantly found on remains of *Betula* and *Fagus* trees (Bondartsev, 1953; Júlich, 1984; Ryvarden and Gilbertson, 1993; Mukhin, 1993; Bondartseva, 1998), whereas in North America - on *Betula* and *Alnus* trees (Gilbertson and Ryvarden, 1986; Farr et al., 1989; McCormick et. al., 2013; Gáper and Gáperová, 2014). Its wide distribution and abundance make it one of the destructors of deciduous tree debris and, accordingly, an important biotic factor in gas exchange (Mukhin et al., 2021).

Tinder fungus is not only an ecologically important fungus in the forest ecosystems, but it is known as a producer of bioactive compounds with medicinal properties (Grienke et al., 2014; Gáper et al., 2016; Badalyan and Shahbazyan, 2015; Badalyan and Gharibyan, 2016, 2017; Badalyan and Borhani, 2019; Badalyan et al., 2019; 2021). The main obstacle associated with the use of this fungus in biotechnology and biomedicine is insufficient knowledge of its geographical and ecological variability, phylogenetic diversity and limitation of species boundaries (Grienke et al., 2014; Dresch et al., 2015; Gáper et al., 2016; Peintner et al., 2019; Badalyan and Zambonelli, 2023).

This fungus is currently regarded as a morphologically homogeneous, monotypic species, the morphological features of which do not show pronounced geographic and ecological variability (Gáperová et al., 2016). However, morphological forms and varieties of this species have been previously described according to *Index Fungorum* database. The results of ITS rDNA analysis showed that *F. fomentarius s. lat.* consists of two genotypes A and B, of which genotype A is predominantly found in fungi growing on *Fagus* tree, while the genotype B – in fungi growing on other substrates (Júdová et al., 2012). Usually, fungi of different genotypes do not differ morphologically (Gáperová et al., 2016) and, according to Júdová et al. (2012), *F. fomentarius s. lat.* consists of at least two sympatric cryptic species.

This point of view was further developed by Pristaš et al. (2013) and Gáper et al. (2016). They confirmed the existence of two genotypes or phylogenetic lineages using molecular markers other than ITS (*efa*, LSU) and showed that the genotype A or phylogenetic lineage A has two A1 and A2 sublineages. The first is found in North America, while the second – in Europe. Phylogenetic lineage B is characteristic for Southern Europe (Greece, Spain, Italy, Portugal, Slovakia, Czech Republic and France), but it is also noted in Iran and China (Gáper et al., 2016).

The study carried out in the Asian part of Russia revealed that in this area lineages A and B are both presented (Mukhin et al., 2018). They are identified by 2 indels and heterozygosity at 8 positions in ITS sequences (Fig. 1).

Sequence	ITS1.1	11 bp	ITS1.2	254 bp	ITS2.1	10 bp	ITS2.2	6 bp	ITS2.3	5 bp	ITS2.4	8 bp	ITS2.5	19 bp	ITS2.6	Sublineage
HM584810	GG--TT		GCCTCGC		ACCTT		TAGCGTTGGA-TGTT		TTTTTGC		C--------AGT		CTTAA		TGTGG	A1
EF155492	GG--TT		GCCTCGC		ACCTT		TAGCGTTGGA-TGTT		TTTT-GC		C--------AGT		CTTAA		TGTGG	A2
MF563974	GG--TT		GCCTCGC		ACCTT		TAGCGTTGGA-TGTT		TTTT-GC		C--------AGT		CTTAA		TGTGG	A2
KJ668550	GGGATT		GCTTTGC		ACCTT		TAGGGTTGGCTTATT		TTTT-GC		CTCGTTTGAGT		CTCAA		TGCGG	B1
FJ865438	GGGATT		GCTTTGC		ACTTT		TAGGGTTGGC-TATT		TTTT-GC		CTCGTTTGAGT		CTCAA		TGCGG	B2
MF563985	GGGATT		GCTTTGC		ACTTT		TAGGGTTGGC-TATT		TTTT-GC		CTCGTTTGAGT		CTCAA		TGCGG	B2

Fig. 1. The alignment of internal transcribed spacer (ITS) sequences of *Fomes fomentarius s. lat.* Shading denotes the variable positions used to determine genotypes.

The ratio of lineage A to lineage B is 2:1 in this area. All isolates of lineage A from the Asian part of Russia belong to sublineage A2, which differs from sublineage A1 by a single nucleotide insertion at ITS2.3. The sublineage A2 is distributed across the Asian part of Russia: the Southern Urals, Altai, Western Sayan, and Baikal regions (Fig. 2A).

The analysis of data showed that lineage B, like lineage A, also possesses 2 sublineages, B1 and B2 which differ by nucleotide sequences in ITS2.1 (Fig. 1). The sublineage B1 represents a small group of samples obtained from Iran, China, Nepal, and South Korea. Their ITS sequences corresponded to lineage B at 9 positions, but the ACCTT sequence in ITS1.2 is characteristic of genotype A. A major proportion of lineage B isolates has ITS sequences typical of genotype B and was designated as sublineage B2 (Fig. 2B). It is represented by samples mainly collected in Europe (Great Britain, Italy, Latvia, Slovakia, and Slovenia). All B lineage isolates found in the Asian part of Russia belong to sublineage B2 which was found only in the Southern Urals – an eastern border between European and Asian subcontinents.

In the Asian part of Russia, sublineages A2 and B2 of *F. fomentarius s. lat.* differ not only by their distribution, but also by substrate characteristics. The fungi of sublineage A2 mainly grow on *Betula* trees, less frequently on *Alnus*, sporadic on *Larix* trees. The spectrum of substrates on which sublineage B2 grows includes *Acer*, *Alnus*, *Prunus*, and *Salix* trees, except *Betula* species. According to GenBank data, the fungi of sublineage A2 have been collected in Europe from *Alnus*, *Quercus*,

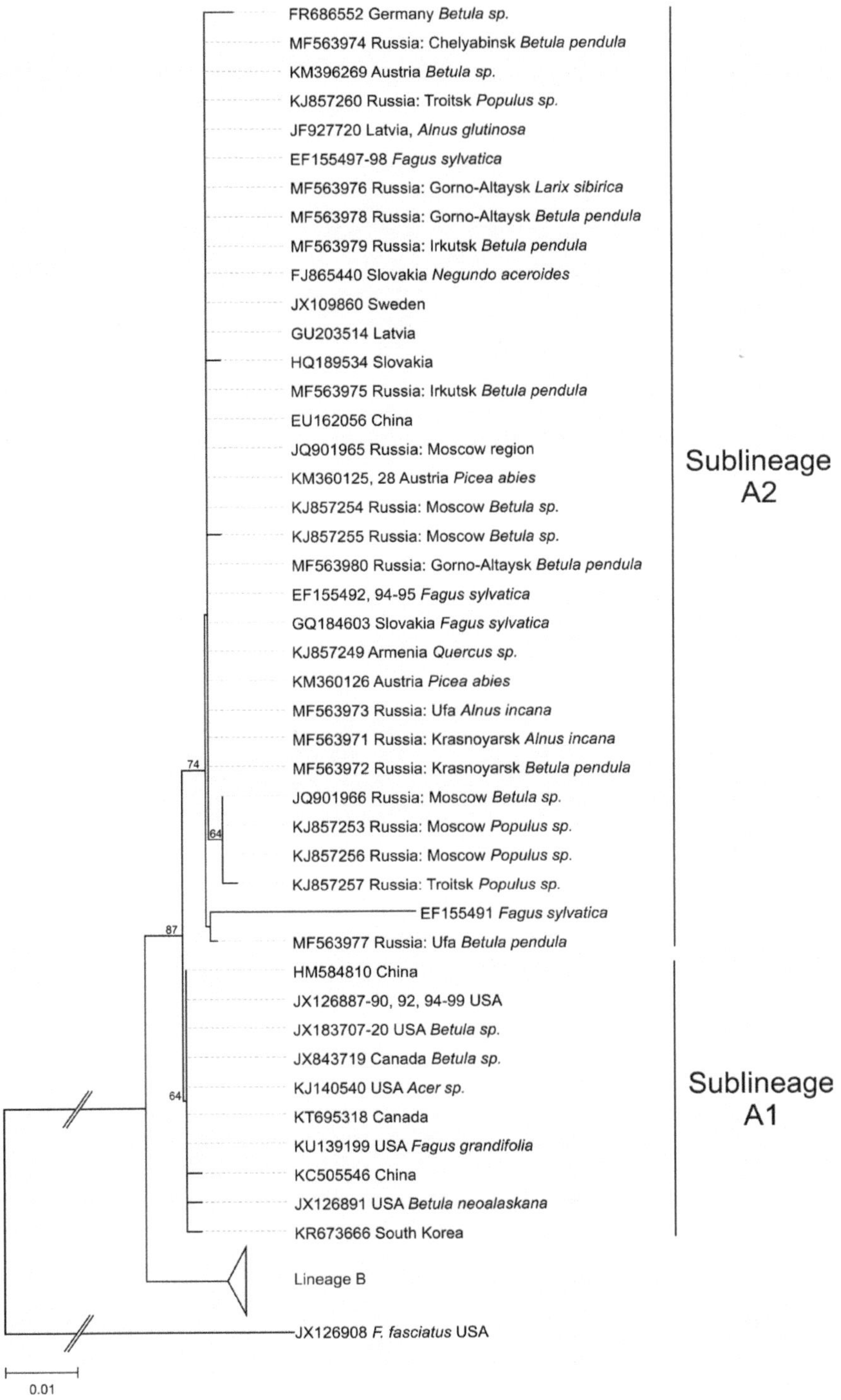
FR686552 Germany *Betula sp.*
MF563974 Russia: Chelyabinsk *Betula pendula*
KM396269 Austria *Betula sp.*
KJ857260 Russia: Troitsk *Populus sp.*
JF927720 Latvia, *Alnus glutinosa*
EF155497-98 *Fagus sylvatica*
MF563976 Russia: Gorno-Altaysk *Larix sibirica*
MF563978 Russia: Gorno-Altaysk *Betula pendula*
MF563979 Russia: Irkutsk *Betula pendula*
FJ865440 Slovakia *Negundo aceroides*
JX109860 Sweden
GU203514 Latvia
HQ189534 Slovakia
MF563975 Russia: Irkutsk *Betula pendula*
EU162056 China
JQ901965 Russia: Moscow region
KM360125, 28 Austria *Picea abies*
KJ857254 Russia: Moscow *Betula sp.*
KJ857255 Russia: Moscow *Betula sp.*
MF563980 Russia: Gorno-Altaysk *Betula pendula*
EF155492, 94-95 *Fagus sylvatica*
GQ184603 Slovakia *Fagus sylvatica*
KJ857249 Armenia *Quercus sp.*
KM360126 Austria *Picea abies*
MF563973 Russia: Ufa *Alnus incana*
MF563971 Russia: Krasnoyarsk *Alnus incana*
MF563972 Russia: Krasnoyarsk *Betula pendula*
74
JQ901966 Russia: Moscow *Betula sp.*
KJ857253 Russia: Moscow *Populus sp.*
64
KJ857256 Russia: Moscow *Populus sp.*
KJ857257 Russia: Troitsk *Populus sp.*
EF155491 *Fagus sylvatica*
87
MF563977 Russia: Ufa *Betula pendula*
HM584810 China
JX126887-90, 92, 94-99 USA
JX183707-20 USA *Betula sp.*
JX843719 Canada *Betula sp.*
KJ140540 USA *Acer sp.*
64
KT695318 Canada
KU139199 USA *Fagus grandifolia*
KC505546 China
JX126891 USA *Betula neoalaskana*
KR673666 South Korea
Lineage B
JX126908 *F. fasciatus* USA
0.01
Sublineage A2
Sublineage A1

Fig. 2 contd. ...

...Fig. 2 contd.

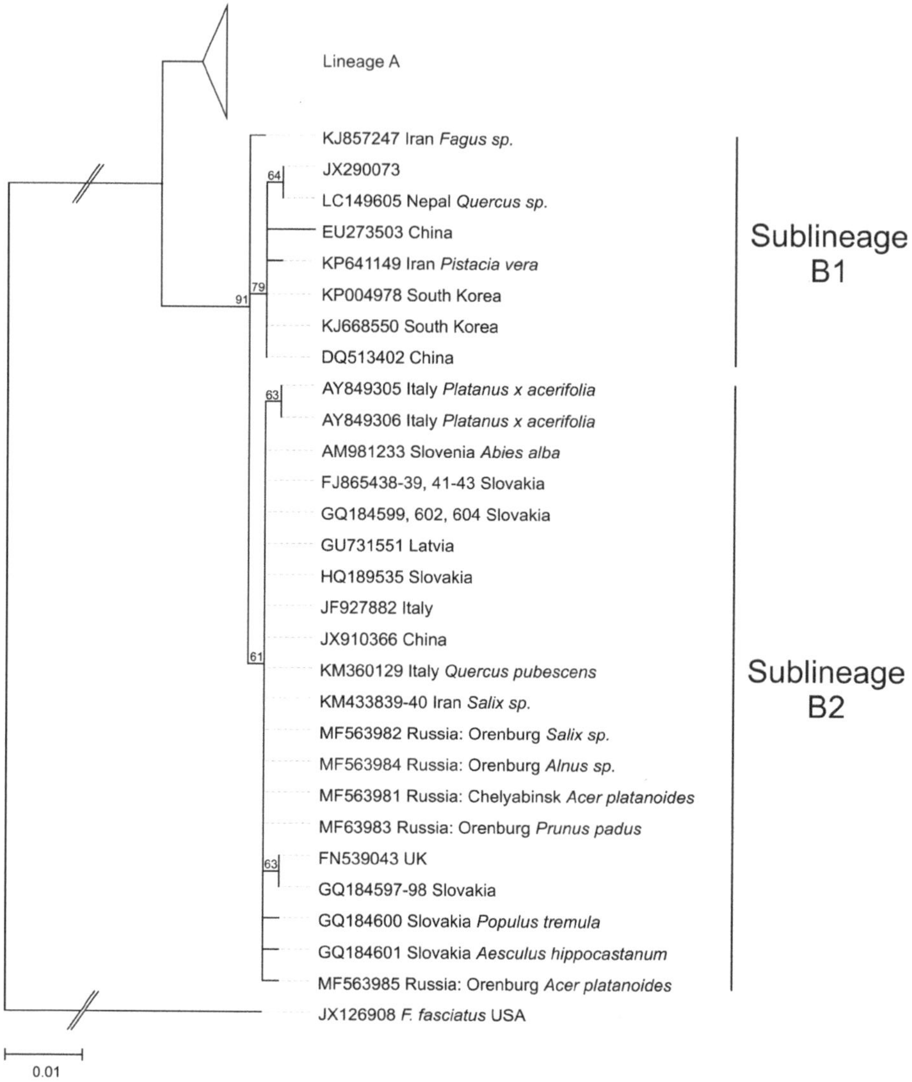

Fig. 2. The phylogenetic tree of *Fomes fomentarius s. lat.* isolates of lineage A and lineage B based on 519 positions of ITS sequences (Mukhin et al., 2018).

Negundo, *Picea* trees, mainly from *Betula*, *Populus* and *Fagus* trees, sublineage B2 - from *Abies*, *Aesculus*, *Quercus*, *Platanus*, *Populus*, and *Salix* trees (Fig. 2). Thus, according to Júdová et al. (2012), Pristaš et al. (2013), Gáper et al. (2016), Mukhin et al. (2018) and Náplavova et al. (2020), *F. fomentarius s. lat.* consists of two phylogenetic lineages, each of which including two sublineages. Based on the analysis of ITS rDNA region, Dresch et al. (2015) and Peintner et al. (2019) have isolated Chinese, North European, North American, East Asian, and South European phylogenetic clades from

F. fomentarius s. lat. They combined into two groups, uniting phylogenetically - related sister North European and North American ITS rDNA isolates on the one hand, and South European, Chinese, on the other. Peintner et al. (2019) have identified the North European group of isolates from Austria, Germany, Italy, Latvia, Poland, Slovakia, and Slovenia as *Fomes fomentarius sensu stricto*, and the South European group, represented by isolates from Italy, Portugal, Slovenia, and France, as *Fomes inzengae* - (Ces. & De Not.) Cooke two sympatric and cryptic species. As it was shown by Badalyan et al. (2022), isolates of sublineage A2 are combined into one group with reference sequences for *F. fomentarius s. str.*, and isolates cluster with sublineage B2 reference sequences are included in the same group with the references for *F. inzengae* (Fig. 3).

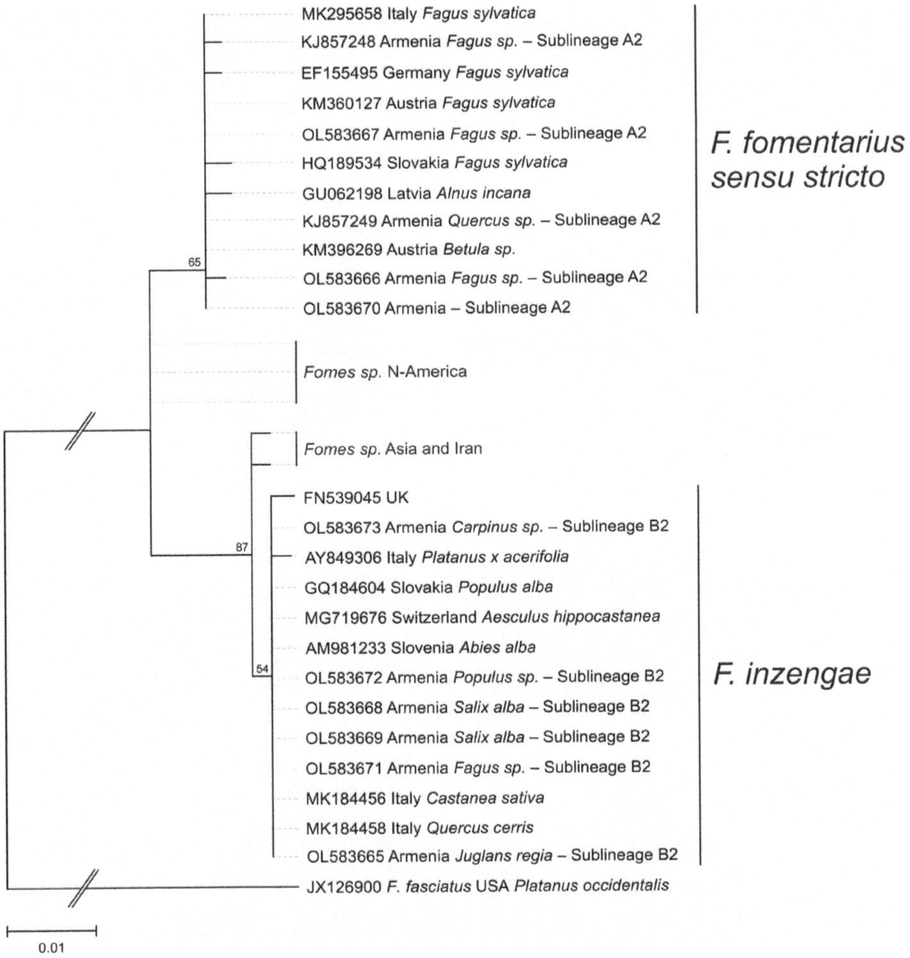

Fig. 3. The phylogenetic tree of *Fomes fomentarius s. lat.* isolates from Armenia and references for *F. fomentarius s. str.* and *F. inzengae* (Peintner et al., 2019) based on 516 positions of ITS sequences (Badalyan et al., 2022).

Fomes fomentarius s. str. and *F. inzengae* are closely related cryptic species regarding basidiocarp morphology (Fig. 4). The basidiocarps of *F. inzengae* differ from *F. fomentarius s. str.* by smaller hymenium pores (0.31 mm *vs* 0.36 mm in *F. fomentarius s. str.*), and by their large number (31–34 pores/cm *vs* 27–30 pores/cm in *F. fomentarius s. str.*). *F. inzengae* also differ by smaller size of basidiospores (9–12.5 × 3–4 μm *vs* 13.5–18 × 4.5–6.5 μm in *F. fomentarius s. str.*) and thicker skeletal hyphae (3.2–6.9 μm) compared to *F. fomentarius s. str.* (3.0–6.4 μm). Basidiocarps of *F. inzengae* also have a greater ability to absorb water from air than basidiocarps of *F. fomentarius* s. str. (Peintner et al., 2019).

Fomes fomentarius s. str. and *F. inzengae* have some specific physiological features. *F. inzengae* strains are fast growing at higher temperature with higher than 30°C growth optimal temperature, whereas mycelia of *F. fomentarius s. str.* grows faster at lower temperatures (10 and 20°C) (Peintner et al., 2019).

The studies of dikaryotic strains isolated from Armenia showed that *F. fomentarius s. str.* strains grow in up to 35°C (only on inoculum) with an optimal temperature of 25°C, in particular 3.0–4.4 mm/d and 2.1–4.2 mm/d on MEA and

Fig. 4. The basidiocarps of *F. fomentarius s. str.*: I – IPAE-*Fomes*-141 (OL549282), II – IPAE-*Fomes*-189 (OL580758), III – IPAE-*Fomes*-143 (OL580756), and *F. inzengae*: IV – IPAE-*Fomes*-B2 (OL579734), V – IPAE-*Fomes*-74 (OL555781), VI – IPAE-*Fomes*-236 (OL555782).

PDA, respectively. Meanwhile, *F. inzengae* strains tolerate temperature of 35°C and above (up to 40 °C only on inoculum) with an optimal temperature of 30°C, as well as higher growth rates (4.8–6.0 mm/d and 4.0–5.3 mm/d on MEA and PDA, respectively). The asexual chlamydospores have been described in mycelial cultures of both species, whereas oidia – mainly in *F. fomentarius s. str.* strains (Badalyan et al., 2015; 2019).

According to Peintner et al. (2019), *F. inzengae* is mainly distributed in Great Britain, Italy, Slovakia, Slovenia, Switzerland, France, Iran and China, particularly in the Mediterranean region. Garrido-Benavent et al. (2020) shown that in the Iberian Peninsula the genus *Fomes* is represented only by *F. inzengae* species. The *F. fomentarius s. str.* Peintner et al. (2019) characterizes as more northern than *F. inzengae* species. According with their distribution *F. fomentarius s. str.* и *F. inzengae* significantly differ in their substrate spectra, in particularly, they are distinguished by the fact that spectrum of the *F. fomentarius s. str.* includes *Betula*, but the *F. inzengae* – not. This provided by *F. fomentarius s. str*. and *F. inzengae* are the possibility of sympatric existence where their ranges overlap.

All these data show that *F. fomentarius s. lat.* is a genetically heterogeneous species consisting of four geographically distinct but morphologically identical phylogenetic lineages or clades. According to Júdová et al. (2012) and Pristaš et al. (2013), lineages A and B are considering as two sympatric cryptic species. As sympatric cryptic species, the North and South European clades are also considered Peintner et al. (2019) and they are not excluding that the "Asia" and "N. America" clades also could be two separate species (Júdová et al., 2012; Pristaš et al., 2013; Peintner et al., 2019).

The species rank of phylogenetic lineages and clades is substantiated by the named authors mainly by the data of ITS analysis of rDNA. ITS-base taxon delimitation or sequence-base taxon delimitation, is the main method in molecular fungal systematics and, according to it, with a nucleotide similarity 97–99%, it can be assumed ("species hypothesis") that they belong to the same species, and to different species, at lower similarity (Kõljag et al, 2013; Taylor and Hibbett, 2014). In Table 1 the data on nucleotide diversity and divergence (Rozas et. al., 2017) of 312 ITS sequences of *F. fomentarius s. lat.* (89 original and 223 from GenBank) calculated are presented. They show that within the phylogenetic sublineages A1, A2, B2, the level of nucleotide diversity is 0.011–0.029%, which indicates their genetic homogeneity and a low level of individual ITS variability. An increased level of the latter (0.46%) is characteristic of sublineage B1, but this is most likely due to its somewhat dubious composition, as pointed out by Pristaš et al. (2013) and Gáper et al. (2016). The level of nucleotide divergence between sublineages A1 and A2, B1 and B2 is 0.46 and 0.6%, respectively, and in this case it indicates that they belong to the same species. Differences between the sequences of *F. fomentarius s. str*. and *F. inzengae* are 1.74%, but also do not go beyond 97–99% nucleotide similarity. At the same time, all four phylogenetic sublineages show significant far beyond 97–99% differences in the ITS rDNA *Fomes fasciatus* (Sw.) Cooke.

Similar results have been described by Júdová et al. (2012), Dresch et al. (2015), Peintner et al. (2019). According to authors, the sequence similarity within

Table 1. Nucleotide diversity (the average number of nucleotide differences per site between two DNA sequences in all possible pairs in the sample population) and nucleotide divergence (the average number of nucleotide substitutions per site between DNA sequences from different populations) of phylogenetic sublineages of *Fomes fomentarius s. lat.* and *Fomes fasciatus.*

Sublineage (number of sequences)	A1	A2	B1	B2	***F. fasciatus***
	1	**2**	**3**	**4**	**5**
1 (57)	0.00029*	0.0046	0.01820	0.01723	0.07578
2 (100)		0.00025*	0.01845	0.01735	0.07247
3 (27)			0.00464*	0.00602	0.09339
4 (115)				0.00011*	0.08734
5 (13)					0.00328*

*, Nucleotide diversity.

A and B genotypes is higher than 99%, between them –97%. Dresch et al. (2015) and Peintner et al. (2019) previous studies have revealed that in Europe the internal sequence divergence between strains of *F. fomentarius s. str.* is 0.02%, *F. inzengae* –0.01%, between them species are 9–18 bp or 2.6% and between of *F. fasciatus* and *F. fomentarius s. str.*, *F. inzengae* 41–62 bp. The same data was obtained as a result of the study of *F. fomentarius s. lat.* in Armenia: the phylogenetic divergence between *F. fomentarius s. str.* and *F. inzengae* is 8–9 bp or 1.84% of nucleotide substitutions per site, but between *F. fomentarius s. str.*, *F. inzengae* and *F. fasciatus* is 33–36 bp (6.73%) and 36–38 bp (7.16%) of nucleotide substitutions per site, respectively (Badalyan et al., 2022).

Thus, based on ITS taxon delimitation methodology, we can draw the following conclusions: (a) the level of nucleotide divergence between phylogenetic lineages of *F. fomentarius s. lat.* allows considering them infraspecific taxa; (b) phylogenetic lineages A and B, *F. fomentarius s. str.* and *F. inzengae* may likely be subspecies of *F. fomentarius s. lat.*; (c) each of the subspecies is represented by two geographically distinct groups of populations corresponding to sublineages A1, A2, B1, B2. Accordingly to the above, *F. fomentarius s. lat.* is a monomorphic, polytypic species consisting of two cryptic subspecies: *F. fomentarius ssp. fomentarius* and *F. fomentarius ssp. inzengae.*

3. Polymorphic Polytypic Species

The polytypic species differs from monomorphic, cryptic polytypic species by phenotypic polymorphism of basidiocarps. As an example, to form morphological species are the species of genus *Daedaleopsis*: *D. confragosa* (Bolton) J. Schröt., *D. tricolor* (Bull.) Bondartsev & Singer, and *D. septentrionalis* (P. Karst.) Niemelä which are differ by morphology of their basidiocarps.

In *D. confragosa*, the hymenophore is tubular, less often daedaloid, the cap color ranges from yellow-brown to light-brown (Fig. 5), whereas the *D. tricolor* and *D. septentrionalis* basidiocarps are with a lamellar hymenophore with a reddish cap with concentric zonation and brown, respectively (Ryvarden and Gilbertson, 1993).

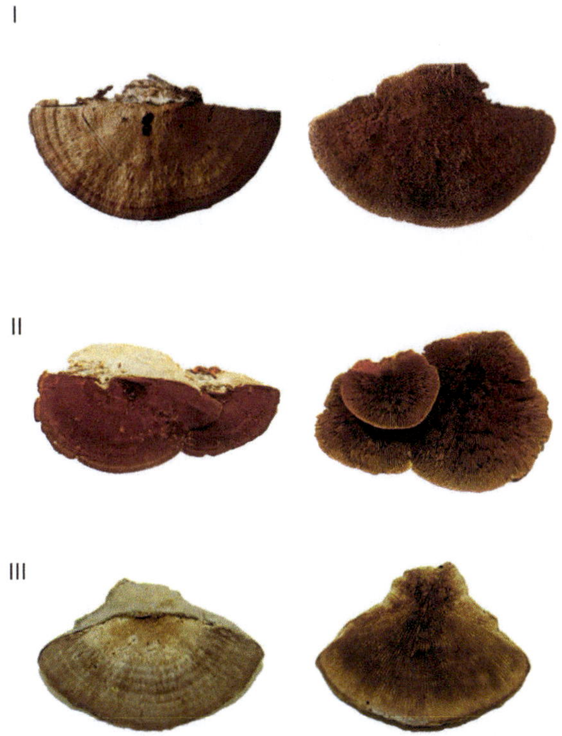

Fig. 5. The basidiocarps of *Daedaleopsis confragosa* (I), *D. tricolor* (II) and *D. septentrionalis* (III).

D. septentrionalis may be also well delimited on the basis of the dichotomously, divided along the edge of the basidiocarp (Niemelä, 1982).

A high morphological variability of basidiocarps and absence of specific anatomical features make their species status debatable. Thus, Bondartsev (1953), Tellería (1980), Jülich (1984) consider *D. tricolor* as a variety of *D. confragosa*, and Ryvarden and Gilbertson (1993) as an independent species, although it is possible that it may be the southern ecotype of *D. confragosa*. To a certain extent, several authors (Ko and Jung, 1999; Bernicchia et al., 2005; Koukol et al., 2014; Mentrida et al., 2015; Galovic et al., 2018) have shown the absence of genetic differences in the ITS region at the species level in *D. confragosa* and *D. tricolor*. However, the ITS rDNA isolate HG973499, a reference for *D. septentrionalis,* is separated from the ITS isolates of *D. confragosa* and *D. tricolor* (Koukol et al., 2014; Mentrida et al., 2015). These results have been obtained due to the phylogenetic analysis for the *tef* region (Koukol et al., 2014). This data allow considering *D. confragosa* and *D. tricolor* as one species different from *D. septentrionalis*.

The phylogenetic proximity of *D. confragosa* and *D. tricolor* in the ITS region has been shown on specimens collected from the Asian part of Russia: the Urals, Siberia and the Far East (Mukhin et al., 2020). Previous study has shown that the cluster analysis combines not only ITS isolates of *D. confragosa* and *D. tricolor*, but also *D. septentrionalis* from the Asian part of its area of distribution into one group.

Fig. 6. The phylogenetic tree isolates of morphospecies *Daedaleopsis* genus based on 553 positions of ITS sequences revealed by Neighbor-Joining method. Numbers above branches indicate Maximum Composite Likelihood bootstrap (Felsenstein, 1985; Kumar et al., 2016; Saitou and Nei, 1987; Tamura et al., 2004).

However, European reference isolate *D. septentrionalis* HG973499 has a separate position (Fig. 6).

Data of individual differences in the nucleotide diversity isolates *D. confragosa*, *D. tricolor*, and *D. septentrionalis* from the Asian part of Russia and their nucleotide divergence (Rozas et. al., 2017) are shown in Table 2. The nucleotide diversity of ITS sequences inside isolates of *D. confragosa*, *D. tricolor* and *D. septentrionalis* varies from 0.59 to 1.22%, and nucleotide divergence – 0.59 to 1.01%. The divergence of the reference European isolate of *D. septentrionalis,* and the same species' isolates from the Asian part of areal is 1.28%, i.e., in the same range as with isolates of *D. confragosa* and *D. tricolor*: 1.11–1.28%. According to the ITS-base taxon delimitation criterion (similarity 99–97%/distinction 1–3%), the monotypic morphological species *D. confragosa*, *D. tricolor* and *D. septentrionalis* are not separate species, but represent infraspecific groups of one polytypic species.

What is their infraspecific taxonomic rank? Some mycologists are considering *D. tricolor* as a variety of *D. confragosa*: *D. confragosa* var. *tricolor* (Pers.: Fr.) Bondartsev (Koukol et al., 2014; Mentrida et al., 2015). In the past, we have also

Table 2. The nucleotide diversity (the average number of nucleotide differences per site between two DNA sequences in all possible pairs in the sample population) and nucleotide divergence (the average number of nucleotide substitutions per site between DNA sequences from different species) of ITS isolates of the morphospecies *Daedaleopsis* genus from the Asian part of Russia.

Species (number of sequences)	*D. confragosa*	*D. tricolor*	*D. septentrionalis*	*D. septentrionalis*, sequence HG973499
	1	2	3	4
1 (8)	0.00596*	0.00958	0.00596	0.01111
2 (7)	0.00958	0.01227*	0.01016	0.01281
3 (4)			0.00663*	0.01278
4 (1)				

* Nucleotide diversity.

shared this opinion and proposed both species *D. tricolor* and *D. confragosa* as varieties of *D. confragosa* (Mukhin et al., 2020). Variety is a taxonomic category that distinguished by variability between individuals and populations is characterized by uncertainty (Mayr, 1970). One of its main characteristics is the lack of geographic confinement to a certain part of the species range among its representatives (Gilyarov, 1989).

Daedaleopsis confragosa, D. tricolor and *D. septentrionalis* are not meet these criteria, since they possess areas that significantly differ in size and geographical confinement. In particular, *D. confragosa* occurs in Eurasia and North America, whereas *D. tricolor* only in Eurasia (Gilbertson and Ryvarden, 1986; Farr et al., 1989; Ryvarden and Gilbertson, 1993). In Europe, *D. confragosa* is found everywhere. The area of distribution of *D. tricolor* is limited to its central and southern parts (Ryvarden and Gilbertson, 1993) and classified as a sub-Mediterranean species (Piątek, 2001). *D. septentrionalis,* like *D. tricolor*, is a Eurasian species found in Northern Europe: Norway, Sweden, Finland (Ryvarden and Gilbertson 1993; Knudsen and Hansen 1997; Kotiranta et al., 2009).

All three morphological species are represented in Russia (Bondartseva, 1998) and their geographical patterns are generally the same, as in Europe. According to Vladykina et al. (2020) and Mukhin and Vladykina (2020) in the Asian part of Russia *D. confragosa* is more common in Western Siberia and North Pacific region, whereas *D. tricolor* is more common in the southern part of forest zone, such as Altai, Western Sayan, Primorsky territory and Sakhalin Island (Fig. 7). *D. septentrionalis* is a relatively rare species compared to other species of the genus in Siberia and the Far East. This fungus is a geographic vicariant species to *D. tricolor* – it is common in the northern part of the forest zone and rare in the southern. However, it does not occur in the Dauria – one of the most southern regions of Siberia (Kotiranta et al., 2016).

Similar character of the geographical distribution of *D. confragosa, D. tricolor*, and *D. septentrionalis* in Europe and Asia indicates that they have stable physiological features. As we have shown, using the examples of *D. confragosa* and *D. tricolor*, their distribution as a whole corresponds to physiological features, such as temperature conditions. According to our data, the range of growth temperature (from

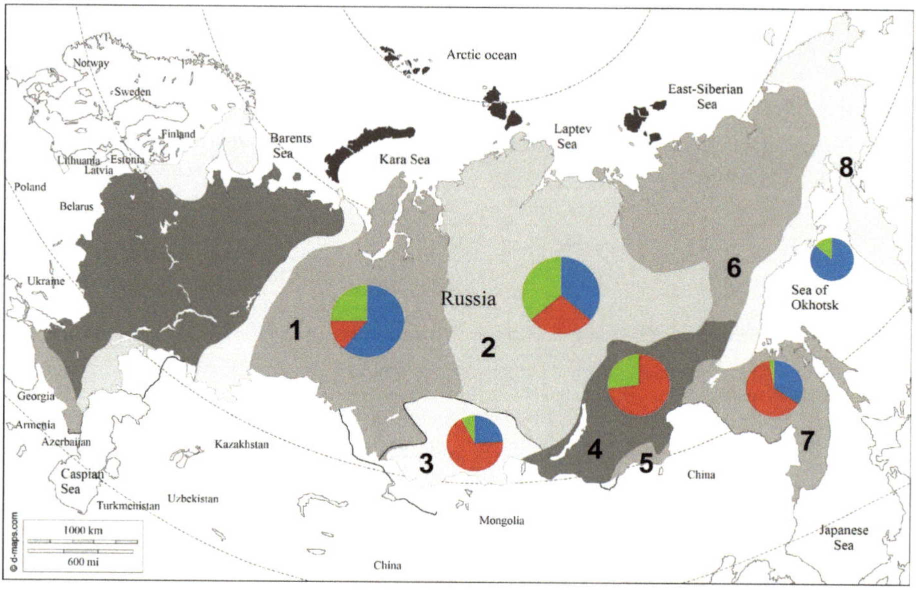

Fig. 7. The composition and relative abundance of morphospecies of *Daedaleopsis* genus in the Asian part of Russia. 1 – Western Siberia, 2 – Central Siberia, 3 – Altai-Sayan Mountainous country, 4 – Baikal area and Transbaikalia Mountainous country, 5 – the Dauria, 6 – North-East Siberia, 7 – Amur-Sakhalin country and 8 – North Pacific Ocean country.

5 to 40°C, optimum 30–35°C), of *D. confragosa* and *D. tricolor* is the same, and they belong to one group of mesophilic fungi. But they exhibit ecological specificity in relation to low and high temperatures. *D. confragosa* grows more intensively at low (5–10°C) temperatures, whereas *D. tricolor* at high temperature (over 30°C), with optimal growth temperature of *D. confragosa* is 30°C and *D. tricolor* is 35°C (Fig. 8). Therefore, *D. confragosa* may be characterized as a psychrophilic mesophilic and *D. tricolor* a mesophilic thermophilic fungi. This is also confirmed by data on the temperature dynamics of their CO_2 gas exchange: in both fungi it is recorded in the same temperature range (5–50°C) with a lower optimal temperature (35°C) in *D. confragosa* and a higher (45°C) - in *D. tricolor* (Fig. 8).

The growth temperature characteristics of *D. confragosa* and *D. tricolor* determine the temperature dynamics of their competitiveness. *D. confragosa* grows more actively at 5–10°C (2.4 ± 0.1 mm/d, p = 0.004) on agar medium compared to *D. tricolor* (1.3 ± 0.2 mm/d). *D. tricolor* also grows more actively at 35°C (p = 0.02) on agar medium than *D. tricolor* (10.1 ± 1.2 mm/day *vs* 5.3 ± 0.3 mm/d). Accordingly, *D. confragosa* and *D. tricolor* have an advantage in the colonization of substrates at low and at high temperatures, respectively (Fig. 9). At 20°C, there are no significant differences (p = 0.5) between the growth rate parameters of *D. confragosa* (7.0 ± 0.2 mm/d) and *D. tricolor* (5.9 ± 0.4 mm/d).

Daedaleopsis confragosa, *D. tricolor*, and *D. septentrionalis* possess different but overlapping ranges. One of the factors of their sympatry is trophic or substrate specialization. In the Asian part of distribution, *D. confragosa*, *D. tricolor* and

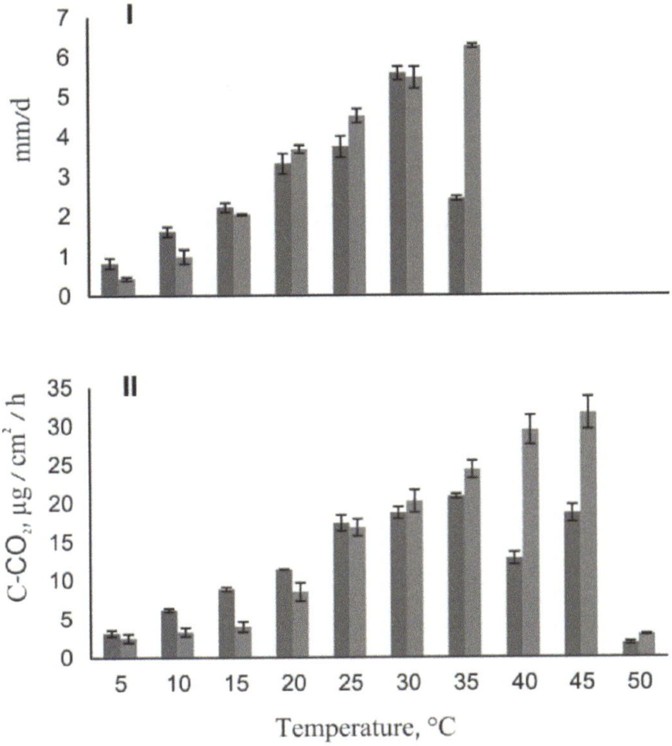

Fig. 8. The temperature dynamics of growth (I) and CO_2 gas exchange (II) of mycelium *D. confragosa* and *D. tricolor* on malts-agar; First columns – *D. confragosa*, second columns – *D. tricolor* (m ± SE).

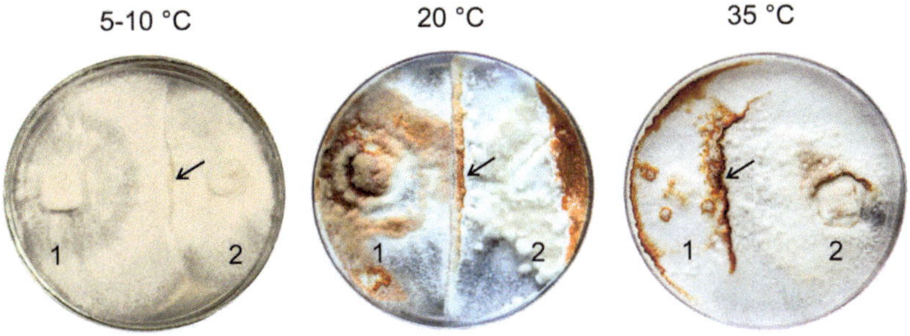

Fig. 9. The barrage test on competitiveness of the *D. confragosa* (1) and *D. tricolor* (2) depends on temperature, arrow – the zone of antagonistic interaction, barrage zone.

D. septentrionalis mainly grow on woody debris of deciduous (*Acer*, *Alnus*, *Betula*, *Carpinus*, *Chosenia*, *Crataegus*, *Quercus*, *Padus*, *Populus*, *Salix*, *Sorbus*, *Tilia*) and are rarely found on coniferous (*Abies*) trees. However, *D. confragosa* is mainly confined to the *Salix* and *Alnus*, whereas *D. tricolor* mostly occurs on *Betula*, and less frequently *Alnus*, *Padus*, and *Salix* wood. *D. septentrionalis* is grows on *Betula*, rarely on *Alnus* and *Salix* woody debris (Fig. 10).

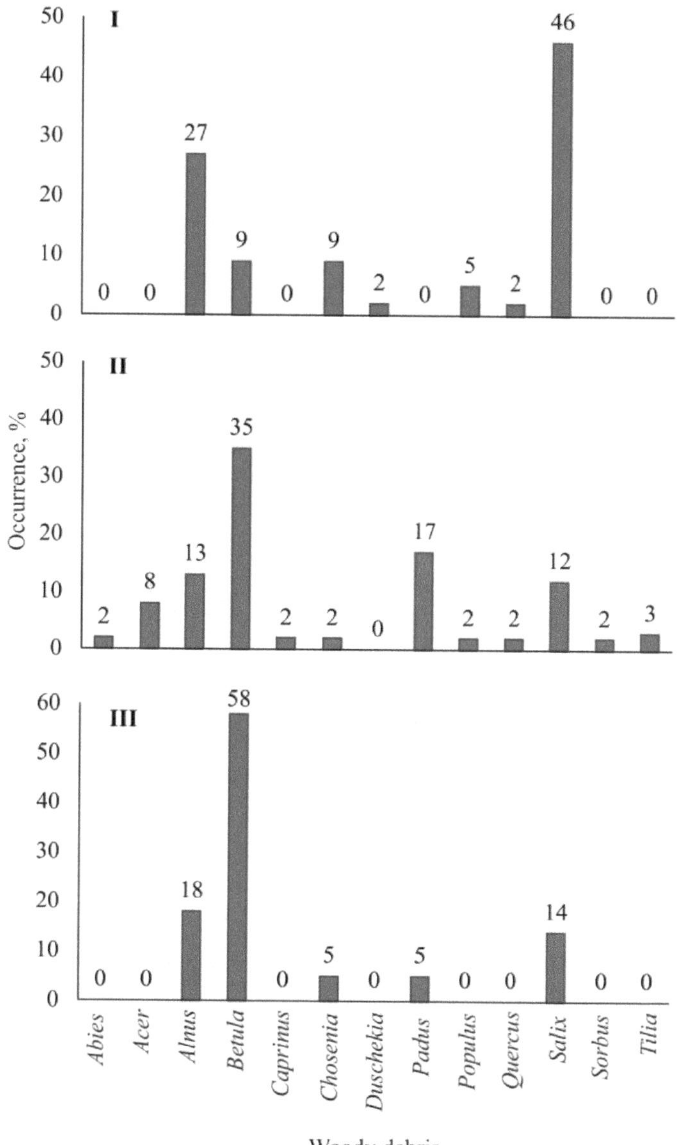

Fig. 10. The substrate spectra of the *D. confragosa* (I), *D. tricolor* (II), *D. septentrionalis* (III) in the Asian part of Russia.

In Europe, the preferred substrates of *D. confragosa* and *D. septentrionalis* are the same (*Salix* and *Betula,* respectively) but for *D. tricolor* they do not match: *Salix* in Europe and *Betula* in Russia. Thus, absence of genetic ITS differences at the species level in *D. confragosa*, *D. tricolor,* and *D. septentrionalis* make them infraspecific groups of the same taxonomic species. According to genetic, geographical and ecological characteristics, their intraspecific taxonomic rank corresponds to a subspecies which are morphologically similar, but geographically and ecologically isolated populations (Mayr, 1970; Gilyarov, 1985). All morphospecies, including *D. confragosa*, *D. tricolor*, and *D. septentrionalis*, together form one polymorphic polytypic taxonomic species *D. confragosa s. lat.* Since different bioactive compounds possessing antifungal, antibacterial and antiviral activities, as well as antioxidant, hypotensive, antitumor, genoprotective, and other therapeutic effects have been described in *D. confragosa* and *D. tricolor*, further studies and elucidation of their taxonomic status has both scientific and practical interests (Table 3).

Table 3. The medicinal properties of *D. confragosa* and *D. tricolor* (Mukhin et al., 2020).

Species	**Type, origin of extract**	**Bioactive compounds**	**Medicinal effects**	**References**
D. confragosa	Dry BC	Phospholipids, glycolipids, fatty acids (7-hydroxy-8,14-dimethyl-9-hexadecenoic and 7-hydroxy-8,16-dimethyl-9-octadecenoic), sterols	—	Dembitsky et al., 1992
	—	Lectins	Anti-H serological specificity	Pemberton, 1994
	Extract, M	Polysaccharides	Antitumor	Ohtsuka et al., 1973
	Water, BC	—	Hypotensive	Melzig et al., 1996
	Methanolic, BC	—	Cytotoxic	Tomasi et al., 2004
	Extract	Phenolics, flavonoids	Antioxidant	Vidović et al., 2011
	Water, M	—	Antiviral	Teplyakova et al., 2012
	Water, petroleum, ether, ethanolic, methanolic, BC, M	Tannis, phenolics, steroids	Antimicrobial, antioxidant, fungistatic	Fakoyal et al., 2013
	Water, BC and M	—	Genoprotective	Knežević et al., 2017
D. tricolor	Extract, BC	Triterpen derivatives (3α-carboxyacetoxyquercinic acid, 3α-carboxyacetoxy-24-methylene-23-oxolanost-8-en-26-oic acid and 5α,8α-epidioxyergosta-6,22-dien-3β-ol)	—	Rösecke and König, 2000
	Extract, BC	Ergosta-5,8,24(28)-trien-3β-ol and other sterols	—	Yaoita et al., 2002
	Petroleum, ether, BC	20(29)-lupen-3-one	Antifungal, antibacterial, antioxidant	Kim et al., 2001

Note: BC, basidiocarps; M, mycelium.

Conclusions and Future Prospects

The widely used high-tech molecular methods and genetic analysis in fungal taxonomy allows the study of transition from a monotypic concept of a species to a polytypic one. One of the primary tasks is to study the genetic and morphological polymorphism of fungi, its geographical patterns and relationships with the environmental factors. The ultimate goal of these works is to elucidate the volume and structure of the species in several wood-decaying fungi, as well as the patterns of speciation in this polymorphic group of organisms. This will also improve the ITS-based taxonomic identification of polypore mushrooms. The main disadvantage of this method is the level of genetic similarity/differences between species and infraspecific taxa for "species hypothesis", and its solution is impossible without data on the individual and inter population level ITS variability, as well as geographical and ecological characters. The study of genetic and morphological polymorphism of Agaricomycetes fungi is important for their further biotechnological usage, since the available data on boundaries of a perspective wood-decaying polypores and its intraspecific variability are required.

Acknowledgments

This chapter arises from a long-standing cooperation between the Yerevan State University, Armenia and Ural Federal University, Russia to study different collections of Agaricomycetes polypore fungi, particularly their taxonomy, morphology, ecology, phylogeny and medicinal properties. The authors are also tthankful to colleagues and collaborators who contributed to the development of molecular taxonomy, fungal biology and biotechnology research to explore wood-decaying fungi as sources of different biotech products, such as pharmaceuticals and cosmeceuticals. This work was supported by the Science Committee of Republic of Armenia, in the frames of the thematic research project 21T-1F228 and Russian Science Foundation, project 22-24-00970. The authors have no conflict of interest to declare.

References

Badalyan, S.M. and Borhani, A. (2019). Medicinal, nutritional and cosmetic values of macrofungi distributed in Mazandaran province of Northern Iran. Review. Int. J. Med. Mushrooms, 21: 1099–1106. https://doi.org/10.1615/IntJMedMushrooms.2019032743.

Badalyan, S.M. and Gharibyan, N.G. (2016). Diversity of polypore bracket mushrooms, Polyporales (Agaricomycetes) recorded in Armenia and their medicinal properties. Int. J. Med. Mushrooms, 18: 347–354. https://doi.org/10.1615/intjmedmushrooms.v18.i4.80.

Badalyan, S.M. and Gharibyan, N.G. (2017). Characteristics of Mycelial Structures of Different Fungal Collections. YSU Press, Yerevan, Armenia, p. 176.

Badalyan, S.M. and Shahbazyan, T.A. (2015). Medicinal properties of two polypore species: *Fomes fomentarius* and *Fomitopsis pinicola*. pp. 277–279. *In*: Dyakov, Y.T. (ed.). Current Mycology in Russia, Proceedings of the 3rd International Mycological Forum. National Academy of Mycology, Moscow.

Badalyan, S.M. and Zambonelli, A. (2023). The potential of mushrooms to develop healthy food and biotech products. pp. 307–344. *In*: Satyanarayana T. and Deshmukh, S.K. (eds.). Fungi and Fungal Products in Human Welfare and Biotechnology. Springer Nature, Singapore. https://doi.org/10.1007/978-19-8853-0_11.

Badalyan, S.M., Barkhudaryan, A. and Rapior, S. (2019). Recent progress in research on the pharmacological potential of mushrooms and prospects for their clinical application. pp. 1–70. *In*: Agrawal, D.C. and Dhanasekaran, M. (eds.). Medicinal Mushrooms: Recent Progress in Research and Development. Springer Nature, Singapore. https://doi.org/10.1007/978-981-13-6382-5_1.

Badalyan, S.M., Shahbazyan, T.A. and Gharibyan, N.G. (2019). The morphological observation of mycelia of several Armenian strains of medicinal bracket fungus *Fomes fomentarius* (L.) Fr. (Polyporales, Agaricomycetes). Proceedings of the YSU, Chemistry and Biology, 53: 92–97.

Badalyan, S.M., Barkhudaryan, A. and Rapior S. (2021). The cardioprotective properties of Agaricomycetes mushrooms growing in the territory of Armenia. Review. Int. J. Med. Mushrooms, 23: 21–31. https://doi.org/10.1615/IntJMedMushrooms.2021038280.

Badalyan, S.M., Shnyreva, A.V., Iotti, M. and Zambonelli, A. (2015). Genetic resources and mycelial characteristics of several medicinal polypore mushrooms (Polyporales, Basidiomycetes). Int. J. Med. Mushrooms, 17: 371–384. https://doi.org/10.1615/IntJMedMushrooms.v17.i4.60.

Badalyan, S.M., Zhuykova E.V. and Mukhin, V.A. (2022). The phylogenetic analysis of Armenian collections of medicinal tinder polypore *Fomes fomentarius* (Agaricomycetes, Polyporaceae). Italian J. Mycology, 51: 23–33. https://doi.org/10.6092/issn.2531-7342/14474.

Bernicchia, A., Fugazzola, M.A., Gemelli, V., Mantovani, B., Lucchetti, A. et al. (2006). DNA recovered and sequenced from an almost 7000 y-old Neolithic polypore, *Daedaleopsis tricolor*. Mycol Res., 110: 14–17. https://doi.org/10.1016/j.mycres.2005.09.012.

Bondartsev, A.S. (1953). Polypore fungi of the European part of the USSR and the Caucasus. AS USSR, Moscow, Leningrad, USSR, p. 1106.

Bondartseva, M.A. (1998). Definitorium Fungorum Rossiae. Ordo Aphyllophorales. Fasc. 2. Familiae Albatrel-laceae, Aporpiaceae, Boletopsidaceae, Bondarzewiaceae, Corticiaceae (genera Tubuliferae), Fistulinaceae, Ganodermataceae, Lachnocladiaceae (genus Tubiliferus), Phaeolaceae, Polyporaceae (genera Tubuliferae), Poriaceae, Rigidoporaceae. Nauka, Saint-Petersburg, Russia, p. 391.

Bridge, P.D., Spooner, B.M. and Roberts, P.J. (2005). The impact of molecular data in fungal systematics. Adv. Bot. Res., 42: 33–67. https://doi.org/10.1016/S0065-2296(05)42002-9

Dembitsky, V.M., Shubina, E.E. and Kashin, A.G. (1992). Phospholipid and fatty acid composition of some Basidiomycetes. Phytochemistry, 313: 845–849.

Dresch, P., D'Aguanno, M., Rosam, K., Grienke, U., Rollinger, J. and Peintner, U. (2015). Fungal strain matters: colony growth and bioactivity of the European medicinal Polypores *Fomes fomentarius*, *Fomitopsis pinicola* and *Piptoporus betulinus*. AMB Express, 5: 1–14. https://doi.org/10.1186/s13568-014-0093-0.

Fakoyal, S., Adegbehingbe, K. and Ogundiimu, A. (2013). Biopharmaceutical assessment of active components of *Deadaleopsis confragosa* and *Ganoderma lucidum*. Open J. Med. Microb., 3: 135–8. https://doi.org/10.4236/ojmm.2013.32020.

Farr, D.F., Bills, G.F., Chamuris, G.P. and Rossman, A.Y. (1989). Fungi on Plants and Plant Products in the United States. St. Paul, Minnesota, The American Phytopathological Society (APS) Press, St. Paul, Minnesota, USA, p. 1252. https://doi.org/10.1002/fedr.19901010703.

Felsenstein, J. (1985). Confidence limits on phylogenies: An approach using the bootstrap. Evolution, 39: 783–791.

Galović, V., Marković, M., Pap, P., Mulett, M., Rakić, M., Vasiljević, A. and Pekeč, S. (2018). Molecular taxonomy and phylogenetics of *Daedaleopsis confragosa* (Bolt.: Fr.) J. Schröt. from wild cherry in Serbia. Genetika, 50: 519–532. https://doi.org/10.2298/GENSR1802519G.

Gáper, J. and Gáperová, S. (2014). A worldwide geographical distribution and host preferences of *Fomes fomentarius*. pp 57–63. *In*: Barta, M., Ferus, P. (eds.). Dendrological days in Mlyňany Arboretum SAV. Mlyňany Arboretum SAV, Vieska nad Žitavou.

Gáper, J., Gáperová, S., Pristas, P. and Náplaková, K. (2016). Medicinal value and taxonomy of the tinder polypore, *Fomes fomentarius* (Agaricomycetes): a review. Int. J. Med. Mushrooms, 18: 851–859. https://doi.org/10.1615/intjmedmushrooms.v18.i10.10.

Gáperová, S., Gáper, J., Gašparcová, T., Náplavová, K. and Pristaš, P. (2016). Morphological variability of *Fomes fomentarius* basidiomata based on literature data. Annales Universitatis Paedagogicae Cracoviensis Studia Naturae, 16: 699–705.

Garrido-Benavent, I., Velasco-Santos, J.M., Pérez-De-Gregorio, M.À. and Pasaban, P.M. (2020). *Fomes inzengae* (Ces. & De Not.) Cooke in the Iberian Peninsula. Butll. Soc. Micol. Valenciana, 24: 151–170.

Gilbertson, R.L. (1980). Wood-rooting fungi of North America. Mycologia, 72: 1–54.

Gilbertson, R.L. and Ryvarden, L. (1986). North American Polypores, vol. 1: Abortiporus–Lindtneria. Fungiflora A/S, Oslo, Norway, p. 433.

Gilbertson, R.L. and Ryvarden, L. (1987). North American Polypores, vol. 2: Megaspoporia–Wrightoporia. Fungiflora A/S, Oslo, Norway, pp. 437–885.

Gilyarov, M.S. (1989). Biological Encyclopedic Dictionary. Ripol Klassik, Moscow, USSR, p. 864.

Grienke, U., Zöll, M., Peintner, U. and Rollingeret, J.M. (2014). European medicinal polypores: A modern view on traditional uses. J. Ethnopharmacol., 154: 564–583. https://doi.org/10.1016/j.jep.2014.04.030.

Horton, T.R. and Bruns, T.D. (2001). The molecular revolution in ectomycorrhizal ecology: Peeking into the black-box. Mol. Ecol., 10: 1855–1871. https://doi.org/10.1046/j.0962-1083.2001.01333.x.

Júdová, J., Dubiková, K., Gáperová, S., Gáper, J. and Pristaš, P. (2012). The occurrence and rapid discrimination of *Fomes fomentarius* genotypes by ITS-RFLP analysis. Fungal Biology, 116: 155–160. https://doi.org/10.1016/j.funbio.2011.10.010.

Jülich, W. (1984). Band IIb/1: Basidiomyceten, 1. Teil. Die Nichtblätterpilze, Gallertpilze und Bauchpilze. Kleine Kryptogamenflora. Jena. VEB Gustav Fischer Verlag, Stuttgart, Germany, p. 626.

Kim, E.M., Jung, H.R. and Min, T.J. (2001). Purification, structure determination and biological activities of 20 (29)-lupen-3-one from *Daedaleopsis tricolor* (Bull. ex Fr.) Bond. et Sing. Bull. Kor. Chem. Soc., 22: 59–62. https://doi.org/10.5012/bkcs.2001.22.1.59.

Kirk, P.M., Cannon, P.F., Minter, D.W. and Stalpers, J.A. (2008). Dictionary of Fungi. Tenth edition. CABI, UK, p. 784.

Knežević, A., Stajić, M., Živković, L., Milovanović, I., Spremo-Potparević, B. and Vukojević, J. (2017). Antifungal, antioxidative, and geno-protective properties of extracts from the blushing bracket mushroom, *Daedaleopsis confragosa* (Agaricomycetes). Int. J. Med. Mushrooms, 19: 509–20. https://doi.org/10.1615/IntJMedMushrooms.v19.i6.30.

Knudsen, H. and Hansen, L. (1997). Nordic macromycetes. Vol. 3. Heterobasidiod, Aphyllophoroid and Gastromycetoid Basidiomycetes. Nordsvamp, Copenhagen, p. 444.

Ko, K.S. and Jung, H.S. (1999). Phylogenetic re-evaluation of *Trametes consors* based on mitochondrial small subunit ribosomal DNA sequences. FEMS Microbiol. Lett., 170: 181–186. https://doi.org/10.1016/S0378-1097(98)00548-5.

Kõljalg, U., Nilsson, R.H., Abarenkov, K., Tedersoo, L., Taylor, A.F. et al. (2013). Towards a unified paradigm for sequence-based identification of fungi. Mol. Ecol., 22: 5271–5277. https://doi.org/10.1111/mec.12481.

Kotiranta, H., Saarenoksa, R. and Kytövuori, I. (2009). Aphyllophoroid fungi of Finland. A check-list with ecology, distribution, and threat categories. Norrlinia, 19: 1–223.

Kotiranta, H., Shiryaev, A.G. and Spirin, V. (2016). Aphyllophoroid fungi (Basidiomycota) of Tuva Republic, southern Siberia, Russia. Folia Cryptogamica Estonica, 53: 51–64. https://doi.org/10.12697/fce.2016.53.07.

Koukol, O., Kotlaba, F. and Pouzar, Z. (2014). Taxonomic evaluation of the polypore *Daedaleopsis tricolor* based on morphology and molecular data. Czech Mycol., 66: 107–119. https://doi.org/10.33585/cmy.66201.

Kües, U. and Badalyan, S.M. (2017). Making use of genomic information to explore the biotechnological potential of medicinal mushrooms. pp. 397–458. *In*: Agrawa, D.C., Tsay, H.S., Shyur, L.F., Wu, Y.C., Wang, S.Y. (eds.). Medicinal Plants and Fungi: Recent Advances in Research and Development, Medicinal and Aromatic Plants of the World, Volume 4. Springer, New York. https://doi.org/10.1007/978-981-10-5978-0_13.

Kumar, S., Stecher, G. and Tamura, K. (2016). MEGA7: Molecular evolutionary genetics analysis version 7.0 for bigger datasets. Mol. Biol. Evol., 33: 1870–1874. https://doi.org/10.1093/molbev/msw054.

Mayr, E. (1970). Population, Species and Evolution. The Belknap Press of Harvard University Press, Cambridge, Massachusetts, USA, p. 453.

McCormick, M.A., Grand, F.L., Post, D.J. and Cubeta, A.M. (2013). Phylogenetic and phenotypic characterization of *Fomes fasciatus* and *Fomes fomentarius* in the United States. Mycologia, 105: 1525–1534. https://doi.org/10.3852/12-336.

Melzig, M.F., Pieper, S., Siems, W.E., Heder, G., Bottger, A. et al. (1996). Screening of selected basidiomycetes for inhibitory activity on neutral endopeptidase (NEP) and angiotensin-converting enzyme (ACE). Pharmazie, 51: 501–503.

Mentrida, S., Krisai-Greilhuber, I. and Voglmayr, H. (2015). Molecular evaluation of species delimitation and barcoding of *Daedaleopsis confragosa* specimens in Austria. Österreichische Zeitschrift für Pilzkund, 24: 173–179.

Mukhin, V. and Vladykina, V. (2020). Distribution of xylotrophic fungi of the genus *Daedaleopsis* in the Asian part of Russia from 1978 to 2019. Occurrence dataset. https://doi.org/10.15468/m4hk49.

Mukhin, V.A. (1978). Structure of the flora of basidial wood-decomposing fungi occurring in the European and Asiatic areas of the Holarctic region (USSR). Mikologiia i fitopatologiia 2(1): 55–60.

Mukhin, V.A. (1993). Biota of Wood-decaying Fungi of West Siberian Plane. Nauka, Ekaterinburg, USSR, p. 231.

Mukhin, V.A., Diyarova, D.K., Gitarskiy, M.L. and Zamolodchikov, D.G. (2021). Carbon and oxygen gas exchange in woody debris: the process and climate-related drivers. Forests, 12: 1156–1171. https://doi.org/10.3390/f12091156.

Mukhin, V.A., Zhuykova, E.V. and Badalyan, S.M. (2018). Genetic variability of the medicinal tinder bracket polypore, *Fomes fomentarius* (Agaricomycetes), from the Asian Part of Russia. Int. J. Med. Mushrooms, 20: 561–568. https://doi.org/10.1615/intjmedmushrooms.2018026278.

Mukhin, V.A., Zhuykova, E.V., Vladykina, V.D. and Badalyan, S.M. (2020). Notes on medicinal polypore species from the genus *Daedaleopsis* (Agaricomycetes), distributed in the Asian part of Russia. Int. J. Med. Mushrooms, 22: 775–780. https://doi.org/10.1615/IntJMedMushrooms.v22.i8.

Náplavová, K., Gáper, J., Gáperová, S., Beck, T., Pristaš, P. et al. (2020). Genetic and plant host differences of *Fomes fomentarius* in selected parts of Southern Europe. Plant Biosystems, 154: 125–127. http://doi.org/10.1080/11263504.2019.1701129.

Niemelä, T. (1982). Taxonomic notes on the polypore genera *Antrodiella, Daedaleopsis, Fibuloporia* and *Phellinus*. Karstenia, 22: 11–12.

Nilsson, R.H., Kristiansson, E., Ryberg, M., Hallenberg, N. and Larsson, K.H. (2008). Intraspecific ITS variability in the kingdom Fungi as expressed in the international sequence databases and its implications for molecular species identification. Evol. Bioinform., 4: 193–201. https://doi.org/10.4137/EBO.S653.

Ohtsuka, S., Ueno, S., Yoshikumi, C., Hirose, F., Ohmura, Y. et al. (1973). Polysaccharides having an anticarcino-genic effect and a method of producing them from species of Basidiomycetes. United Kingdom patent # 1331513.

Peintner, U., Kuhnert-Finkernagel, R., Wille, V., Biasioli, F., Shiryaev, A. and Perini, C. (2019). How to resolve cryptic species of polypores: An example in *Fomes*. IMA Fungus, 10: 1–21. https://doi.org/10.1186/s43008-019-0016-4.

Pemberton, R.T. (1994). Agglutinins (lectins) from some British higher fungi. Mycol Res., 98: 277–90. https://doi.org/10.1016/S0953-7562(09)80455-3.

Piątek, M. (2001). New discovery of *Daedaleopsis tricolor* (Fungi, Foriales) and a review of its distribution in Poland. Pol. Bot. J., 46: 277–279.

Pristaš, P., Gáperová, S., Gáper, J. and Júdová, J. (2013). Genetic variability in *Fomes fomentarius* reconfirmed by translation elongation factor 1-α DNA sequences and 25S LSU rRNA sequences. Biologia, 68: 816–820. https://doi.org/10.2478/s11756-013-0228-9.

Rogers, S.O., Holdenrieder, O. and Sieber, T.N. (1999). Intraspecific comparisons of *Laetiporus sulphureus* isolates from broadleaf and coniferous trees in Europe. Mycological Research, 103: 1245–1251. https://doi.org/10.1017/S0953756299008564.

Rösecke, J. and König, W.A. (2000). Constituents of the fungi *Daedalea quercina* and *Daedaleopsis confragosa* var. *tricolor*. Phytochemistry, 54: 757–762. https://doi.org/10.1016/S0031-9422(00)00130-8.

Rozas, J., Ferrer-Mata, A., Sánchez-Del Barrio, J.C., Guirao-Rico, S., Librado, P. et al. (2017). DnaSP 6: DNA Sequence polymorphism analysis of large datasets. Mol. Biol. Evol., 34: 3299–3302. https://doi.org/10.1093/molbev/msx248.

Ryvarden, L. and Gilbertson, R.L. (1993). European Polypores, Part. 1. (*Abortiporus–Lindtneria*). Fungiflora A/S, Oslo, Norway, p. 387.

Ryvarden, L. and Gilbertson, R.L. (1994). European Polypores, Part. 2. (*Inonotus–Tyromyces*). Fungiflora A/S, Oslo, Norway, pp. 388–743.

Saitou, N. and Nei, M. (1987). The neighbor-joining method: A new method for reconstructing phylogenetic trees. Mol. Biol. Evol., 4: 406–425. https://doi.org/10.1093/oxfordjournals.molbev.a040454.

Stengel, A., Stanke, K.M., Quattrone, A.C. and Herr, J.R. (2022). Improving taxonomic delimitation of fungal species in the age of genomics and phenomics. Front. Microbiol., 13: 847067. https://doi.org/10.3389/fmicb.2022.847067.

Stepanova, N.T. and Mukhin, V.A. (1979). Fundamentals of Wood-destroying Fungi Ecology. Nauka, Moscow, USSR, p. 99.

Tamura, K., Nei, M. and Kumar, S. (2004). Prospects for inferring very large phylogenies by using the neighbor-joining method. Proceedings of the National Academy of Sciences (USA), 101: 11030–11035. https://doi.org/10.1073/pnas.04042061.

Taylor, J.W. and Hibbett, D.S. (2013). Toward sequence-based classification of fungal species. IMA Fungus, 4: A33–A34. https://doi.org/10.1007/BF03449308.

Taylor, J.W., Turner, E., Townsend, J.P., Dettman, J.R. and Jacobson, D. (2006). Eukaryotic microbes, species recognition and the geographic limits of species: examples from the kingdom Fungi. Phil. Trans. R. Soc. B. 361: 1947–1963. https://doi.org/10.1098/rstb.2006.1923.

Tellería, M.T. (1980). Contribución al estudio de los Aphyllophorales españoles. Bibl. Mycol., 74: 1–464.

Teplyakova, T.V., Psurtseva, N.V., Kosogova, N.V., Mazurkova, T.A., Khanin, V.A. and Viasenko, V.A. (2012). Antiviral activity of polyporoid mushrooms (higher Basidiomycetes) from Altai Mountains (Russia). Int. J. Med. Mushrooms, 14: 37–45. https://doi.org/10.1615/intjmedmushr.v14.i1.40.

Tomasi, S., Lohezic-Le Devehat, F., Sauleau, P., Bezivin, C. and Boustie, J. (2004). Cytotoxic activity of methanol extracts from Basidiomycete mushrooms on murine cancer cell lines. Pharmazie, 59: 290–293.

Vidović, S., Zeković, Z., Mujić, I., Lepojević, Ž., Radojković, M. and Živković, J. (2011). The antioxidant properties of polypore mushroom *Daedaleopsis confragosa*. Cent. Eur. J. Biol., 6: 575–582. https://doi.org/10.2478/s11535-011-0029-5.

Vladykina, V.D., Mukhin, V.A. and Badalyan, S.M. (2020). *Daedaleopsis* Genus in Siberia and the Far East of Russia. ARPHA Proceedings (Proceedings BDI-2020: III Russian National Conference "Information Technology in Biodiversity Research"), 2: 17–26. https://doi.org/10.3897/ap.2.e58134.

Yaoita, Y., Ebina, K., Kakuda, R., Machida, K. and Kikuchi, M. (2002). Sterol constituents from *Daedaleopsis tricolor*. J. Nat. Med., 56: 117–119.

Index

A

Agaricomycetes 311, 328
Animals 124, 125, 127, 129, 134, 148–152, 154, 155
Associations 123–125, 127, 128, 134, 149, 154, 155

B

Basidiomycota 180–182, 186
Beneficial symbionts 17
Bioremediation 3, 10, 79, 94, 95
Brown rot 180, 182–186

C

Climatic regions 57, 61
Coastal sand dunes 103–105, 107, 115
Coniophora 280–284, 289, 296–305
Conservation 53
Coprophilous fungi 109
Core-group fungi 103, 107, 108

D

Daedaleopsis 320–324
Decomposition 17, 19, 22, 24–26, 28, 29, 36, 41
Dipterocarpaceae 5
Dispersal 79–81, 91
Disturbance 2, 3, 9
Diversity 222–225, 227, 233, 234
Domestic fungi 280

E

Ecological niches 101, 102, 104, 111, 115
Ecology 222, 223, 234
Ecosystem services 79, 94, 101, 102, 114, 165, 174
Ectomycorrhizae 109, 114
Ectomycorrhizal fungi 57
Entomopathogens 108
Environmental factors 79

F

Faunal association 5
Fomes 313, 314, 316–320
Foothills 166, 173
Forests 159–165, 174

G

Geographic distribution 245, 260–262

H

Human interference 9
Humid tropical forest 17, 18, 23, 24, 26, 30, 35, 39, 41, 42
Hymenochaetales 223, 232

I

Inonotus 244, 251
Inorganic-nitrogen 54, 61, 62, 67, 70, 72
Interactions 129, 148–152, 154, 155
ITS 312–314, 316, 317, 319–323, 326, 328

J

Jarrah forest 57, 59–64, 66, 67, 69, 70, 72, 73

M

Macrofungi 15, 17, 21, 22, 24–26, 30, 32, 34–36, 39–42
Mangroves 103–105, 107, 115
Maritime habitats 173
Metabolomics 296, 300–302
Microhabitats 79, 84, 92
Microsuccessions 203
Mushrooms 109, 114, 115
Mutualism 4, 6, 79, 89, 90

Mycogeography 123, 125
Mycorrhizae 19–22

N

Nutrient cycling 3, 4, 17, 19, 26, 29, 42
Nutrient elements 70–72

P

Parasites 38–40
Parasitism 79, 89, 90
Pathogenicity 89
Pathogens 3, 4, 7, 8
pH 54, 57, 61, 62, 64–68, 70, 72
Phellinus 244, 252, 256, 258, 259, 261, 267, 270, 271
Phylogenetic lineages 313, 314, 316, 319, 320
Pinaceae 245, 248
Polymorphism 309, 311, 320, 328
Polyporaceae 208–214
Polyporales 223, 230–232
Polypores 246, 312, 313, 328

S

Saprotrophic fungi 72
Scrub jungles 161, 166, 169, 172
Serpula 280–283, 286–289, 296–305
Similarity 125, 135, 147, 148
Substrate ecology 179, 197, 208, 214, 215
Substrates 2–6, 10
Successional species assemblage 54, 61

T

Tapinella 280–283, 290, 305
Taxonomy 312, 313, 328
Termitomycetes 112

U

Urea application 54, 56–61, 63, 65, 67–72

W

West coast 172
Western Australia 53, 59, 61, 73
Western Ghats 101–103, 105–109, 111, 114, 115, 159, 160, 166, 173
White rot 180, 182–187, 214, 244, 247
Wood decay 183, 207

About the Editors

Prof. Kandikere R. Sridhar is an adjunct faculty at Mangalore University, India. His primary area of study is aquatic fungi in freshwaters and marine waters. He has research collaborations in the USA, Canada, Germany, Switzerland and Portugal. He was president of the Mycological Society of India (2018), a distinguished Asian Mycologist (2015) and one of the world's top 2% scientists in the field of mycology (2019–21). He has published over 500 research articles and edited 10 books. He is in the editorial board of several national and international journals and reviewed over 150 research papers, 25 project proposals and 10 book proposals.

Dr. Sunil Kumar Deshmukh is Scientific Advisor to Greenvention Biotech, Uruli-Kanchan, Pune, India and Agpharm Bioinnovations LLP, Patiala, Punjab, India. Being a veteran industrial mycologist, he spent a substantial part of his career in drug discovery at Hoechst Marion Roussel Limited [now Sanofi India Ltd.], Mumbai, and Piramal Enterprises Limited, Mumbai. He has also served TERI-Deaken Nano Biotechnology Centre, TERI, New Delhi, and as an Adjunct Associate Professor at Deakin University, Australia. He has to his credit 8 patents, 145 publications, and 19 books on various aspects of fungi and natural products of microbial origin. He is a president of the Association of Fungal Biologists (AFB) and a past president of the Mycological Society of India (MSI). Dr. Deshmukh is Associate editor of Frontiers in Microbiology and the series editor of Progress in Mycological Research published by CRC press. Dr. Deshmukh serves as a referee for more than 20 national and international journals. He has approximately four decades of research experience in getting bioactives from fungi and keratinophilic fungi.